Python

可以这样学

董付国 ◎ 著

清华大学出版社
北京

内 容 简 介

全书共分 16 章,对 Python 内部工作原理进行了一定深度的剖析,99％以上的案例代码使用 Python 3.5.1 实现,也适用于 Python 3.4.x(除少数几个新特性之外)和最新版本 Python 3.5.2 以及 Python 3.6.0,极个别案例使用 Python 2.7.11 实现(同样适用于其他版本 Python 2.7.x,包括最新的 Python 2.7.12),适当介绍了 Python 代码优化、系统编程和安全编程的有关知识,满足不同层次读者的需要。另外,书中通过小提示、小技巧、注意拓展知识等形式介绍了更多的内容,全部内容远比章节目录所显示的要多,需要认真阅读才能真正领会其中的奥妙。

本书适合作为 Python 程序员的开发指南,也可以作为高等院校计算机专业、软件工程专业等专业的 Python 教材,还可以作为 Python 爱好者的指导用书。

本书封面贴有清华大学出版社防伪标签,无标签者不得销售。
版权所有,侵权必究。侵权举报电话:010-62782989 13701121933

图书在版编目(CIP)数据

Python 可以这样学/董付国著. ——北京:清华大学出版社,2017(2018.8重印)
 ISBN 978-7-302-45646-9

Ⅰ. ①P… Ⅱ. ①董… Ⅲ. ①软件工具—程序设计 Ⅳ. ①TP311.56

中国版本图书馆 CIP 数据核字(2016)第 283896 号

责任编辑:白立军
封面设计:杨玉兰
责任校对:梁 毅
责任印制:杨 艳

出版发行:清华大学出版社
  网　　址:http://www.tup.com.cn,http://www.wqbook.com
  地　　址:北京清华大学学研大厦 A 座   邮　编:100084
  社 总 机:010-62770175   邮　购:010-62786544
  投稿与读者服务:010-62776969,c-service@tup.tsinghua.edu.cn
  质量反馈:010-62772015,zhiliang@tup.tsinghua.edu.cn
  课件下载:http://www.tup.com.cn,010-62795954

印　刷　者:北京富博印刷有限公司
装　订　者:北京市密云县京文制本装订厂
经　　　销:全国新华书店
开　　　本:185mm×260mm   印　张:33.25   字　数:787 千字
版　　　次:2017 年 2 月第 1 版   印　次:2018 年 8 月第 6 次印刷
定　　　价:69.00 元

产品编号:069714-01

前言

FOREWORD

Python 并不是一门新语言，它由 Guido van Rossum 于 1989 年年底开始设计，并于 1991 年推出第一个公开发行版本，比 Java 早 4 年。Python 推出不久就迅速得到各行业人士的青睐，经过 20 多年的发展，已经渗透到统计分析、移动终端开发、科学计算可视化、系统安全、逆向工程、软件测试与软件分析、图形图像处理、人工智能、机器学习、游戏设计与策划、网站开发、数据爬取与大数据处理、密码学、系统运维、音乐编程、影视特效制作、计算机辅助教育、医药辅助设计、天文信息处理、化学、生物信息处理、神经科学与心理学、自然语言处理、电子电路设计、电子取证、树莓派等几乎所有专业和领域，在黑客领域更是多年来一直拥有霸主地位。

作为一个非常不完整列表，这里给出几个 Python 应用案例：著名搜索引擎 Google 的核心代码使用 Python 实现，迪士尼公司的动画制作与生成采用 Python 实现，大部分 UNIX 和 Linux 操作系统都内建了 Python 环境支持，豆瓣网使用 Python 作为主体开发语言进行网站架构和有关应用的设计与开发，网易大量网络游戏的服务器端代码超过 70% 采用 Python 进行设计与开发，易度的 PaaA 企业应用云端开发平台和百度云计算平台 BAE 也都大量采用了 Python 语言，eBay 已经使用 Python 超过 15 年以上（在 eBay 官方宣布支持 Python 之前就已经有程序员在使用了），美国宇航局使用 Python 实现了 CAD/CAE/PDM 库及模型管理系统，微软集成开发环境 Visual Studio 2015 开始默认支持 Python 语言而不需要像之前的版本一样再单独安装 PTVS 和 IronPython，开源 ERP 系统 Odoo 完全采用 Python 语言开发，树莓派使用 Python 作为官方编程语言，引力波数据是用 Python 进行处理和分析的，YouTube、美国银行等也在大量使用 Python 进行开发，类似的案例数不胜数。

早在多年前 Python 就已经成为卡耐基梅隆大学、麻省理工学院、加州大学伯克利分校、哈佛大学、多伦多大学等国外很多大学计算机专业或非计算机专业的程序设计入门教学语言，近几年来国内也有不少学校的多个专业陆续开设了 Python 程序设计课程。Python 语言连续多年在 TIOBE 网站的编程语言排行榜上排名七八位，2011 年 1 月 Python 被 TIOBE 网站评为 2010 年年度语言；在 2014 年 12 月份 IEEE Spectrum 推出的编程语言排行榜中，Python 取得了第 5 位的好名次；2015 年 12 月份 TIOBE 编程语言排行榜上 Python 跃居第 4 位，仅次于 Java、C 和 C++，已经成为脚本语言的标准；Top developer Languages of 2015 更是把 Python 排到了第 3 位。

Python 是一门免费、开源的跨平台高级动态编程语言，支持命令式编程、函数式编程，完全支持面向对象程序设计，拥有大量功能强大的内置对象、标准库、涉及各行业领域的扩展库以及众多狂热的支持者，使得各领域的工程师、科研人员、策划人员甚至管理人员能够快速实现和验证自己的思路、创意或者推测。在有些编程语言中需要编写大量代码才能实现的功能，在Python中直接调用内置函数或标准库方法即可实现，大幅度简化了代码的编写和维护。Python用户只需要把主要精力放在业务逻辑的设计与实现上，在开发速度和运行效率之间达到了完美的平衡，其精妙之处令人击节赞叹。

如何学习Python

要想改变世界，首先要改变自己的世界。要想学好Python，首先要从内心认识到Python的强大与美，树立起学好、用好Python的信念并坚持不懈的努力，然后才有可能攀登Python高手之巅。

很多人从内心很恐惧Python，曾经有不少人问我："Python功能那么强大，肯定很难学吧？"其实，从编程语言发展史来看，人类语言和机器语言之间的鸿沟越来越小，人机交互越来越方便，越高级的编程语言越接近人类自然语言，越容易学习、掌握和运用，所以请不要有丝毫的恐惧和犹豫，放手去学便是。

以我个人20年的经验，不管学习和使用哪种编程语言，大概都需要经历4个阶段：第一阶段，能看懂和调试别人的代码；第二阶段，能在别人的代码基础上进行适当改写；第三阶段，能把多段已有的代码拼凑起来实现自己需要的功能；第四阶段，自己动手编写代码实现特定功能需求。一般而言，如果每天坚持3个小时学习Python，两周左右应该就能入门，3个月后就可以展示出不错的成果。当然，随着学习和开发时间越来越长，功力会越来越深厚，能够掌控的代码行数（LOC）会越来越多，能够控制的业务逻辑越来越复杂。

毫无疑问，Python是一门快乐、优雅的语言，上手非常容易，稍加接触就会喜欢上Python并能够在短时间内写出几个小程序。与C语言系列和Java等语言相比，Python大幅度降低了学习与使用的难度。Python易学易用，语法简洁清晰，代码可读性强，编程模式非常符合人类思维方式和习惯。尽管如此，但这并不代表可以毫不费力地学会和熟练运用Python，在学习Python的路上没有秘籍，也没有哈利波特魔法杖，唯一的捷径就是勤学苦练。

多看。读书破万卷，下笔如有神。学习Python也是这样，不仅要多看书，还要看很多遍。很多知识点是互相关联的，单独一个知识点也无法实现稍微复杂一点的功能，书中很多案例代码用到了后面章节的知识点，而后面章节的案例代码又用到了前面章节的内容，这实在是无法避免的一件事。因此，不能奢望看一遍就能熟悉和掌握书中的内容，更不能奢望只看一本书就能学会Python的全部知识。以我个人而言，除了Python官方帮助文档和一些社区的资料之外，已经阅读了超过25本Python方面的书，并且还在不断地购买和阅读别人编写的Python书籍来提高自己对Python的理解。一书一茶一下午，这样平淡的快乐或许有人难以体会，于我却是乐此不疲。

多练。太极拳论曰"由招熟而渐悟懂劲，由懂劲而阶及神明，然非用力之久，不能豁然贯通焉"。陆游的教子诗《冬夜读书示子聿》也认为"纸上得来终觉浅，绝知此事要躬行"。掌握了正确的方法以后，多练是唯一的捷径。一定要动手编写和调试书上的代码，踏踏实实把基础掌握好再有针对性地学习自己需要的扩展库,切忌只看不练。很多人眼高手低,一看就会,一编就错,根本原因就是练得太少了。子曰"学而时习之",也是这个道理,充分说明练习的重要性。一层功夫一层天。欲穷千里目,更上一层楼。多练,可以说是攀登Python高手之巅最重要的途径。

多想。学而不思则罔,思而不学则殆。一味地看书和埋头苦练是不行的,还要多想、多总结、多整理,争取把学到的知识和技术彻底理解。理解得越多,需要记忆的就越少。

多交流。独学而无友,则孤陋而寡闻。除了重视基础知识的学习和练习之外,还要多交流。除了Python官方网站和在线帮助文档之外,经常浏览一些Python论坛并阅读和调试其中的优秀代码,汲取他人代码中的精华。子曰"三人行必有我师焉,择其善者而从之",也是相同的道理。遇到不懂的问题也可以去一些论坛发帖提问,或者请教身边的朋友和老师,但是提问之前一定要充分思考,有针对性地请教别人,经过充分思考以后再请教别人不仅是对别人的尊重,也能让自己有更大的收获。百思不得其解的问题经过高手一点拨才能有茅塞顿开和恍然大悟的效果,这样的交流更加有效,不浪费彼此的时间。曾经有学生拿一个十几页代码的程序来问我问题,我问哪里看不懂,他说整个程序都看不懂,想让我帮忙看看然后把整篇代码给他讲讲。正如大家所想的一样,我直接拒绝了他。也曾经有读者问我"怎么用Python做图像处理?"真的很抱歉,这样没营养的问题我实在没法回答。

俗话说,心急吃不了热豆腐。控制好学习的进度和节奏才能获得最好的效果,每天学习一点、进步一点、提高一点,时间久了会突然有一天发现自己已经成为Python高手,很有零存整取的感觉。欲速则不达,把下面这个图送给各位读者朋友共勉。

学拳容易改拳难。不是所有慢悠悠的拳都是太极拳,也不是所有用Python语言写出来的代码都足够Pythonic。很多人认为编程语言都是一通百通,无非是语法不一样而已,认为"没吃过猪肉也见过猪跑",简单看看语法就能立刻使用另外一种语言编写程序。这样的想法确实有一定的道理,但实际上就算天天看猪跑也没法知道猪肉是啥味道,按照C语言的思路用Python写出来的代码绝对不是好的Python程序,会显得不伦不类,代码会非常啰唆,不得Python精髓,完全不能发挥Python的优势。应该在熟悉Python编程

模式的基础上,尽量尝试从最自然、最简洁的角度出发去思考和解决问题,这样才能写出更加优雅、更加 Pythonic 的代码,像诗一样美。

汝果欲学诗,功夫在诗外。没有丰富的人生阅历很难写出优美并且有内涵、有灵魂的诗,学习 Python 也是这样。归根到底,Python 是用来表达我们思想、算法或帮我们解决某个问题的语言和工具而已,idea 才是一个程序的灵魂。切不可把全部精力放到 Python 语言本身的学习上,而是要把主要精力放到自己的专业知识学习上,最终再用 Python 把自己的思想或算法准确地表达出来。本书从不同领域选取了一些有代表性的案例,同时还结合自己多年的项目开发和教学经验整理和设计了一些案例,希望能够起到抛砖引玉的作用。

内容组织与阅读建议

对于 Python 程序员来说,熟练运用优秀、成熟的扩展库可以快速实现业务逻辑和创意,而 Python 语言基础知识和基本数据结构的熟练掌握则是理解和运用其他扩展库的必要条件。并且,在实际开发中建议优先使用 Python 内置对象和标准库对象实现预定功能,这样可以获得更高的执行效率。本书前 7 章使用大量篇幅介绍 Python 编程基础知识,通过大量案例演示 Python 语言的精妙与强大。然后从第 8 章开始介绍大量标准库和扩展库在 GUI 编程、网络编程、数据库编程、大数据处理、多线程与多进程编程、系统运维、图形图像编程、科学计算可视化、密码学编程、移动终端编程等多个领域的应用。最后一章通过一个完整的系统演示了 Python 在实际系统开发中的应用。全书共 16 章,读者在熟练掌握前 7 章之后,可以结合自己的专业领域或兴趣爱好,在其他章节中有选择地进行阅读。

第 1 章 Python 基础。介绍如何选择 Python 版本和开发环境,Python 对象模型,数字、字符串等基本数据类型,运算符与表达式,常用内置函数,基本输入输出函数,扩展库管理与使用。

第 2 章 Python 序列。讲解序列常用方法和基本操作,列表基本操作与常用方法,切片操作,列表推导式,元组与生成器推导式,序列解包,字典、集合基本操作与常用方法,字典推导式与集合推导式。

第 3 章 程序控制结构与函数设计。讲解 Python 选择结构、for 循环与 while 循环、带 else 子句的循环结构,break 与 continue 语句,循环代码优化,函数定义与使用,关键参数、默认值参数、长度可变参数等不同参数类型,全局变量与局部变量,参数传递时的序列解包,return 语句,lambda 表达式以及 map()、reduce()、filter(),生成器与可调用对象。

第 4 章 面向对象程序设计。讲解类的定义与使用,self 与 cls 参数,类成员与实例成员,私有成员与公有成员,继承与派生,属性,特殊方法与运算符重载等内容,以及自定义类实现数组、矩阵、队列、栈、二叉树、有向图、集合等数据结构。

第 5 章 字符串与正则表达式。讲解字符串编码格式,字符串格式化、替换、分割、连接、查找、排版等基本操作,正则表达式语法、正则表达式对象、子模式与子模式扩展语法、match 对象,以及 Python 正则表达式模块 re 的应用。

第 6 章 文件与文件夹操作。讲解文件操作基本知识,Python 文件对象,文本文件读写操作、二进制文件读写与对象序列化、文件复制、移动、重命名、文件类型检测、文件完整性检查、压缩与解压缩、文件夹大小统计、文件夹增量备份、删除指定类型的文件，以及word、excel、zip、apk、rar等常见文件类型的操作。

第7章 异常处理结构、代码测试与调试。讲解Python异常类层次结构与自定义异常类，多种不同形式的异常处理结构，使用IDLE和pdb模块调试Python程序，Python单元测试相关知识。

第8章 数据库应用开发。介绍SQLite数据库及其相关概念，Connection对象、Cursor对象、Row对象，使用Python操作Access、MS SQL Server、MySQL等关系型数据库以及使用Python操作NoSQL数据库MongoDB。

第9章 网络应用开发。讲解计算机网络基础知识，TCP、UDP协议编程，网络嗅探器与端口扫描器设计，域名解析与网页爬虫设计原理，代理服务器与FTP软件原理与实现，使用Python编写CGI程序，使用Flask和django框架开发Web应用，以及使用C#与Python混合开发Web应用。

第10章 多线程与多进程。讲解Python标准库threading和multiprocessing在多线程编程与多进程编程中的应用，以及多线程与多进程之间的数据共享与同步控制。

第11章 大数据处理。介绍大数据处理框架MapReduce、Hadoop和Spark基本概念，重点介绍MapReduce和Spark应用。

第12章 图形编程与图像处理。讲解扩展库PyOpenGL在计算机图形学编程中的应用，扩展库pillow在图像编程中的应用。

第13章 数据分析与科学计算可视化。讲解扩展库numpy、scipy、matplotlib在科学计算与可视化领域的应用，以及标准库statistics与扩展库pandas在统计与分析、数据处理中的应用。

第14章 密码学编程。介绍恺撒密码、维吉尼亚密码等经典密码算法的Python实现，以pycrypto、rsa、hashlib等模块为主讲解安全哈希算法、对称密钥密码算法DES与AES以及非对称密钥密码算法RSA与DSA的应用。

第15章 tkinter编程精彩案例。讲解如何使用Python标准库tkinter进行GUI编程，通过大量实际案例演示基本组件的用法，包括用户登录界面设计、选择类组件应用、简单文本编辑器、画图程序设计与实现、电子时钟、简单动画、屏幕任意区域截图、音乐播放器、远程桌面监控程序等。

第16章 课堂教学管理系统设计与实现。通过一个综合案例来演示前面章节知识的应用，提供了学生名单和题库的导入、在线点名、在线提问、在线答疑、在线收作业、在线自测与考试、数据导出、防作弊与服务器自动发现、信息汇总、试卷生成等功能。

本书的最大特点是信息量大、知识点紧凑、案例丰富、注释量大、实用性强，把书中一些代码进行简单拼凑就可以满足实际工作中需要的很多功能。全书近200个涉及不同行业领域的实用案例和上千个代码片段并且配有大量注释以方便理解，没有插入多余的程序输出结果或软件安装截图，只保留了必要的代码运行结果或截图以供读者参考和对比，充分利用宝贵的篇幅来介绍和演示尽可能多的知识，绝对物超所值。本书作者具有16年程序设计教学经验，先后讲授过汇编语言、C/C++/C#、Java、PHP、Python等多门程设计语言,并且编写过大量的应用程序,其中有几套系统已投入使用多年并一直在使用。本书内容结合了作者多年教学与开发过程中积累的许多经验和案例,并巧妙地糅合进了相应的章节。

本书对 Python 内部工作原理进行了一定深度的剖析,书中 99%以上的案例均使用 Python 3.5.1 实现,这些代码同样也适用于 Python 3.4.x(除少数几个新特性之外,如矩阵运算符@)和最新版本 Python 3.5.2 以及马上就要正式面世的 Python 3.6.0,极个别案例使用 Python 2.7.11 实现(同样适用于其他版本 Python 2.7.x,包括最新的 Python 2.7.12),并适当介绍了 Python 代码优化、系统编程和安全编程的有关知识,可以满足不同层次读者的需要。另外,书中通过小提示、小技巧、注意、拓展知识等形式介绍了更多的内容,所以全部内容远比章节目录所显示的要多,需要认真阅读才能真正领会其中的奥妙。

配套资源

本书提供所有案例源代码,可以登录清华大学出版社网站(www.tup.com.cn)下载,或加入本书读者群(QQ 群号为 282819961)下载最新配套资源并与作者直接交流,作者微信号 Python_dfg 也随时期待您的反馈和交流,当然也欢迎关注微信公众号"Python 小屋"及时阅读作者写的最新案例代码。

本书适用读者

本书可以作为(但不限于):
- 本科、专科或研究生程序设计课程教材。
- Python 培训用书。
- 具有一定 Python 基础的读者进阶首选学习资料。
- 涉及 Python 开发的工程师、策划人员、科研人员和管理人员阅读书目。
- 打算利用业余时间学习一门快乐的程序设计语言并编写几个小程序来娱乐的读者首选学习资料。
- 少数对编程具有浓厚兴趣和天赋的中学生课外阅读资料。

感谢

首先感谢父母的养育之恩,在当年那么艰苦的条件下还坚决支持我读书,没有让我像其他同龄的孩子一样辍学。感谢姐姐、姐夫多年来对我的爱护以及在老家对父母的照顾,感谢善良的弟弟、弟媳在老家对父母的照顾,正是有了你们,远离家乡的我才能安心工作。当然,最应该感谢的是妻子和孩子对我这个技术狂人的理解,这些年来她们已经习惯了正在吃饭的我突然想起个思路然后就跑到计算机前面去写代码了,习惯了我每个周末和假期都在教研室看书或写代码而不陪她们,也习惯了周末的中午和晚上做好饭以后再打电话让我回家。为了表示对我的支持,她们还阅读了本书定稿前的版本并发现了几个错别字。

感谢每一位读者,感谢您在茫茫书海中选择了本书,衷心祝愿您能够从本书中受益,学到真正需要的知识!同时也期待每一位读者的热心反馈,随时欢迎您指出书中的不足!

本书的出版获2014年山东省普通高校应用型人才培养专业发展支持计划项目资助。我校专业共建合作伙伴——浪潮优派科技教育有限公司总裁邵长臣先生审阅了全书,并提出很多宝贵的意见,在此致以诚挚的谢意。本书在编写出版过程中也得到清华大学出版社的大力支持和帮助,在此表示衷心的感谢。

董付国于山东烟台

2016 年 7 月

目 录

CONTENTS

第1章 Python 基础 ··· 1
  1.1 Python 是一种什么样的语言 ··································· 1
  1.2 Python 开发环境 ··· 2
      1.2.1 百家争鸣的繁荣景象 ····································· 2
      1.2.2 IDLE 简单使用 ·· 6
  1.3 变量、运算符与表达式 ··· 9
      1.3.1 Python 变量与内置数据类型 ························· 9
      1.3.2 常用内置函数 ·· 15
      1.3.3 运算符与表达式 ·· 21
      1.3.4 人机对话基本接口 ······································· 25
  1.4 模块安装与使用 ·· 28
      1.4.1 安装 Python 扩展库 ···································· 28
      1.4.2 模块导入与使用 ·· 30
      1.4.3 编写自己的模块和包 ···································· 32

第2章 Python 序列 ··· 35
  2.1 列表与列表推导式 ·· 36
      2.1.1 列表创建与删除 ·· 36
      2.1.2 列表常用方法 ·· 38
      2.1.3 列表推导式 ··· 46
      2.1.4 切片 ··· 50
  2.2 元组与生成器推导式 ·· 54
      2.2.1 元组 ··· 54
      2.2.2 生成器推导式 ·· 55
  2.3 字典 ·· 57
      2.3.1 字典创建和元素添加、修改与删除 ················ 57
      2.3.2 访问字典对象的数据 ··································· 59
      2.3.3 案例精选 ··· 61
  2.4 集合 ·· 63

2.4.1　集合基础知识 ·· 63
　　2.4.2　集合操作与运算 ·· 64
　　2.4.3　案例精选 ·· 67
2.5　序列解包 ·· 69

第3章　程序控制结构与函数设计 ······································ 71

3.1　选择结构 ·· 71
　　3.1.1　条件表达式 ·· 71
　　3.1.2　选择结构的几种形式 ·· 73
　　3.1.3　案例精选 ·· 79
3.2　循环结构 ·· 82
　　3.2.1　for 循环与 while 循环的基本语法 ······························ 82
　　3.2.2　break 与 continue 语句 ·· 83
　　3.2.3　循环代码优化技巧 ·· 84
　　3.2.4　案例精选 ·· 86
3.3　函数设计与使用 ·· 90
　　3.3.1　基本语法 ·· 91
　　3.3.2　函数参数不得不说的几件事 ···································· 94
　　3.3.3　变量作用域 ·· 99
　　3.3.4　lambda 表达式 ·· 102
　　3.3.5　案例精选 ·· 104

第4章　面向对象程序设计 ·· 122

4.1　基础知识 ·· 122
　　4.1.1　类的定义与使用 ·· 122
　　4.1.2　私有成员与公有成员 ·· 123
　　4.1.3　数据成员 ·· 125
　　4.1.4　方法 ·· 126
　　4.1.5　属性 ·· 129
　　4.1.6　继承 ·· 131
　　4.1.7　特殊方法与运算符重载 ·· 134
4.2　案例精选 ·· 135
　　4.2.1　自定义数组 ·· 135
　　4.2.2　自定义矩阵 ·· 141
　　4.2.3　自定义队列 ·· 147
　　4.2.4　自定义栈 ·· 151
　　4.2.5　自定义二叉树 ·· 154
　　4.2.6　自定义有向图 ·· 157

4.2.7　自定义集合 ··· 158
第5章　字符串与正则表达式 ·· **165**
　5.1　字符串 ·· 165
　　　5.1.1　字符串格式化的两种形式 ··· 168
　　　5.1.2　字符串常用方法 ··· 171
　　　5.1.3　案例精选 ·· 186
　5.2　正则表达式 ··· 190
　　　5.2.1　正则表达式语法与子模式扩展语法 ···························· 190
　　　5.2.2　re模块方法与正则表达式对象 ··································· 193
　　　5.2.3　案例精选 ·· 199
第6章　文件与文件夹操作 ·· **206**
　6.1　文件对象常用方法与属性 ··· 207
　6.2　文本文件操作案例精选 ·· 209
　6.3　二进制文件操作案例精选 ··· 217
　　　6.3.1　使用pickle模块读写二进制文件 ································ 217
　　　6.3.2　使用struct模块读写二进制文件 ································· 219
　　　6.3.3　使用shelve模块操作二进制文件 ································ 220
　　　6.3.4　使用marshal模块操作二进制文件 ····························· 220
　6.4　文件与文件夹操作 ·· 221
　　　6.4.1　标准库os、os.path与shutil简介 ································ 221
　　　6.4.2　案例精选 ·· 227
第7章　异常处理结构、代码测试与调试 ··· **252**
　7.1　异常处理结构 ·· 252
　　　7.1.1　异常是什么 ·· 252
　　　7.1.2　Python内置异常类层次结构 ····································· 254
　　　7.1.3　常见异常处理结构形式 ·· 255
　7.2　代码测试 ·· 262
　　　7.2.1　doctest ·· 263
　　　7.2.2　单元测试 ·· 264
　7.3　代码调试 ·· 270
　　　7.3.1　使用IDLE调试 ··· 270
　　　7.3.2　使用pdb调试 ··· 273
阶段性寄语 ·· **278**

第 8 章 数据库应用开发 ... 279

8.1 使用 Python 操作 SQLite 数据库 ... 279
8.1.1 Connection 对象 ... 280
8.1.2 Cursor 对象 ... 281
8.1.3 Row 对象 ... 284

8.2 使用 Python 操作其他关系型数据库 ... 285
8.2.1 操作 Access 数据库 ... 285
8.2.2 操作 MS SQL Server 数据库 ... 286
8.2.3 操作 MySQL 数据库 ... 288

8.3 操作 MongoDB 数据库 ... 290

第 9 章 网络应用开发 ... 293

9.1 计算机网络基础知识 ... 293

9.2 Socket 编程 ... 295
9.2.1 UDP 编程 ... 296
9.2.2 TCP 编程 ... 298
9.2.3 网络嗅探器 ... 300
9.2.4 多进程端口扫描器 ... 302
9.2.5 代理服务器端口映射功能的实现 ... 305
9.2.6 自己编写 FTP 通信软件 ... 308

9.3 域名解析与网页爬虫 ... 313
9.3.1 网页内容读取与域名分析 ... 313
9.3.2 网页爬虫 ... 315
9.3.3 scrapy 框架 ... 316
9.3.4 BeautifulSoup4 ... 318

9.4 网站开发 ... 323
9.4.1 使用 IIS 运行 Python CGI 程序 ... 323
9.4.2 Python 在 ASP.NET 中的应用 ... 325
9.4.3 Flask 框架简单应用 ... 327
9.4.4 django 框架简单应用 ... 328

第 10 章 多线程与多进程 ... 333

10.1 多线程编程 ... 334
10.1.1 线程创建与管理 ... 336
10.1.2 线程同步技术 ... 339

10.2 多进程编程 ... 346
10.2.1 进程创建与管理 ... 347

		10.2.2 进程间数据交换	347
		10.2.3 进程同步技术	350

第 11 章 大数据处理 — 351

11.1 大数据简介 — 351
11.2 MapReduce 框架 — 352
11.3 Spark 应用开发 — 356

第 12 章 图形编程与图像处理 — 361

12.1 图形编程 — 361
    12.1.1 绘制三维图形 — 361
    12.1.2 绘制三次贝塞尔曲线 — 364
    12.1.3 纹理映射 — 365
    12.1.4 响应键盘事件 — 368
    12.1.5 光照模型 — 369
12.2 图像处理 — 372
    12.2.1 pillow 模块基本用法 — 372
    12.2.2 计算椭圆中心 — 375
    12.2.3 动态生成比例分配图 — 376
    12.2.4 生成验证码图片 — 377
    12.2.5 gif 动态图像分离与生成 — 379
    12.2.6 材质贴图 — 380
    12.2.7 图像融合 — 381
    12.2.8 棋盘纹理生成 — 383

第 13 章 数据分析与科学计算可视化 — 384

13.1 扩展库 numpy 简介 — 384
13.2 科学计算扩展库 scipy — 393
    13.2.1 数学、物理常用常数与单位模块 constants — 394
    13.2.2 特殊函数模块 special — 395
    13.2.3 信号处理模块 signal — 395
    13.2.4 图像处理模块 ndimage — 397
13.3 扩展库 pandas 简介 — 401
13.4 统计分析标准库 statistics 用法简介 — 405
13.5 matplotlib — 407
    13.5.1 绘制正弦曲线 — 408
    13.5.2 绘制散点图 — 408
    13.5.3 绘制饼状图 — 410

13.5.4 绘制带有中文标签和图例的图 ……………………………………… 410
13.5.5 绘制图例标签中带有公式的图 ……………………………………… 411
13.5.6 使用 pyplot 绘制,多个图形单独显示 …………………………… 412
13.5.7 绘制三维参数曲线 ………………………………………………… 413
13.5.8 绘制三维图形 ……………………………………………………… 414
13.5.9 使用指令绘制自定义图形 ………………………………………… 416
13.5.10 在 tkinter 中使用 matplotlib ……………………………………… 417
13.5.11 使用 matplotlib 提供的组件实现交互式图形显示 ……………… 419
13.5.12 根据实时数据动态更新图形 ……………………………………… 421
13.5.13 使用 Slider 组件调整曲线参数 …………………………………… 422

第 14 章 密码学编程 ……………………………………………………………… 425

14.1 经典密码算法 …………………………………………………………………… 425
14.1.1 恺撒密码算法 ……………………………………………………… 425
14.1.2 维吉尼亚密码 ……………………………………………………… 427
14.1.3 换位密码算法 ……………………………………………………… 428
14.2 安全哈希算法 …………………………………………………………………… 429
14.3 对称密钥密码算法 DES 和 AES ……………………………………………… 432
14.4 非对称密钥密码算法 RSA 与数字签名算法 DSA …………………………… 433
14.4.1 RSA ………………………………………………………………… 433
14.4.2 DSA ………………………………………………………………… 435

第 15 章 tkinter 编程精彩案例 ………………………………………………… 436

15.1 用户登录界面 …………………………………………………………………… 436
15.2 选择类组件应用 ………………………………………………………………… 438
15.3 简单文本编辑器 ………………………………………………………………… 441
15.4 简单画图程序 …………………………………………………………………… 445
15.5 电子时钟 ………………………………………………………………………… 449
15.6 简单动画 ………………………………………………………………………… 451
15.7 多窗口编程 ……………………………………………………………………… 454
15.8 屏幕任意区域截图 ……………………………………………………………… 456
15.9 音乐播放器 ……………………………………………………………………… 458
15.10 远程桌面监控系统 …………………………………………………………… 462

第 16 章 课堂教学管理系统设计与实现 …………………………………… 466

16.1 功能简介 ………………………………………………………………………… 466
16.1.1 教师端功能 ………………………………………………………… 466
16.1.2 学生端功能 ………………………………………………………… 467

16.2 数据库设计 ··· 468
16.3 系统总框架与通用功能设计 ·· 469
16.4 数据导入功能 ··· 473
    16.4.1 学生名单导入 ·· 473
    16.4.2 题库导入 ··· 473
16.5 点名与加分功能 ·· 475
    16.5.1 在线点名 ··· 475
    16.5.2 离线点名与加分 ·· 477
16.6 随机提问功能 ··· 480
16.7 在线收作业功能 ·· 482
    16.7.1 学生端 ·· 482
    16.7.2 教师端 ·· 483
16.8 在线自测与在线考试功能 ·· 485
    16.8.1 学生端 ·· 486
    16.8.2 教师端 ·· 488
16.9 信息查看功能 ··· 490
16.10 数据导出功能 ·· 492
16.11 其他辅助功能 ·· 493
    16.11.1 防作弊功能 ·· 493
    16.11.2 服务器自动发现功能 ·· 493
    16.11.3 Word 版试卷生成功能 ·· 494

结束语 ··· 496

附录 A 本书中例题清单 ·· 497

附录 B 本书中插图清单 ·· 503

附录 C 本书中表格清单 ·· 507

附录 D 本书中拓展知识摘要清单 ·· 508

参考文献 ··· 513

第 1 章 Python 基础

1.1 Python 是一种什么样的语言

小时不识月,呼作白玉盘。很多人习惯地(甚至还会配合着鄙夷的眼神和表情)说 Python 不过是一种脚本语言而已,实际上这种说法是非常不准确的,完全不能体现出 Python 的强大。严格来说,Python 是一门跨平台、开源、免费的解释型高级动态编程语言。除了解释执行,Python 还支持伪编译将源代码转换为字节码来优化程序提高运行速度和对源代码进行保密,并且支持使用 py2exe、pyinstaller、cx_Freeze 或其他类似工具将 Python 程序及其所有依赖库打包为扩展名为 exe 的可执行程序,从而可以脱离 Python 解释器环境和相关依赖库而在 Windows 平台上独立运行;Python 支持命令式编程、函数式编程,完全支持面向对象程序设计,语法简洁清晰,并且拥有大量的几乎支持所有领域应用开发的成熟扩展库;也有人喜欢把 Python 称为"胶水语言",因为它可以把多种不同语言编写的程序融合到一起实现无缝拼接,更好地发挥不同语言和工具的优势,满足不同应用领域的需求。

Python 官方网站同时发行并维护着 Python 2.x 和 Python 3.x 两个不同系列的版本,目前的最新版本分别是 Python 2.7.12、Python 3.5.2 和 Python 3.6.0a3。Python 2.x 和 Python 3.x 这两个系列的版本之间很多用法是不兼容的(让人欣慰的是,除了一些新特性、运算符和标准库对象之外,同一个系列的不同版本之间用法大多是一致的),除了输入输出方式有所不同,很多内置函数及标准库的内部实现和返回值类型也有较大的区别,不同 Python 版本的扩展库之间更是差别巨大。众多的 Python 及其扩展库版本让很多新手眼花缭乱,恐怕还吓跑了一些人,不同版本之间的不兼容也让不少人感到很苦恼,甚至痛苦。不过稍加了解和熟悉之后就会发现这并不是什么问题,实际上功能超级强大的 Python 自诞生不久就迅速得到了各行各业的人士的喜爱。

我从哪里来,要到哪里去,这是每个人都应该经常思考的问题,人生必须有个明确的、高大上的目标并且不停地为之而奋斗。同样,在选择 Python 的时候,一定要先考虑清楚自己学习 Python 的目的是什么,打算做哪方面的开发,有哪些扩展库可用,这些扩展库最高支持哪个版本的 Python。这些问题全部确定以后,再做出自己的选择,这样才能事半功倍,而不至于把太多时间浪费在 Python 以及各种扩展库的反复安装和卸载上。另外,当较新的 Python 版本推出之后,不要急于赶时髦,看到别人用高版本的 Python 也不用自卑,而是应该在确定自己所必须使用的扩展库也推出了与之匹配的新版本之后再一起进行更新。

目前来看,Python 3.x 的设计理念更加人性化,全面普及和应用已经是大势所趋,越来越多的扩展库以最快的速度推出了与最新 Python 版本相适应的版本。如果暂时还没想到要做什么行业领域的应用开发,或者仅仅是为了尝试一种新的、好玩的语言,那么请毫不犹豫地选择 Python 3.x 系列的最高版本。

拓展知识:Python 名字的来源:虽然在英语中 Python 是大蟒蛇的意思,但是 Python 语言却和大蟒蛇没有任何关系。Python 语言的名字来自于一个著名的电视剧(Monty Python's Flying Circus),Python 之父 Guido van Rossum 是这个电视剧的狂热爱好者,故此把他发明的语言命名为 Python。

小提示:由于历史原因,短期内还无法完全放弃 Python 2.x,不过预计在 2020 年将会退出历史舞台,Python 3.x 的全面普及是个大趋势。

小提示:除了在 IDLE 主界面上可以直接看到当前使用的 Python 版本号(见图 1-1),还可以使用下面的方法来查看当前 Python 的版本。

```
>>>import platform                        #导入Python模块platform
>>>platform.python_version()              #调用模块中的函数
'3.5.1'
>>>import sys                             #导入Python模块sys
>>>sys.version
'3.5.1(v3.5.1:37a07cee5969, Dec  6 2015, 01:54:25) [MSC v.1900 64 bit(AMD64)]'
>>>sys.winver
'3.5'
>>>sys.version_info
sys.version_info(major=3, minor=5, micro=1, releaselevel='final', serial=0)
>>>sys.executable                         #查看Python主程序文件
'C:\\Python 3.5\\pythonw.exe'
```

小提示:另外,sys 模块还提供了大量与系统编程有关的接口,后面章节将根据需要进行展开介绍。platform 模块还提供了一些查看操作系统信息的函数,例如下面的代码:

```
>>>platform.win32_ver()
('7', '6.1.7601', 'SP1', 'Multiprocessor Free')
>>>platform.version()
'6.1.7601'
>>>platform.machine()
'AMD64'
>>>platform.python_compiler()
'MSC v.1900 64 bit(AMD64)'
```

1.2 Python 开发环境

1.2.1 百家争鸣的繁荣景象

工欲善其事,必先利其器。学习编程也是同样的道理,对开发环境的熟悉应该是学习一门编程语言的第一步。

IDLE 是 Python 的官方标准开发环境,从官方网站 www.python.org 下载并安装合适的 Python 版本之后,同时就安装了 IDLE。相对于其他 Python 开发环境而言,IDLE 相对来说确实有点简陋,但已经具备了 Python 应用开发的几乎所有功能(例如,语法智能提示、使用不同颜色显示不同类型的内容等),并且也不需要过于复杂的配置,因此得到很多人的喜爱。Python 3.5.1 IDLE 的界面如图 1-1 所示。

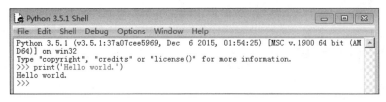

图 1-1　Python 3.5.1 IDLE 的界面

是的,你没有看错,输出了 Hello world,这是我们用 Python 编写的第一段代码,意味着我们进入到了 Python 语言的世界,先向新世界的朋友们打个招呼。据说某程序员退休之后喜欢上了书法,买回来笔墨纸砚后沉思良久,然后在纸上重重地写下了 Hello world,宣告自己正式进军书法界。

除了默认安装的 IDLE,还有大量的其他 IDE 开发环境,例如 wingIDE、PyCharm、PythonWin、Eclipse、Spyder、IPython、Komodo 等。不要总去问别人到底哪个开发环境好,因为你得到的答案取决于你问的是谁。每个 IDE 都有不同的风格,也分别得到了不同开发人员的喜爱,但万变不离其宗,严格来说那些开发环境都是对 Python 解释器 python.exe 的封装,核心是完全一样的,只是加了一个"外挂"而已,使用起来更加方便,减少了出错率,尤其是拼写错误,而这恰恰是很多程序员最容易犯的一个错误。wingIDE、PyCharm 和 Eclipse＋PyDev 等 Python 开发环境的运行界面分别如图 1-2～图 1-4 所示。

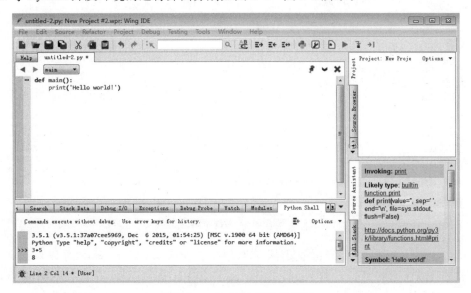

图 1-2　wingIDE 的运行界面

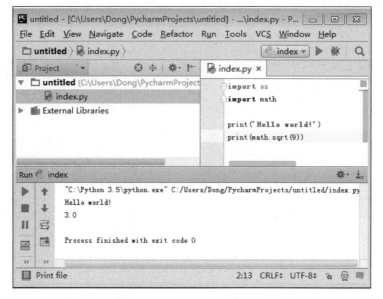

图 1-3　PyCharm 的运行界面

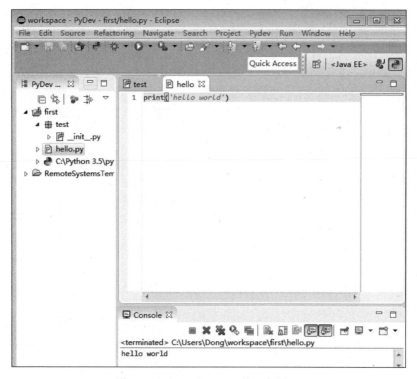

图 1-4　Eclipse＋PyDev 的运行界面

如果暂时什么都不想安装，只是简单地想试试 Python 好不好玩，可以试试 Python 官方网站（www.python.org）提供的 Interactive Shell，登录 Python 官方网站之后，单击图 1-5 中方框内的那个图标（一个大于号和一个下画线），然后稍等片刻即可进入图 1-6

所示的界面。

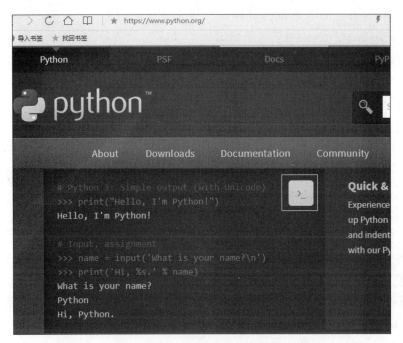

图 1-5　Python 官方网站提供的 Interactive Shell 入口

图 1-6　Python 官方网站提供的 Interactive Shell 界面

有不少太极拳爱好者刚接触太极拳时面对众多的太极拳"版本"也是一样的困惑,杨式、陈式、孙式、吴式、武式、赵堡等等,杨式太极拳有 85 式、103 式、老六路等等,陈式又有洪传陈式、混元太极等等。到底该学哪个呢?正所谓万变不离其宗,不管选择哪一个,符合太极拳理才是最根本的。学好的话,哪一个都可以成为拳中之王;如果只学个花架子而不得拳术精髓,学再好的拳也是徒有虚表。与其在众多的 Python 开发环境面前纠结选择哪一个,不如马上行动起来,尽快地确定一个,然后熟练掌握其用法。

拓展知识:如果有读者想尝试一下在安卓手机上编写 Python 程序,可以安装支持 Python 3.x 的 QPython3 或者支持 Python 2.x 的 QPython,关于 SL4A 和安卓类库调用的相关知识可以查阅相关资料。

1.2.2 IDLE 简单使用

本书采用标准的 IDLE 作为开发环境来演示 Python 的强大功能,几乎所有代码都可以直接拿到其他开发环境中运行,不需要任何修改。有时候可能需要同时安装多个不同的版本,例如,同时安装 Python 2.7.11 和 Python 3.5.1,并根据不同的开发需求在两个版本之间进行切换。多版本并存一般不影响在 IDLE 环境中直接运行程序,只需要启动相应版本的 IDLE 即可。在命令提示符环境中运行 Python 程序时,在调用 Python 主程序时指定其完整路径,或者通过修改系统 Path 变量来实现不同版本之间的切换。在 Win 7 系统下修改系统 Path 变量的步骤如下:单击"开始"菜单,右击"计算机"并选择"属性",单击"高级系统设置"切换至"高级"选项卡,再单击"环境变量"按钮,然后修改系统 Path 变量中 Python 安装路径,如图 1-7 所示。

图 1-7　Win 7 环境中系统 Path 变量的修改方法

小提示:有些 Python 程序需要在命令提示符环境中运行。在 Win 7 系统中,可以依次单击"开始"→"所有程序"→"附件"→"命令提示符"进入命令提示符环境,或者单击"开始"菜单然后在"搜索程序和文件"文本框内输入 cmd 回车确定直接进入命令提示符环境,最后再通过 cd 命令切换至相应的文件夹。也可以参考 1.4.1 节的内容直接进入命令提示符环境并切换至相应的文件夹。

如果能够熟练使用开发环境提供的一些快捷键,将会大幅度提高开发效率。在 IDLE 环境中,除了撤销(Ctrl+Z)、全选(Ctrl+A)、复制(Ctrl+C)、粘贴(Ctrl+V)、剪切(Ctrl+X)等常规快捷键之外,其他比较常用的快捷键如表 1-1 所示。

表 1-1　IDLE 中的常用快捷键

快 捷 键	功 能 说 明
Tab	补全单词,列出全部可选单词供选择
Alt+P	浏览历史命令(上一条)
Alt+N	浏览历史命令(下一条)
Ctrl+F6	重启 Shell,之前定义的对象和导入的模块全部失效
F1	打开 Python 帮助文档
Alt+/	自动补全前面曾经出现过的单词,如果之前有多个单词具有相同前缀,则在多个单词中循环切换
Ctrl+]	缩进代码块
Ctrl+[	取消代码块缩进
Alt+3	注释代码块
Alt+4	取消代码块注释

启动 IDLE 之后默认为交互模式,直接在 Python 提示符">>>"后面输入相应的命令并回车执行即可,如果执行顺利的话,马上就可以看到执行结果,否则会提示错误或者抛出异常。

```
>>>3+5                              #井号之后的内容是注释,不会被执行
8
>>>import math                      #导入 Python 标准库 math
>>>math.sqrt(9)                     #使用标准库函数计算平方根
3.0
>>>9 ** 0.5                         #使用运算符**计算平方根
3.0
>>>3 * (2+6)
24
>>>2 / 0                            #除 0 错误,抛出异常,详见第 7 章
Traceback(most recent call last):
  File "<pyshell#18>", line 1, in <module>
    2/0
ZeroDivisionError: integer division or modulo by zero
>>>x='Hello world                   #语法错误,字符串结尾缺少一个单引号
```

```
SyntaxError: EOL while scanning string literal
```

小提示：Python非常追求代码的可读性。很明显，如果代码密密麻麻地挤成一团肯定不可能有好的可读性，所以好的Python代码在形式上一般是比较松散的（形散神不散）。一般来说，建议在运算符两侧和逗号后面增加一个空格，在不同功能的代码块之间增加一个空行，这样看起来会更舒服一些。

交互模式一般用来实现一些简单的业务逻辑，或者验证某些功能。复杂的业务逻辑更多的是通过编写Python程序来实现，同时也方便代码的不断完善和重复利用。在IDLE界面中使用菜单File→New File创建一个程序文件，输入代码并保存为文件（务必要保证扩展名为.py，如果是GUI程序可以保存扩展名为pyw的文件。如果保存为其他扩展名的文件，一般并不影响在IDLE中直接运行，但是在"命令提示符"环境中运行时需要显式调用Python主程序，并且在资源管理器中直接双击该文件时可能会无法关联Python主程序从而导致无法运行）后，使用菜单Run→Check Module来检查程序中是否存在语法错误，或者使用菜单Run→Run Module运行程序，程序运行结果将直接显示在IDLE交互界面上。除此之外，也可以通过在资源管理器中双击扩展名为.py或.pyc的Python程序文件直接运行；在有些情况下，可能还需要在命令提示符环境中运行Python程序文件。在"开始"菜单的"附件"中单击"命令提示符"，然后执行Python程序。例如，假设有程序HelloWorld.py内容如下：

```
def main():
    print('Hello world')
main()
```

在IDLE环境中运行该程序结果如图1-8所示。

图1-8　在IDLE中运行程序

在命令提示符环境中运行该程序的方法与结果如图1-9所示，该图中演示了两种执行Python程序的方法，虽然第二种方法看上去更加简单，但是请尽量使用第一种方法来运行Python程序，否则可能会影响某些程序的正确运行。

图1-9　在命令提示符中运行程序

👉**注**：全书涉及命令提示符环境的插图都尽量采用白底黑字而不是默认的黑底白字，其他插图也尽量采用浅色背景，这样可以减少印刷时的用墨量。

👉**小技巧**：为提高代码运行速度，同时也对 Python 源代码进行保密，可以在命令提示符环境中使用"python -OO -m py_compile file.py"将 Python 程序 file.py 伪编译成为.pyc 文件，选项-OO 表示优化编译。

👉**拓展知识**：自定义 IDLE 清屏快捷键。有不少读者问我 Python IDLE 有没有像命令提示符环境中 cls 那样的清屏命令，能不能一下子删除交互模式中所有已执行的命令和输出结果，毕竟每次都关掉再重新打开 IDLE 确实有点啰唆，显得也没有技术含量。IDLE 本身并没有提供清屏命令，但是可以通过扩展来实现。可以从网上下载 ClearWindow.py(配套资源里已经提供了这个文件)并放到 Python 安装路径中的 lib\idlelib 文件夹中，然后用记事本打开文件 lib\idlelib\config-extensions.def，在最后添加以下几行：

```
[ClearWindow]
enable=1
enable_editor=0
enable_shell=1
[ClearWindow_cfgBindings]
clear-window=<Control-Key-;>
```

保存 lib\idlelib\config-extensions.def 文件后重启一下 IDLE，会发现菜单 Options 下面多了一个可以清屏的菜单项，如图 1-10 所示。另外，也可以随时使用"Ctrl＋;"快捷键实现清屏。当然，这个快捷键是可以任意修改的，修改的地方就在上面配置代码最后一行，看到了吗？那就试试吧，把它改成自己喜欢的快捷键。

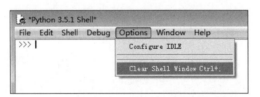

图 1-10　配置 IDLE 并增加清屏菜单和快捷键

1.3　变量、运算符与表达式

1.3.1　Python 变量与内置数据类型

对象是 Python 语言中最基本的概念之一，在 Python 中的一切都是对象，常用的内置对象如表 1-2 所示。除此之外，还有大量的标准库对象和扩展库对象，标准库是 Python 默认安装的，但需要导入之后才能使用其中的对象，扩展库对象则需要首先安装扩展库然后再导入并使用其中的对象。在 Python 中可以创建任意类型的变量，一般情况下与对象的概念不作严格的区分。

表 1-2 Python 内置对象

对象类型	示　例	简　要　说　明
数字	1234, 3.14, 1.3e5, 3+4j	数字大小没有限制,且支持复数及其运算
字符串	'swfu', "I'm student", '''Python '''	使用单引号、双引号、三引号作为界定符
列表	[1, 2, 3],['a', 'b', ['c', 2]]	所有元素放在一对方括号中,元素之间使用逗号分隔
字典	{1:'food' ,2:'taste', 3:'import'}	所有元素放在一对大括号中,元素之间使用逗号分隔,元素形式为"键:值"
元组	(2, −5, 6)	所有元素放在一对圆括号中,元素之间使用逗号分隔
文件	f=open('data.dat', 'rb')	open 是 Python 内置函数,使用指定的模式打开文件
集合	set('abc'), {'a', 'b', 'c'}	所有元素放在一对大括号中,元素之间使用逗号分隔,元素不允许重复
布尔型	True, False	
空类型	None	
编程单元	函数(使用 def 定义) 类(使用 class 定义)	类和函数都属于可调用对象

在 Python 中,不需要事先声明变量名及其类型,直接赋值即可创建各种类型的对象变量,并且变量的类型是可以随时发生改变的。例如,语句

```
>>>x=3
```

创建了整型变量 x,并赋值为 3,语句

```
>>>x='Hello world.'
```

创建了字符串变量 x,并赋值为'Hello world.',语句

```
>>>x=[1, 2, 3]
```

创建了列表对象 x,并赋值为[1, 2, 3]。这一点同样适用于元组、字典、集合以及其他 Python 任意类型的对象,以及自定义类型的对象。

虽然不需要在使用之前显式地声明变量及其类型,但是 Python 仍属于强类型编程语言,Python 解释器会根据赋值或运算来自动推断变量类型。每种类型的对象支持的运算也不完全一样,因此在使用变量时需要程序员自己确定所进行的运算是否合适,以免出现异常或者意料之外的结果。另外,Python 还是一种动态类型语言,也就是说,变量的类型是可以随时变化的,下面的代码演示了 Python 变量类型的变化。

```
>>>x=3
>>>print(type(x))                    #内置函数 type()用来返回变量类型
<class 'int'>
>>>x='Hello world.'
>>>print(type(x))
```

```
<class 'str'>
>>>x=[1,2,3]
>>>print(type(x))
<class 'list'>
>>>isinstance(3, int)              #内置函数isinstance()用来测试对象是否为指定类型的实例
True
>>>isinstance('Hello world', str)
True
```

代码中首先创建了整型变量x,然后又分别创建了字符串和列表类型的变量x。当创建了字符串类型的变量x之后,之前创建的整型变量x自动失效;创建列表对象x之后,之前创建的字符串变量x自动失效。可以将该模型理解为"状态机",除非显式修改变量类型或删除变量,否则变量将一直保持之前的类型。

在大多数情况下,如果变量出现在赋值运算符或复合赋值运算符(如+=、*=等)的左边则表示创建变量或修改变量的值,否则表示引用该变量的值,这一点同样适用于使用下标来访问列表、字典等可变序列以及其他自定义对象中元素的情况。例如:

```
>>>x=3                            #创建整型变量
>>>print(x**2)                    #访问变量的值
9
>>>x +=6                          #修改变量的值
>>>print(x)                       #读取变量值并输出显示
9

>>>x=[1,2,3]                      #创建列表对象
>>>print(x)
[1, 2, 3]
>>>x[1]=5                         #修改列表元素值
>>>print(x)                       #输出显示整个列表
[1, 5, 3]
>>>print(x[2])                    #输出显示列表指定元素
3
```

小提示:在Python中可以使用变量表示任意大的数字,不用担心范围的问题,但是对于浮点数的计算由于精确度的问题偶尔可能会出现略显奇葩的结果。例如:

```
>>>9999 ** 99                     #这里**是幂乘运算符
99014835352672348760226312475328262557055952889579105732432652912179483789405351346442217682691643393258692438667776624403200162375682140043297505120882020498009873555270384136230466997051069124380021820284037432937880069492030979195418511779843432959121215910629869993866990806757337472433120894242554489391091007320504903165678922088956073296292622630586570659359491789627675639684851490098999
>>>0.3+0.2                        #实数相加
0.5
```

```
>>>0.3+0.3
0.6
>>>0.4-0.1                              #实数相减,结果稍微有点偏差
0.30000000000000004
>>>0.4-0.1==0.3
False
>>>0.9-0.5
0.4
```

另外,Python内置支持复数运算,例如:

```
>>>x=3+4j                               #使用j或J表示复数虚部
>>>y=5+6j
>>>x+y                                  #复数之间的加、减、乘、除
(8+10j)
>>>x-y
(-2-2j)
>>>x * y
(-9+38j)
>>>x / y
(0.6393442622950819+0.03278688524590165j)
>>>abs(x)                               #复数的模
5.0
>>>x.imag                               #虚部
4.0
>>>x.real                               #实部
3.0
>>>x.conjugate()                        #共轭复数
(3-4j)
```

拓展知识:Python标准库fractions中的Fraction对象支持分数运算。

```
>>>from fractions import Fraction
>>>x=Fraction(3, 5)                     #创建分数
>>>y=Fraction(3, 7)
>>>x
Fraction(3, 5)
>>>x.numerator                          #分子
3
>>>x.denominator                        #分母
5
>>>x+y                                  #分数之间的四则运算,支持通分
Fraction(36, 35)
>>>x * y
Fraction(9, 35)
```

```
>>>x / y
Fraction(7, 5)
>>>x-y
Fraction(6, 35)
>>>x * 2                                    #分数与数字之间的运算
Fraction(6, 5)
```

字符串和元组属于不可变序列,不能通过下标的方式来修改其中的元素值,例如:

```
>>>x = (1,2,3)
>>>print(x)
(1, 2, 3)
>>>x[1]=5                                   #元组是不可变序列,不支持元素值的修改
Traceback(most recent call last):
  File "<pyshell#7>", line 1, in <module>
    x[1]=5
TypeError: 'tuple' object does not support item assignment
```

在 Python 中,允许多个变量指向同一个值,例如:

```
>>>x=3
>>>id(x)
1786684560
>>>y=x                                      #现在 y 和 x 是同一个对象
>>>id(y)                                    #所以内存地址是一样的
1786684560
```

然而,继续上面的示例代码,当为其中一个变量修改值以后,其内存地址将会变化,但这并不影响另一个变量,例如:

```
>>>x +=6
>>>id(x)                                    #变量 x 已经不再是之前的 x
1786684752
>>>y
3
>>>id(y)                                    #变量 y 的地址和值都没有变
1786684560
```

在这段代码中,内置函数 id()用来返回变量所指值的内存地址。可以看出,在 Python 中修改变量值的操作,并不是直接修改变量的值,而是修改了变量指向的内存地址(引用)。这是因为 Python 解释器首先读取变量 x 原来的值,然后将其加 6,并将结果存放于内存中,最后将变量 x 指向该结果的内存空间,如图 1-11 所示。

Python 采用的是基于值的内存管理方式,如果为不同变量赋值为相同值,这个值在内存中只有一份,多个变量指向同一块内存地址,前面的几段代码也说明了这个特点。再例如:

```
>>>tempList=[1, 1, 1]
```

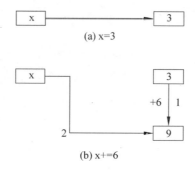

图 1-11 Python 内存管理模式

```
>>>id(tempList[0])==id(tempList[1])==id(tempList[2])    #内存地址一样
True
```

当一个变量不再使用时,可以使用 del 命令将其删除。Python 具有自动内存管理功能,对于没有任何变量指向的值,Python 自动将其删除。Python 会跟踪所有的值,并自动删除不再有变量指向的值。

最后,在定义变量名的时候,需要注意以下问题。

(1) 变量名必须以字母(以小写字母居多,但是也可以用大写字母)或下画线开头,但以下画线开头的变量在 Python 中有特殊含义,参考第 4 章内容。

(2) 变量名中不能有空格或标点符号(括号、引号、逗号、斜线、反斜线、冒号、句号、问号等)。

(3) 不能使用关键字作为变量名,可以导入 keyword 模块后使用 print(keyword.kwlist)查看 Python 的所有关键字。要注意的是,随着 Python 版本的变化,关键字列表可能也会有所变化。

(4) 不建议使用系统内置的模块名、类型名或函数名以及已导入的模块名及其成员名作为变量名,这会改变其类型和含义,甚至会导致其他代码无法正常执行。可以通过 dir(__builtins__)查看所有内置对象名称。

(5) 变量名对英文字母的大小写敏感,例如 student 和 Student 是不同的变量。

小提示:Python 变量不直接存储值,而是存储对象的引用,正是这个特点使得 Python 变量类型可以动态改变,随时可以指向另一个完全不同类型的对象。为变量赋值时,首先在内存中寻找一块合适的区域并把值存于其中,然后把这个内存地址赋值给变量。

拓展知识:Python 字符串对象提供了一个方法 isidentifier()可以用来判断指定字符串是否可以作为变量名、函数名、类名等标识符。例如:

```
>>>'abc'.isidentifier()                #abc 可以作为 Python 变量名
True
>>>'3abc'.isidentifier()               #变量名不能以数字开头
False
>>>'_3abc'.isidentifier()
```

```
True
>>>'__3abc'.isidentifier()
True
>>>',__3abc'.isidentifier()                #变量名不能以标点符号开头
False
>>>'__3abc('.isidentifier()                #变量名中也不能含有标点符号
False
>>>'a,bc'.isidentifier()
False
```

1.3.2 常用内置函数

内置函数(Built-In Functions,BIF)不需要导入任何模块即可直接使用,执行下面的命令可以列出所有内置函数和内置对象:

```
>>>dir(__builtins__)
```

可以使用 help(函数名)可以查看某个函数的用法,不需要导入模块就可以直接使用 help(模块名)查看该模块的帮助文档,例如 help('math')。常用的内置函数及其功能简要说明如表 1-3 所示,其中方括号内的参数可以省略。

表 1-3　Python 常用内置函数

函　　数	功能简要说明
abs(x)	返回数字 x 的绝对值或复数 x 的模
all(iterable)	如果对于可迭代对象 iterable 中所有元素 x 都有 bool(x)为 True,则返回 True。对于空的可迭代对象也返回 True
any(iterable)	只要可迭代对象 iterable 中存在元素 x 使得 bool(x)为 True,则返回 True。对于空的可迭代对象,返回 False
bin(x)	把数字 x 转换为二进制串
bool(x)	返回与 x 等价的布尔值 True 或 False
callable(object)	测试对象 object 是否可调用。类和函数是可调用的,包含__call__()方法的类的对象也是可调用的
compile()	用于把 Python 代码编译成可被 exec()或 eval()函数执行的代码对象
chr(x)	返回 Unicode 编码为 x 的字符
dir(obj)	返回指定对象 obj 或模块 obj 的成员列表
eval(s[, globals[, locals]])	计算并返回字符串 s 中表达式的值
exec(x)	执行代码或代码对象 x
filter(func, seq)	返回 filter 对象,其中包含序列 seq 中使得单参数函数 func 返回值为 True 的那些元素,如果函数 func 为 None 则返回那些值等价于 True 的元素

续表

函　　数	功能简要说明
float(x)	把数字或字符串 x 转换为浮点数并返回
hasattr(obj, name)	测试对象 obj 是否具有成员 name
hash(x)	返回对象 x 的哈希值,如果 x 不可哈希则抛出异常
help(obj)	返回对象 obj 的帮助信息
hex(x)	把数字 x 转换为十六进制串
id(obj)	返回对象 obj 的标识(内存地址)
input([提示内容字符串])	接收键盘输入的内容,返回字符串
int(x[, d])	返回数字 x 的整数部分,或把 d 进制的字符串 x 转换为十进制并返回,d 默认为十进制
isinstance(object, class-or-type-or-tuple)	测试对象 object 是否属于指定类型(如果有多个类型的话需要放到元组中)的实例
len(obj)	返回对象 obj 包含的元素个数,适用于列表、元组、集合、字典、字符串以及 range 对象和其他可迭代类型的对象
list([x])、set([x])、tuple([x])、dict([x])	把对象 x 转换为列表、集合、元组或字典并返回,或生成空列表、空集合、空元组、空字典
map(func, seq)	将函数 func 映射至序列 seq 中每个元素,返回包含函数值的 map 对象
max(x)、min(x)	返回序列 x 中的最大值、最小值,要求序列 x 中的所有元素之间可比较大小
next(x)	返回可迭代对象 x 中的下一个元素
sum(x)	返回序列 x 中所有元素之和,要求序列 x 中所有元素必须为数字
oct(x)	把数字 x 转换为八进制串
open(name[, mode])	以指定模式 mode 打开文件 name 并返回文件对象
ord(x)	返回 1 个字符 x 的 Unicode 编码
pow(x, y)	返回 x 的 y 次方,等价于 x**y
print(value, …, sep=' ', end='\n', file=sys.stdout, flush=False)	基本输出函数
range([start,] end [, step])	返回 range 对象,其中包含[start,end)区间内以 step 为步长的整数
reduce(func, seq)	将双参数的函数 func 以迭代的方式从左到右依次应用至序列 seq 中每个元素,最终返回单个值作为结果。在 Python 2.x 中该函数为内置函数,在 Python 3.x 中需要从 functools 中导入 reduce 函数再使用
reversed(seq)	返回 seq(可以是列表、元组、字符串、range 以及其他可迭代对象)中所有元素逆序后的迭代器对象
round(x [,小数位数])	对 x 进行四舍五入,若不指定小数位数,则返回整数

续表

函　数	功能简要说明
str(obj)	把对象 obj 直接转换为字符串
sorted(iterable，key＝None，reverse＝False)	返回排序后的列表,其中 iterable 表示要排序的序列或迭代对象,key 用来指定排序规则或依据,reverse 用来指定升序或降序。该函数不改变 iterable 内任何元素的顺序
type(obj)	返回对象 obj 的类型
zip(seq1 [，seq2 […]])	返回 zip 对象,其中元素为(seq1[i]，seq2[i]，…)形式的元组

内置函数数量众多且功能强大,很难一下子全部解释清楚,下面先简单介绍其中几个,后面的章节中将根据内容组织的需要逐步展开并演示更多函数更加巧妙的用法。

小提示:①可以通过内置函数 help()查看某个函数的使用帮助;②编写程序时应优先考虑使用内置函数,因为内置函数不仅成熟、稳定,而且速度相对较快;③可以导入 sys 模块后使用 print(sys.builtin_module_names)查看 Python 所有内置模块名称;④可以使用 help('modules')查看本机所有可用模块的名称。

拓展阅读:除了可以在交互模式下使用 help()函数直接查看某个对象的使用帮助,例如 help(id)可以查看内置函数 id()的帮助,还可以在交互模式中直接执行 help()函数进入 Python 的内置帮助系统,如图 1-12 所示。如果想退出这个帮助系统,可以执行 quit 命令。

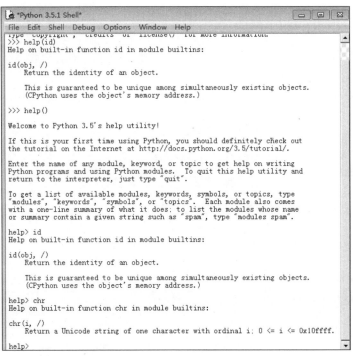

图 1-12　Python 内置帮助系统

内置函数 bin()、oct()、int()、hex()用来将数字转换为二进制、八进制、十进制和十六进制形式。

```
>>>bin(555)                              #把数字转换为二进制串
'0b1000101011'
>>>oct(555)                              #转换为八进制串
'0o1053'
>>>hex(555)                              #转换为十六进制串
'0x22b'
>>>int(_, 16)                            #把十六进制数转换为十进制数
555
>>>int(bin(54321), 2)                    #二进制与十进制之间的转换
54321
```

ord()和chr()是一对功能相反的函数，ord()用来返回单个字符的Unicode码，而chr()则用来返回Unicode编码对应的字符，str()则直接将其任意类型参数转换为字符串。下面的代码演示了这几个函数的用法：

```
>>>ord('a')                              #查看指定字符的Unicode编码
97
>>>chr(65)
'A'
>>>chr(ord('A')+1)
'B'
>>>chr(ord("3")+1)
'4'
>>>chr(ord('国')+1)                       #支持中文
'图'
>>>ord('董')                              #这个用法仅适用于Python 3.x
33891
>>>ord('付')
20184
>>>ord('国')
22269
>>>''.join(map(chr,(33891, 20184, 22269)))
'董付国'
>>>str(1234)                             #直接变成字符串
'1234'
>>>str([1,2,3])
'[1, 2, 3]'
>>>str((1,2,3))
'(1, 2, 3)'
>>>str({1,2,3})
'{1, 2, 3}'
```

max()、min()、sum()这 3 个内置函数分别用于计算列表、元组或其他可迭代对象中所有元素最大值、最小值以及所有元素之和,sum()只支持数值型元素的序列或可迭代对象,max()和 min()则要求序列或可迭代对象中的元素之间可比较大小。例如下面的示例代码,首先使用列表推导式生成包含 10 个随机数的列表,然后分别计算该列表的最大值、最小值和所有元素之和。列表推导式的介绍详见 2.1.3 节。

```
>>>from random import randint
>>>a=[randint(1,100)for i in range(10)]          #包含 10 个[1,100]之间随机数的列表
>>>print(max(a),min(a),sum(a))                   #最大值、最小值、所有元素之和
```

很显然,如果需要计算该列表中所有元素的平均值,可以直接使用下面的方法:

```
>>>sum(a)/len(a)
```

小技巧:max()函数还有个 key 参数,可以指定比较大小的依据,例如,表达式 max(['2', '111'], key=len)的值为'111',即返回长度最大的字符串。

对于初学者而言,也许 dir()和 help()这两个内置函数是最有用的,使用 dir()函数可以查看指定模块中包含的所有成员或者指定对象类型所支持的操作,而 help()函数则返回指定模块或函数的说明文档,这对于了解和学习新的模块与知识是非常重要的,能够熟练使用这两个函数也是学习能力的重要体现。例如,下面的代码首先导入数学模块 math,然后查看该模块的常量和函数,并查看指定函数的使用帮助:

```
>>>import math
>>>dir(math)                                     #查看模块中可用对象
['__doc__', '__loader__', '__name__', '__package__', '__spec__', 'acos', '
acosh', 'asin', 'asinh', 'atan', 'atan2', 'atanh', 'ceil', 'copysign', 'cos',
'cosh', 'degrees', 'e', 'erf', 'erfc', 'exp', 'expm1', 'fabs', 'factorial', 'floor',
'fmod', 'frexp', 'fsum', 'gamma', 'hypot', 'isfinite', 'isinf', 'isnan', 'ldexp',
'lgamma', 'log', 'log10', 'log1p', 'log2', 'modf', 'pi', 'pow', 'radians', 'sin', '
sinh', 'sqrt', 'tan', 'tanh', 'trunc']
>>>help(math.sqrt)                               #查看指定方法的使用帮助
Help on built-in function sqrt in module math:
sqrt(…)
    sqrt(x)
    Return the square root of x.
>>>help(math.sin)                                #查看指定方法的使用帮助
Help on built-in function sin in module math:
sin(…)
    sin(x)
Return the sine of x(measured in radians).
>>>dir(3+4j)                                     #查看复数类型成员
['__abs__', '__add__', '__class__', '__coerce__', '__delattr__', '__div__',
'__divmod__', '__doc__', '__eq__', '__float__', '__floordiv__', '__format__',
'__ge__', '__getattribute__', '__getnewargs__', '__gt__', '__hash__', '__init
```

__', '__int__', '__le__', '__long__', '__lt__', '__mod__', '__mul__', '__ne__', '__neg__', '__new__', '__nonzero__', '__pos__', '__pow__', '__radd__', '__rdiv__', '__rdivmod__', '__reduce__', '__reduce_ex__', '__repr__', '__rfloordiv__', '__rmod__', '__rmul__', '__rpow__', '__rsub__', '__rtruediv__', '__setattr__', '__sizeof__', '__str__', '__sub__', '__subclasshook__', '__truediv__', 'conjugate', 'imag', 'real']

```
>>>dir('')                                            #查看字符串类型成员
```
['__add__', '__class__', '__contains__', '__delattr__', '__doc__', '__eq__', '__format__', '__ge__', '__getattribute__', '__getitem__', '__getnewargs__', '__getslice__', '__gt__', '__hash__', '__init__', '__le__', '__len__', '__lt__', '__mod__', '__mul__', '__ne__', '__new__', '__reduce__', '__reduce_ex__', '__repr__', '__rmod__', '__rmul__', '__setattr__', '__sizeof__', '__str__', '__subclasshook__', '_formatter_field_name_split', '_formatter_parser', 'capitalize', 'center', 'count', 'decode', 'encode', 'endswith', 'expandtabs', 'find', 'format', 'index', 'isalnum', 'isalpha', 'isdigit', 'islower', 'isspace', 'istitle', 'isupper', 'join', 'ljust', 'lower', 'lstrip', 'partition', 'replace', 'rfind', 'rindex', 'rjust', 'rpartition', 'rsplit', 'rstrip', 'split', 'splitlines', 'startswith', 'strip', 'swapcase', 'title', 'translate', 'upper', 'zfill']

> **小提示**：在 Python 中带有下画线的成员有特殊含义，详见本书第 4 章。

内置函数 type() 和 isinstance() 可以用来判断数据类型，常用来对函数参数进行检查，可以避免错误的参数类型导致函数崩溃或返回意料之外的结果。不过，从另一方面来讲，过多地使用 type() 和 isinstance() 函数会一定程度上影响多态，建议谨慎使用。

```
>>>type(3)                                            #查看 3 的类型
<class 'int'>
>>>type([3])                                          #查看[3]的类型
<class 'list'>
>>>type({3})in(list, tuple, dict)                     #判断{3}是否为 list、tuple 或 dict
False
>>>type({3})in(list, tuple, dict, set)                #判断{3}是否为 list、tuple、dict 或 set
True
>>>isinstance(3, int)
True
>>>isinstance(3j, int)
False
>>>isinstance(3j,(int, float, complex))               #判断 3 是否为 int、float 或 complex 类型的实例
True
```

> **拓展知识**：Python 之禅。《道德经》云：人法地，地法天，天法道，道法自然。截拳道创始人、武术大家李小龙也说过"武术的非凡之处在于它的简单"。禅不是一种学说或者理论，而是一种境界，是一种简简单单的自然。禅不是用来专门修的，功夫到了，代码写

好了,禅自然就有了。下面的"Python 之禅"不用去翻译,更不要试图从这段话里学到什么实质性的技术和知识,只需要用心体会并在编写代码时多想想这几句话,每写完一段代码之后要反复阅读,检查是否符合 Python 之禅。

```
>>>import this
The Zen of Python, by Tim Peters

Beautiful is better than ugly.
Explicit is better than implicit.
Simple is better than complex.
Complex is better than complicated.
Flat is better than nested.
Sparse is better than dense.
Readability counts.
Special cases aren't special enough to break the rules.
Although practicality beats purity.
Errors should never pass silently.
Unless explicitly silenced.
In the face of ambiguity, refuse the temptation to guess.
There should be one—and preferably only one—obvious way to do it.
Although that way may not be obvious at first unless you're Dutch.
Now is better than never.
Although never is often better than * right *  now.
If the implementation is hard to explain, it's a bad idea.
If the implementation is easy to explain, it may be a good idea.
Namespaces are one honking great idea—let's do more of those!
```

1.3.3　运算符与表达式

Python 是纯面向对象的编程语言,在 Python 中一切都是对象。而熟悉面向对象编程的读者应该知道,对象由数据和行为两部分组成,而行为主要通过方法来实现,通过一些特殊方法的重写,可以实现运算符重载。从这个角度来讲,运算符也是表现对象行为的一种形式,不同类的对象支持的运算符会有区别,而同一种运算符作用于不同的对象时也可能会表现出不同的行为和结果。

除了算术运算符、关系运算符、逻辑运算符以及位运算符等常见运算符之外,Python 还支持一些特有的运算符,例如成员测试运算符、集合运算符、同一性测试运算符等。使用时需注意,Python 很多运算符具有多种不同的含义,作用于不同操作数的含义并不相同,非常灵活。常用的 Python 运算符如表 1-4 所示,优先级遵循的规则为:算术运算符优先级最高,其次是位运算符/集合运算符、关系运算符、逻辑运算符,算术运算符之间遵循"先乘除,后加减"的基本运算原则。虽然 Python 运算符有一套严格的优先级规则,但是强烈建议在编写复杂的表达式时尽量使用圆括号来明确说明其中的逻辑来提高代码可读性。记住,圆括号是明确和改变表达式运算顺序的利器。

表 1-4 Python 运算符

运算符示例	功能说明
x+y	算术加法，列表、元组、字符串合并
x－y	算术减法，集合差集
x*y	乘法，序列重复
x/y	真除法
x//y	求整商
－x	相反数
x%y	求余数，字符串格式化
x**y	幂运算
x<y;x<=y;x>y;x>=y	大小比较，集合的包含关系比较
x==y;x!=y	相等(值)比较，不等(值)比较
x or y	逻辑或(只有 x 为假才会计算 y)
x and y	逻辑与(只有 x 为真才会计算 y)
not x	逻辑非
x in y;x not in y	成员测试运算符
x is y;x is not y	对象实体同一性测试(地址)
|、^、&、<<、>>、~	位运算符
&、|、^	集合交集、并集、对称差集
@	矩阵相乘运算符

＋运算符除了用于算术加法，还可以用于列表、元组和字符串的合并或连接，生成新对象，例如：

```
>>>3 + (3+4j)                    #整数和复数相加
(6+4j)
>>>[1, 2, 3]+[4, 5, 6]           #连接两个列表
[1, 2, 3, 4, 5, 6]
>>>(1, 2, 3)+(4,)                #连接两个元组
(1, 2, 3, 4)
>>>'abcd'+'1234'                 #连接两个字符串
'abcd1234'
```

下面的代码演示了集合运算符的用法：

```
>>>{1, 2, 3, 4, 5}-{3}           #差集
{1, 2, 4, 5}
>>>{1, 2, 3, 4, 5} | {6}         #并集
{1, 2, 3, 4, 5, 6}
```

```
>>>{1, 2, 3, 4, 5} & {3}                    #交集
{3}
>>>{1, 2, 3, 4, 5, 6} ^ {5, 6, 7, 8}        #对称差集
{1, 2, 3, 4, 7, 8}
```

*运算符除了表示算术乘法，还可用于序列与整数的乘法，表示序列元素的重复，生成新的序列对象，例如：

```
>>>[1, 2, 3] * 3
[1, 2, 3, 1, 2, 3, 1, 2, 3]
>>>(1, 2, 3) * 3
(1, 2, 3, 1, 2, 3, 1, 2, 3)
>>>'abc' * 3
'abcabcabc'
```

由于Python列表中存储的是地址而不是元素值，当包含子列表的列表进行元素重复的时候，情况会复杂一些，例如：

```
>>>x=[[1]] * 3
>>>x
[[1], [1], [1]]
>>>id(x[0])==id(x[1])==id(x[2])
True
>>>x[0].append(3)
>>>x
[[1, 3], [1, 3], [1, 3]]
>>>id(x[0])==id(x[1])==id(x[2])
True
```

运算符/和//在Python中分别表示算术除法和算术求整商，例如：

```
>>>3 / 2
1.5
>>>15 // 4
3
```

小提示：/运算符在Python 3.x中表示真除法，而在Python 2.x中含义略有不同，其运算规则为"整数相除的结果还是整数，实数与整数相除的结果是实数"。学过C语言的读者估计会对这样的运算规则感到亲切，但请记住这仅限于Python 2.x。也就是说，avg=sum(lst) * 1.0 / len(lst)这样的写法在Python 3.x中就没必要了，直接写avg=sum(lst)/ len(lst)就可以了。

%运算符可以用于整数或实数的求余数，还可以用于字符串格式化，例如：

```
>>>123.45 %3.2
1.849999999999996
>>>789 %23
```

```
7
>>>'%c, %d'%(65, 65)
'A, 65'
```

Python关系运算符最大的特点是可以连用,并且其含义与我们日常的理解完全一致。当然使用关系运算符的一个最重要的前提是,操作数之间必须可比较大小。例如,把一个字符串和一个数字进行大小比较是毫无意义的,所以 Python 也不支持这样的运算。例如:

```
>>>1<3<5                            #等价于 1<3 and 3<5
True
>>>3<5>2
True
>>>1>6<8
False
>>>'Hello'>'world'                  #比较字符串大小
False
>>>[1, 2, 3]<[1, 2, 4]              #比较列表大小
True
>>>'Hello'>3                        #字符串和数字不能比较
Traceback(most recent call last):
  File "<pyshell#4>", line 1, in <module>
    'Hello'>3
TypeError: unorderable types: str()>int()
```

成员测试运算符 in 用于成员测试,即测试一个对象是否是另一个对象的成员,例如:

```
>>>3 in [1, 2, 3]                   #测试 3 是否为列表[1, 2, 3]的成员
True
>>>5 in range(1, 10, 1)             #range()是用来生成指定范围数字的内置函数
True
>>>'abc' in 'abcdefg'
True
>>>for i in(3, 5, 7):               #循环,成员遍历
    print(i, end='\t')
3    5    7
```

位运算符只能用于整数,内部执行过程是:首先将整数转换为二进制数,然后右对齐,必要的时候左侧补0,按位进行运算,最后再把计算结果转换为十进制数字返回。位运算符用法如下面的代码所示:

```
>>>3 <<2                            #把 3 左移 2 位
12
>>>3 & 7                            #位与运算
3
>>>3 | 8                            #位或运算
```

```
11
>>>4 | 6
6
>>>3 ^ 5                          #位异或运算
6
```

拓展知识：位运算规则为 1&1=1、1&0=0&1=0&0=0,1|1=1|0=0|1=1,0|0=0,1^1=0^0=0、1^0=0^1=1,左移位时右侧补 0,右移位时左侧补 0。

Python 3.5 增加了一个新的矩阵相乘运算符 @,下面的代码演示了该运算符的用法：

```
>>>import numpy                   #numpy 是用于科学计算的 Python 扩展库
>>>x=numpy.ones(3)                #ones()函数用于生成全 1 矩阵,参数表示矩阵大小
>>>m=numpy.eye(3) * 3             #eye()函数用于生成单位矩阵
>>>m[0,2]=5                       #设置矩阵指定位置上元素的值
>>>m[2, 0] =3
>>>x @ m                          #矩阵相乘
array([ 6.,   3.,   8.])
```

在 Python 中,单个任何类型的对象或常数属于合法表达式,使用表 1-4 中运算符连接的变量和常量以及函数调用的任意组合也属于合法的表达式。

拓展知识：复合赋值运算符。除了表 1-4 中列出的运算符,Python 还支持类似于 +=、-=、/=、*=、&=、|= 等复合赋值运算符,表示在自身的基础上进行运算。例如,表达式 x+=3 等价于 x=x+3,其他复合赋值运算符的功能类似。

1.3.4 人机对话基本接口

所谓对话,最简单的形式就是你问我答,或者我问你答。而人机对话或人机交互最基本的功能应该是机器能够接收用户的输入,并且能够把处理结果通过一定的形式展示给用户。在 Python 中,内置函数 input()用来接收用户输入,print()则用于把处理结果或其他信息展示给用户。对于 input()而言,不论用户输入什么内容,一律作为字符串对待,必要的时候可以使用内置函数 eval()对用户输入的内容进行类型转换。例如：

```
>>>x=input('Please input: ')
Please input: 345
>>>x
'345'
>>>type(x)                        #把用户的输入作为字符串对待
<class 'str'>
>>>int(x)
345
>>>eval(x)                        #对字符串求值
345
```

```
>>>x=input('Please input: ')
Please input: [1, 2, 3]
>>>x
'[1, 2, 3]'
>>>type(x)
<class 'str'>
>>>eval(x)
[1, 2, 3]
>>>x=input('Please input:')            #不论用户输入什么,都作为一个字符串来对待
Please input:'hello world'
>>>x                                    #如果本来就想输入字符串,就不用再输入引号了
"'hello world'"
>>>eval(x)
'hello world'
```

内置函数 print()用于输出特定信息,语法格式为

```
print(value, …, sep=' ', end='\n', file=sys.stdout, flush=False)
```

其中,sep 参数之前为需要输出的内容(可以有多个);sep 参数用于指定数据之间的分隔符,默认为空格;file 参数用于指定输出位置,默认为标准控制台,也可以重定向输出到文件。例如:

```
>>>print(1, 3, 5, 7, sep='\t')          #修改默认分隔符
1    3    5    7
>>>for i in range(10):                  #修改默认行尾结束符,不换行
    print(i, end=' ')
0 1 2 3 4 5 6 7 8 9
>>>fp=open('D:\\test.txt', 'a+')
>>>print('Hello world!', file=fp)       #重定向,将内容输出到文件中
>>>fp.close()
```

小提示:本书全部以 Python 3.5.1 进行讲解和演示。在 Python 2.x 中内置输入函数 input()和输出语句 print 的用法与 Python 3.x 不同,非常喜欢 Python 2.x 的读者可以查阅官方帮助文档,或阅读我的另一本书《Python 程序设计》(清华大学出版社 2015 年 8 月第一版)。

拓展知识:Python 标准库 sys 还提供了 read()和 readline()方法用来从键盘接收指定数量的字符。例如:

```
>>>import sys
>>>x=sys.stdin.read(5)                  #读取 5 个字符,如果输入不足 5 个,等待继续输入
asd
s
>>>x
'asd\ns'
```

```
>>>x=sys.stdin.read(5)              #读取5个字符,如果超出5个,截断
abcdefghijklmnop
>>>x
'abcde'
>>>x=sys.stdin.read(5)              #从缓冲区内继续读取5个字符
>>>x
'fghij'
>>>x=sys.stdin.read(5)
>>>x
'klmno'
>>>x=sys.stdin.read(5)              #缓冲区内不足5个字符,就等待用户继续输入
1234
>>>x
'p\n123'
>>>x=sys.stdin.readline()           #从缓冲区内读取字符,遇到换行符就结束
>>>x
'4\n'
>>>x=sys.stdin.readline()
abcd
>>>x
'abcd\n'
>>>x=sys.stdin.readline(13)         #如果缓冲区内容比需要的少,就遇到换行符结束
abcdefg
>>>x
'abcdefg\n'
>>>x=sys.stdin.readline(13)         #如果缓冲区内容比需要的多,就截断
abcdefghijklmnopqrst
>>>x
'abcdefghijklm'
>>>x=sys.stdin.readline(13)         #从缓冲区继续读取
>>>x
'nopqrst\n'
```

拓展知识:Python 标准库 pprint 还提供了更加友好的输出函数(pretty printer) pprint(),可以更好地控制输出格式,如果要输出的内容多于一行则会自动添加换行和缩进来更好地展示内容的结构。例如:

```
>>>import pprint
>>>t=[[[['black', 'cyan'], 'white', ['green', 'red']], [['magenta', 'yellow'],
'blue']]]
>>>pprint.pprint(t)                 #默认width=80
[[[['black', 'cyan'], 'white', ['green', 'red']],
  [['magenta', 'yellow'], 'blue']]]
>>>pprint.pprint(t, width=50)
```

```
[[[['black', 'cyan'], 'white', ['green', 'red']],
  [['magenta', 'yellow'], 'blue']]]
>>>pprint.pprint(t, width=30)          #根据需要进行换行和缩进
[[[['black', 'cyan'],
   'white',
   ['green', 'red']],
  [['magenta', 'yellow'],
   'blue']]]
```

1.4 模块安装与使用

1.4.1 安装 Python 扩展库

Python 之所以得到各行业领域工程师、策划师以及管理人员的青睐,与涉及各行业各领域开发的扩展库也有很大关系,不仅数量众多、功能强大,关键是用起来很方便。虽然 Python 标准库已经拥有了非常强大的功能,但很多时候如果我们能够熟练运用扩展库,会大幅度提高软件开发速度。

可以把 Python 模块看作一个个用来存放积木的收纳箱,每个收纳箱里放着很多特定类型的积木(函数或类),用的时候我们把收纳箱从仓库里拖出来然后打开它,选择合适的积木来搭建自己的房子、汽车、轮船、飞机等作品(程序)就可以了。

当前,pip 已经成为管理 Python 扩展库(或模块,一般不做区分)的主流方式,使用 pip 不仅可以实时查看本机已安装的 Python 扩展库列表,还支持 Python 扩展库的安装、升级和卸载等操作。使用 pip 工具管理 Python 扩展库只需要在保证计算机联网的情况下输入几个命令即可完成,极大方便了用户。常用 pip 命令的使用方法如表 1-5 所示。

表 1-5　常用 pip 命令使用方法

pip 命令示例	说　　明
pip install SomePackage	安装 SomePackage 模块
pip list	列出当前已安装的所有模块
pip install—upgrade SomePackage	升级 SomePackage 模块
pip uninstall SomePackage	卸载 SomePackage 模块

在 https://pypi.python.org/pypi 中可以获得一个 Python 扩展库的综合列表,本书编写完成时,该网站已经收纳了 88 348 个涉及各领域的扩展库,并且每天都以几十个的速度在增加。有些扩展库安装时要求本机已安装相应版本的 C++ 编译器,或者有些扩展库暂时还没有与本机 Python 版本对应的官方版本,这时可以从 http://www.lfd.uci.edu/~gohlke/pythonlibs/下载对应的 whl 文件,然后在命令提示符环境中使用 pip 命令进行安装,例如:

```
pip install pygame-1.9.2a0-cp35-none-win_amd64.whl
```

小技巧：使用 pip 命令安装 Python 扩展库需要在命令提示符环境中进行，并且需要切换至 pip 命令所在目录，这个步骤很多人并不熟悉。可以在"资源管理器"或"计算机"中通过单击进入 Python 安装文件夹中的 scripts 文件夹，然后按住 Shift 键再右击空白处，选择"在此处打开命令窗口"直接进入命令提示符环境（见图 1-13），然后即可使用 pip 命令管理 Python 扩展库。然而，有些读者根本不知道 Python 安装到什么地方了（尤其是学校某些机房采用默认安装模式时会出现这种情况），此时可以从"开始"菜单依次展开到 Python 启动程序的快捷方式，然后右击并选择"属性"，在"属性"窗口中选择"打开文件位置"按钮就可以直接进入 Python 安装文件夹，如图 1-14 和图 1-15 所示。

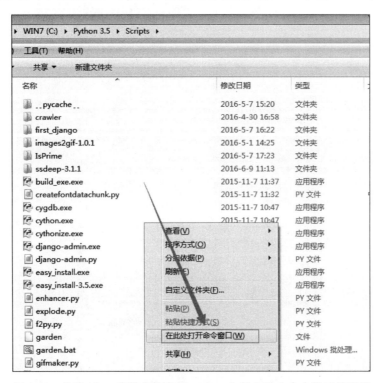

图 1-13　从 Python 安装文件夹的 scripts 文件夹进入命令提示符环境

拓展知识：Python 支持创建多个虚拟环境，每个虚拟环境都是包含 Python 和相应扩展库的一个目录，多个虚拟环境（文件夹）之间互相不干扰。如果有可能根据需要使用不同版本的扩展库，这就需要使用 Python 提供的虚拟环境了。下面我们通过一个实际的例子来演示如何创建和使用 Python 虚拟环境，首先进入命令提示符环境并切换至 Python 安装目录的 tools\Scripts 文件夹，然后执行下面的命令：

..\..\python pyvenv.py ..\..\Python_docx

然后稍等片刻，当再次出现命令提示符时就表明 Python 虚拟环境创建成功了，接下来使用 cd 命令切换至 Python 安装目录的 Python_docx\Scripts 文件夹中，执行 activate 命令，成功的话会发现前面的提示符有些变化，这时就可以使用 pip 工具来安装需要使用的

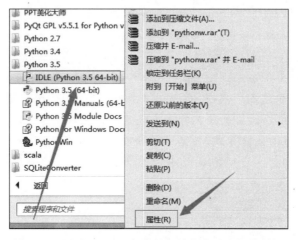

图 1-14　展开 Python 启动程序并右击后选择"属性"

图 1-15　选择"打开文件位置"按钮进入 Python 安装文件夹

扩展库了，安装完成后输入并执行 python 命令，就可以进入 Python 开发环境并使用已安装的扩展库了。

1.4.2　模块导入与使用

　　Python 默认安装仅包含部分基本或核心模块，启动时也仅加载了基本模块，在需要时再显式地加载（有些模块可能需要先安装）其他模块，这样可以减小程序运行的压力，且具有很强的可扩展性，这样的设计与系统安全配置时遵循的"最小权限"原则的思想是一致的，有助于提高系统安全性。可以使用 sys.modules.items() 显示所有预加载模块的相关信息。

内置对象可以直接使用,而标准库和扩展库需要导入之后才能使用其中的对象,当然,扩展库还需要先正确安装才能导入。Python 中导入模块的方法主要有两个。

- import 模块名[as 别名]

使用这种方式导入以后,使用时需要在对象之前加上模块名作为前缀,也就是必须以"模块名.对象名"的方式进行访问。也可以为导入的模块设置一个别名(或者外号),然后就可以使用"别名.对象名"的方式来使用其中的对象了。

```
>>>import math
>>>math.sin(0.5)                #求 0.5(单位是弧度)的正弦值
0.479425538604203
>>>import random
>>>x=random.random()            #获得[0,1)内的随机小数
>>>n=random.randint(1,100)      #获得[1,100]区间上的随机整数
>>>import numpy as np           #导入模块并设置别名
>>>a=np.array((1,2,3,4))        #通过模块的别名来访问其中的对象
>>>print(a)
[1 2 3 4]
```

- from 模块名 import 对象名[as 别名]

使用这种方式仅导入明确指定的对象,并且可以为导入的对象起一个别名。这种导入方式可以减少查询次数,提高访问速度,同时也减少了程序员需要输入的代码量,而不需要使用模块名作为前缀。例如:

```
>>>from math import sin         #只导入模块中的指定对象
>>>sin(3)
0.1411200080598672
>>>from math import sin as f    #给导入的对象起个别名
>>>f(3)
0.141120008059867
```

比较极端的情况是一次导入模块中的所有对象,例如:

```
>>>from math import *
>>>sin(3)                       #求正弦值
0.1411200080598672
>>>gcd(36, 18)                  #求最大公约数
18
```

这种方式简单粗暴,虽然写起来比较省事,可以直接使用模块中的所有函数和对象而不需要再使用模块名作为前缀,但一般并不推荐使用。如果多个模块中有同名的对象,这种方式将会导致只有最后一个导入的模块中的同名对象是有效的,而之前导入的模块中该对象无法访问。例如,a.py 文件中内容如下:

```
def test():
```

```
        print('test in a.py')
```
b.py 文件中的内容如下：
```
def test():
    print('test in b.py')
```

那么导入 a 模块以后，test()方法是可用的，而导入 b 模块之后 a 模块中的 test()方法就无法使用了。例如：

```
>>>from a import *
>>>test()
test in a.py
>>>from b import *                    #这会导致 a 模块中的 test()方法无法使用
>>>test()
test in b.py
```

拓展知识：重新导入模块。在自己编写模块时，可能需要反复修改代码然后重新导入模块进行测试，此时可以使用 imp 模块或 importlib 模块的 reload()函数，重新加载模块时要求该模块已经被正确加载。也就是说，第一次导入和加载模块时不能使用 reload()方法。另外，在 Python 2.x 中 reload()是内置函数，可以直接使用。

小技巧：使用 from…import…的方式明确导入指定的对象而不导入整个模块，可以适当提高程序加载和运行速度。

拓展知识：导入模块时文件搜索的顺序。在导入模块时，Python 首先在当前目录中查找需要导入的模块文件，如果没有找到则从 sys 模块的 path 变量所指定的目录中查找，如果仍没有找到模块文件则抛出异常提示模块不存在。可以查看 sys 模块中 path 变量的值来获知 Python 导入模块时搜索模块的路径，也可以使用 append()方法向其中添加自定义的文件夹以扩展搜索路径。另外，在导入模块时，会优先导入相应的.pyc 文件，如果相应的.pyc 文件与.py 文件时间不相符或不存在对应的.pyc 文件，则导入.py 文件。

建议：①每个 import 语句只导入一个标准库或扩展库；②按照标准库、扩展库、自定义库的先后顺序进行导入。

1.4.3 编写自己的模块和包

Python 程序除了可以直接运行，还可以作为模块导入并使用其中的对象。通过__name__属性可以识别程序的使用方式。每个 Python 脚本在运行时都有一个__name__属性，如果脚本作为模块被导入，则其__name__属性的值被自动设置为模块名；如果脚本独立运行，则其__name__属性值被自动设置为字符串__main__。例如，假设程序 hello.py 中代码如下：

```
def main():                           #def 是用来定义函数的 Python 关键字
```

```
    if __name__=='__main__':              #选择结构,识别当前的运行方式
        print('This program is run directly.')
    elif __name__=='hello':                #冒号、换行、缩进表示一个语句块的开始
        print('This program is used as a module.')

main()                                     #调用上面定义的函数
```

那么在 IDLE 中直接运行该程序时得到结果如下:

```
This program is run directly.
```

而在 IDLE 交互模式中使用 import hello 导入该模块时,得到结果如下:

```
This program is used as a module.
```

小提示:Python 没有类似 C/C++/C♯或者 Java 中用来限定语句块的大括号,而是使用缩进来体现代码之间的从属关系和逻辑关系,同一个级别的代码应具有相同的缩进量。

对于大型软件的开发,不可能把所有代码都存放到一个文件中,那样会使得代码很难维护。对于复杂的大型系统,可以使用包来管理多个模块。包是 Python 用来组织命名空间和类的重要方式,可以看作是包含大量 Python 程序模块的文件夹。在包的每个目录中都必须包含一个__init__.py 文件,该文件可以是一个空文件,用于表示当前文件夹是一个包。__init__.py 文件的主要用途是设置__all__变量以及执行初始化包所需的代码,其中__all__变量中定义的对象可以在使用"from…import *"时全部被正确导入。

假设有如下结构的包:

```
sound/                         Top-level package
    __init__.py                Initialize the sound package
    formats/                   Subpackage for file format conversions
        __init__.py
        wavread.py
        wavwrite.py
        aiffread.py
        aiffwrite.py
        auread.py
        auwrite.py
        …
    effects/                   Subpackage for sound effects
        __init__.py
        echo.py
        surround.py
        reverse.py
        …
    filters/                   Subpackage for filters
        __init__.py
```

```
equalizer.py
vocoder.py
karaoke.py
...
```

那么，可以在自己的程序中使用下面的代码导入其中一个模块：

```
import sound.effects.echo
```

然后使用完整名字来访问或调用其中的成员，例如：

```
sound.effects.echo.echofilter(input, output, delay=0.7, atten=4)
```

如果 sound\effects__init__.py 文件中有下面一行的代码：

```
__all__ = ['echo', 'surround', 'reverse']
```

那么就可以使用这样的方式来导入：

```
from sound.effects import *
```

然后使用下面的方式来使用其中的成员：

```
echo.echofilter(input, output, delay=0.7, atten=4)
```

第 2 章 Python 序列

Python 序列类似于 C 或 Basic 中的一维、多维数组等，但功能要强大很多，使用也更加灵活、方便，*Head First Python* 一书中就戏称列表是"打了激素"的数组。

Python 中常用的序列结构有列表、元组、字典、字符串、集合等，大部分可迭代对象也支持类似于序列的用法，如图 2-1 所示。列表、元组、字符串等序列类型以及 range 对象均支持双向索引，第一个元素下标为 0，第二个元素下标为 1，以此类推；如果使用负数作为索引，则最后一个元素下标为 -1，倒数第二个元素下标为 -2，以此类推。可以使用负整数作为索引是 Python 序列的一大特色，熟练掌握和运用它可以大幅度提高开发效率。

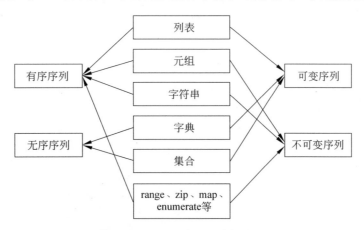

图 2-1　Python 序列分类示意图

《道德经》有云："合抱之木，生于毫末；九层之台，起于累土；千里之行，始于足下。"简单地说就是一句话，没法一口吃成个胖子，做什么都要一步步来，尤其不要忽视基础知识积累的重要性。《庄子·逍遥游》也认为"且夫水之积也不厚，则其负大舟也无力。"没有坚实的基础知识做支撑和后盾，就很难有大的上升空间。练拳不练功，到老一场空。大量实际开发经验表明，熟练掌握 Python 基本数据结构（尤其是序列）的用法可以更加快速有效地解决实际问题。大家慢慢会发现，实际工作中遇到的每个问题最终都会用到一些基本数据结构的方法或内置函数来解决，恰如《道德经》所说"玄之又玄，众妙之门"。这一章通过大量案例介绍了列表、元组、字典、集合等几种基本数据结构的用法，同时还有 range 对象、zip 对象以及 enumerate 对象的巧妙应用，以及在实际应用中非常有用的列表推导式、切片操作、生成器推导式等。

2.1 列表与列表推导式

列表是重要的 Python 内置可变序列之一,是包含若干元素的有序连续内存空间。在形式上,列表的所有元素放在一对中括号"["和"]"中,相邻元素之间使用逗号分隔开。当列表增加或删除元素时,列表对象自动进行内存的扩展或收缩,从而保证元素之间没有缝隙。Python 列表的这个内存自动管理功能可以大幅度减少程序员的负担,但插入和删除非尾部元素时会涉及列表中大量元素的移动,效率较低,并且对于某些操作可能会导致意外的错误结果。因此,除非确实有必要,否则应尽量从列表尾部进行元素的增加与删除操作,这不仅可以大幅度提高列表处理速度,并且总是可以保证得到正确的结果。

在 Python 中,同一个列表中元素的数据类型可以各不相同,可以同时分别为整数、实数、字符串等基本类型,也可以是列表、元组、字典、集合以及其他自定义类型的对象。例如,下面几个都是合法的列表对象:

```
[10, 20, 30, 40]
['crunchy frog', 'ram bladder', 'lark vomit']
['spam', 2.0, 5, [10, 20]]
[['file1', 200,7], ['file2', 260,9]]
[{3}, {5:6},(1, 2, 3)]
```

小提示:Python 采用的是基于值的内存管理模式,Python 变量不直接存储值,而是存储值的引用,Python 列表中元素也是存储值的引用,所以列表中各元素可以是不同类型的数据。

2.1.1 列表创建与删除

使用"="直接将一个列表赋值给变量即可创建列表对象,例如:

```
>>>a_list=['a', 'b', 'mpilgrim', 'z', 'example']
>>>a_list=[]                              #创建空列表
```

也可以使用 list() 函数将元组、range 对象、字符串、字典、集合或其他类型的可迭代对象类型的数据转换为列表。需要注意的是,把字典转换为列表时默认是将字典的"键"转换为列表,而不是把字典的元素转换为列表,如果想把字典的元素转换为列表,需要使用字典对象的 items() 方法明确说明。例如:

```
>>>a_list=list((3,5,7,9,11))              #将元组转换为列表
>>>a_list
[3, 5, 7, 9, 11]
>>>list(range(1,10,2))                    #将 range 对象转换为列表
[1, 3, 5, 7, 9]
>>>list('hello world')                    #将字符串转换为列表
['h', 'e', 'l', 'l', 'o', ' ', 'w', 'o', 'r', 'l', 'd']
```

```
>>>list({3,7,5})                          #将集合转换为列表
[3, 5, 7]
>>>list({'a':3, 'b':9, 'c':78})           #将字典的"键"转换为列表
['a', 'c', 'b']
>>>list({'a':3, 'b':9, 'c':78}.items())   #将字典的"键:值"对转换为列表
[('b', 9),('c', 78),('a', 3)]
>>>x=list()                               #创建空列表
```

> **小知识**：在 Python 社区中，习惯把 list() 还有后面很快就会学到的 tuple()、set()、dict() 这样的函数称为"工厂函数"，因为这些函数可以生成新的数据类型。

创建列表之后，可以使用整数作为下标来访问其中的元素，其中 0 表示第 1 个元素，1 表示第 2 个元素，2 表示第 3 个元素，以此类推；列表还支持使用负整数作为下标，其中 −1 表示最后 1 个元素，−2 表示倒数第 2 个元素，−3 表示倒数第 3 个元素，以此类推，如图 2-2 所示。

```
>>>x=list(range(10))                      #创建列表
>>>import random
>>>random.shuffle(x)                      #把列表中的元素打乱顺序
>>>x
[6, 8, 7, 1, 0, 9, 2, 4, 3, 5]
>>>x[0]                                   #访问第 1 个元素
6
>>>x[1]
8
>>>x[-1]                                  #访问最后一个元素
5
>>>x[-2]
3
```

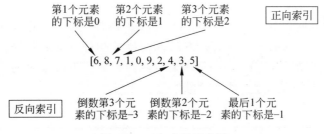

图 2-2 双向索引示意图

当一个列表不再使用时，可以使用 del 命令将其删除，这一点适用于所有类型的 Python 对象。另外，也可以使用 del 命令删除列表、字典等可变序列中的部分元素，而不能删除元组、字符串等不可变序列中的部分元素。例如：

```
>>>x=[1, 2, 3]
>>>del x[1]                               #删除列表中指定位置的元素
```

```
>>>x
[1, 3]
>>>del x                                        #删除列表对象
>>>x                                            #对象删除后无法再访问,抛出异常
Traceback(most recent call last):
  File "<pyshell#58>", line 1, in <module>
    x
NameError: name 'x' is not defined
>>>x={'a':3, 'b':6, 'c':9}
>>>del x['a']                                   #删除字典中部分元素
>>>x
{'c': 9, 'b': 6}
>>>x = (1, 2, 3)                                #创建元组对象
>>>del x[0]                                     #不允许删除元组中的元素
Traceback(most recent call last):
  File "<pyshell#63>", line 1, in <module>
    del x[0]
TypeError: 'tuple' object doesn't support item deletion
>>>x[0]=4                                       #也不能修改元组中的元素值
Traceback(most recent call last):
  File "<pyshell#1>", line 1, in <module>
    x[0]=4
TypeError: 'tuple' object does not support item assignment
```

拓展知识：垃圾回收机制。一般来说，使用 del 删除对象之后 Python 会在恰当的时机调用垃圾回收机制来释放内存，我们也可以在必要的时候导入 Python 标准库 gc 之后调用 gc.collect()函数立刻启动垃圾回收机制来释放内存。

2.1.2 列表常用方法

《象》曰"君子以同而异"。对于 Python 中不同的序列类型而言，有很多方法是通用的，而不同类型的序列又有一些特有的方法或支持某些特有的运算符。常用的列表对象方法如表 2-1 所示。另外，Python 的很多内置函数和命令也可以对列表和其他序列对象进行操作，后面章节中将通过一些案例陆续进行介绍。

表 2-1 常用的列表对象方法

方　　法	说　　明
lst.append(x)	将元素 x 添加至列表 lst 尾部
lst.extend(L)	将列表 L 中所有元素添加至列表 lst 尾部
lst.insert(index，x)	在列表 lst 指定位置 index 处添加元素 x，该位置后面的所有元素后移一个位置

续表

方　法	说　明
lst.remove(x)	在列表 lst 中删除首次出现的指定元素,该元素之后的所有元素前移一个位置
lst.pop([index])	删除并返回列表 lst 中下标为 index(默认为−1)的元素
lst.clear()	删除列表 lst 中的所有元素,但保留列表对象
lst.index(x)	返回列表 lst 中第一个值为 x 的元素的下标,若不存在值为 x 的元素则抛出异常
lst.count(x)	返回指定元素 x 在列表 lst 中的出现次数
lst.reverse()	对列表 lst 所有元素进行逆序
lst.sort(key=None, reverse=False)	对列表 lst 中的元素进行排序,key 用来指定排序依据,reverse 决定升序(False)还是降序(True)
lst.copy()	返回列表 lst 的浅复制,浅复制的介绍参考 2.1.4 节

1. append()、insert()、extend()

这 3 个方法都可以用于向列表对象中添加元素,其中,append()用于向列表尾部追加一个元素,insert()用于向列表任意指定位置插入一个元素,extend()用于将另一个列表中的所有元素追加至当前列表的尾部,这 3 个方法都属于原地操作,不影响列表对象在内存中的起始地址。

```
>>>x=[1, 2, 3]
>>>id(x)                        #查看对象的内存地址
50159368
>>>x.append(4)                  #在尾部追加元素
>>>x
[1, 2, 3, 4]
>>>x.insert(0, 0)               #在指定位置插入元素
>>>x
[0, 1, 2, 3, 4]
>>>x.extend([5, 6, 7])          #在尾部追加多个元素
>>>x
[0, 1, 2, 3, 4, 5, 6, 7]
>>>id(x)                        #列表在内存中的地址不变
50159368
```

另外,运算符+和*也可以实现列表增加元素的目的,但这两个运算符不属于原地操作,而是返回新列表。

```
>>>x=[1, 2, 3]
>>>id(x)
50232456
```

```
>>>x=x+[4]                          #连接2个列表
>>>x
[1, 2, 3, 4]
>>>id(x)                            #内存地址发生改变
50159368
>>>x=x * 2                          #列表元素重复
>>>x
[1, 2, 3, 4, 1, 2, 3, 4]
>>>id(x)
50226952
```

2. pop()、remove()、clear()

这3个方法用于删除列表中的元素，其中，pop()用于删除并返回指定位置(默认是最后一个)上的元素，remove()用于删除列表中第一个值与指定值相等的元素，clear()用于清空列表，这3个方法属于原地操作，不影响列表对象的内存地址。另外，也可以使用del命令删除列表中指定位置的元素，也属于原地操作。

```
>>>x=[1, 2, 3, 4, 5, 6, 7]
>>>x.pop()                          #弹出并返回尾部元素
7
>>>x.pop(0)                         #弹出并返回指定位置的元素
1
>>>x.clear()                        #删除所有元素
>>>x
[]
>>>x=[1, 2, 1, 1, 2]
>>>x.remove(2)                      #删除首个值为2的元素
>>>x
[1, 1, 1, 2]
>>>del x[3]                         #删除指定位置上的元素
>>>x
[1, 1, 1]
```

小提示：在列表中间位置插入或删除元素时，会影响该位置后面所有元素的下标，要尽量避免在列表中间位置进行元素的插入和删除操作。

3. count()、index()

列表方法count()用于返回列表中指定元素出现的次数，index()用于返回指定元素在列表中首次出现的位置，如果该元素不在列表中则抛出异常。除此之外，成员测试运算符in也可以测试列表中是否存在某个元素。

```
>>>x=[1, 2, 2, 3, 3, 3, 4, 4, 4, 4]
>>>x.count(3)                       #元素3在列表x中出现的次数
```

```
3
>>>x.count(5)
0
>>>x.index(2)                               #元素 2 在列表 x 中首次出现的索引
1
>>>x.index(4)
6
>>>x.index(5)                               #列表 x 中没有 5,抛出异常
Traceback(most recent call last):
  File "<pyshell#102>", line 1, in <module>
    x.index(5)
ValueError: 5 is not in list
>>>5 in x                                   #5 不是列表 x 的元素
False
>>>3 in x
True
```

4. sort( )、reverse( )

一提到排序,有其他语言编程经验的读者脑海里立刻会涌现出冒泡法、选择法、插入法、快速排序等名词,以及烦琐的代码和让人头痛的循环边界条件,而逆序也涉及大量元素的交换操作和循环边界的控制。虽然学习和掌握各种排序算法的过程会锻炼我们严谨的逻辑思维能力,但实际运用中我们更希望简单一些。试想,如果只需要把排序的要求明确地告诉计算机,计算机就能立刻满足我们的要求,那将会是一种什么样的体验!至于计算机内部是如何实现排序的,谁会关心!

列表对象的 sort()方法用于按照指定的规则对所有元素进行排序,默认规则是直接比较元素大小;reverse()方法用于将列表所有元素逆序排列。

```
>>>x=list(range(11))                        #包含 11 个整数的列表
>>>import random
>>>random.shuffle(x)                        #随机乱序
>>>x
[6, 0, 1, 7, 4, 3, 2, 8, 5, 10, 9]
>>>x.sort(key=lambda item:len(str(item)), reverse=True)
                                            #按指定规则排序
>>>x
[10, 6, 0, 1, 7, 4, 3, 2, 8, 5, 9]
>>>x.reverse()                              #逆序
>>>x
[9, 5, 8, 2, 3, 4, 7, 1, 0, 6, 10]
>>>x.sort(key=str)                          #按转换为字符串后的大小排序
>>>x
[0, 1, 10, 2, 3, 4, 5, 6, 7, 8, 9]
```

```
>>>x.sort()                              #按默认规则排序
>>>x
[0, 1, 2, 3, 4, 5, 6, 7, 8, 9, 10]
```

需要注意的是，列表对象的 sort() 和 reverse() 是对列表进行原地排序（in-place sorting）和逆序。所谓"原地"，意思是用处理后的数据替换原来的数据，列表中元素原来的顺序全部丢失。如果原来的顺序不想丢失，可以使用内置函数 sorted() 和 reversed() 分别得到新的列表和迭代对象。其中，内置函数 sorted() 返回新列表，内置函数 reversed() 返回一个逆序排列后的迭代对象，这两个函数都不对原列表做任何修改。

```
>>>x=list(range(11))
>>>import random
>>>random.shuffle(x)                     #打乱顺序
>>>x
[2, 4, 0, 6, 10, 7, 8, 3, 9, 1, 5]
>>>sorted(x)                             #以默认规则排序
[0, 1, 2, 3, 4, 5, 6, 7, 8, 9, 10]
>>>sorted(x, key=lambda item:len(str(item)), reverse=True)
                                         #以指定规则排序
[10, 2, 4, 0, 6, 7, 8, 3, 9, 1, 5]
>>>sorted(x, key=str)                    #以指定规则排序
[0, 1, 10, 2, 3, 4, 5, 6, 7, 8, 9]
>>>x
[2, 4, 0, 6, 10, 7, 8, 3, 9, 1, 5]
>>>reversed(x)                           #逆序
<list_reverseiterator object at 0x0000000003089E48>
>>>list(reversed(x))
[5, 1, 9, 3, 8, 7, 10, 6, 0, 4, 2]
```

小提示：上面的几段代码中用到了 lambda 表达式，有关知识请参考 3.3.4 节。

拓展知识：排序方法的 key 参数。充分利用列表对象的 sort() 方法和内置函数 sorted() 的 key 参数，可以实现更加复杂的排序，以内置函数 sorted() 为例：

```
>>>gameresult=[['Bob', 95.0, 'A'],
               ['Alan', 86.0, 'C'],
               ['Mandy', 83.5, 'A'],
               ['Rob', 89.3, 'E']]
>>>from operator import itemgetter
>>>sorted(gameresult, key=itemgetter(2))      #按子列表第 3 个元素进行升序排序
[['Bob', 95.0, 'A'], ['Mandy', 83.5, 'A'], ['Alan', 86.0, 'C'], ['Rob', 89.3, 'E']]
>>>sorted(gameresult, key=itemgetter(2, 0))   #按第 3 个元素升序，然后按第 1 个升序
[['Bob', 95.0, 'A'], ['Mandy', 83.5, 'A'], ['Alan', 86.0, 'C'], ['Rob', 89.3, 'E']]
>>>sorted(gameresult, key=itemgetter(2, 0), reverse=True)
[['Rob', 89.3, 'E'], ['Alan', 86.0, 'C'], ['Mandy', 83.5, 'A'], ['Bob', 95.0, 'A']]
```

```
>>>list1=["what", "I'm", "sorting", "by"]      #以一个列表内容为依据
>>>list2=["something", "else", "to", "sort"]   #对另一个列表内容进行排序
>>>pairs=zip(list1, list2)                     #把两个列表中的对应位置元素配对
>>>[item[1] for item in sorted(pairs, key=lambda x:x[0], reverse=True)]
['something', 'to', 'sort', 'else']
>>>x=[[1, 2, 3], [2, 1, 4], [2, 2, 1]]         #以第2个元素升序、第3个元素降序排序
>>>sorted(x, key=lambda item:(item[1], -item[2]))
[[2, 1, 4], [1, 2, 3], [2, 2, 1]]
>>>x=['aaaa', 'bc', 'd', 'b', 'ba']            #先按长度排序,长度一样的正常排序
>>>sorted(x, key=lambda item:(len(item), item))
['b', 'd', 'ba', 'bc', 'aaaa']
```

5．内置函数对列表的操作

除了列表对象自身方法之外,很多 Python 内置函数也可以对列表进行操作。例如,max()、min()函数用于返回列表中所有元素的最大值和最小值,sum()函数用于返回数值型列表中所有元素之和,len()函数用于返回列表中元素个数,zip()函数用于将多个列表中元素重新组合为元组并返回包含这些元组的 zip 对象,enumerate()函数返回包含若干下标和值的迭代对象。

```
>>>x=list(range(11))                   #生成列表
>>>import random
>>>random.shuffle(x)                   #打乱列表中元素顺序
>>>x
[0, 6, 10, 9, 8, 7, 4, 5, 2, 1, 3]
>>>max(x)                              #返回最大值
10
>>>max(x, key=str)                     #按指定规则返回最大值
9
>>>min(x)
0
>>>sum(x)                              #所有元素之和
55
>>>len(x)                              #列表元素个数
11
>>>list(zip(x, [1] * 11))              #多列表元素重新组合
[(0, 1),(6, 1),(10, 1),(9, 1),(8, 1),(7, 1),(4, 1),(5, 1),(2, 1),(1, 1),(3, 1)]
>>>list(zip(range(1,4)))               #zip()函数也可以用于一个序列或迭代对象
[(1,),(2,),(3,)]
>>>list(zip(['a', 'b', 'c'], [1, 2]))  #如果两个列表不等长,以短的为准
[('a', 1),('b', 2)]
>>>enumerate(x)                        #枚举列表元素,返回 enumerate 对象
<enumerate object at 0x00000000030A9120>
>>>list(enumerate(x))                  #enumerate 对象可迭代
```

[(0, 0),(1, 6),(2, 10),(3, 9),(4, 8),(5, 7),(6, 4),(7, 5),(8, 2),(9, 1),(10, 3)]

在Python中，map()、reduce()、filter()是函数式编程的重要体现形式。内置函数map()可以将一个函数依次作用（或映射）到序列或迭代器对象的每个元素上，并返回一个map对象作为结果，其中，每个元素是原序列中元素经过该函数处理后的结果，不对原序列或迭代器对象做任何修改。下面的代码用到后面章节的知识，大家如果看不懂的话也不要焦虑，一本好书本来就是需要前后反复翻看很多遍才能融会贯通的，正所谓"书读百遍，其义自见"。

```
>>>list(map(str, range(5)))              #转换为字符串
['0', '1', '2', '3', '4']
>>>def add5(v):                          #单参数函数
    return v+5
>>>list(map(add5, range(10)))            #把单参数函数映射到一个序列的所有元素
[5, 6, 7, 8, 9, 10, 11, 12, 13, 14]
>>>def add(x, y):                        #可以接收2个参数的函数
    return x+y
>>>list(map(add, range(5), range(5,10))) #把双参数函数映射到两个序列上
[5, 7, 9, 11, 13]
>>>list(map(lambda x, y: x+y, range(5), range(5,10)))
[5, 7, 9, 11, 13]
>>>[add(x,y)for x, y in zip(range(5), range(5,10))]
[5, 7, 9, 11, 13]
```

标准库functools中的函数reduce()可以将一个接收2个参数的函数以累积的方式从左到右依次作用到一个序列或迭代器对象的所有元素上。

```
>>>from functools import reduce
>>>seq=[1, 2, 3, 4, 5, 6, 7, 8, 9]
>>>reduce(add, range(10))                #add是上一段代码中定义的函数
45
>>>reduce(lambda x, y: x+y, seq)         #使用lambda表达式实现相同功能
45
```

上面的代码运行过程如图2-3所示。

💡小提示：在Python 2.x中，reduce()是内置函数，不需要导入任何模块即可使用。

内置函数filter()将一个单参数函数作用到一个序列上，返回该序列中使得该函数返回值为True的那些元素组成的filter对象，如果指定函数为None，则返回序列中等价于True的元素。

```
>>>seq=['foo', 'x41', '?!', '***']
>>>def func(x):
    return x.isalnum()
>>>filter(func, seq)                     #返回filter对象
```
#测试是否为字母或数字

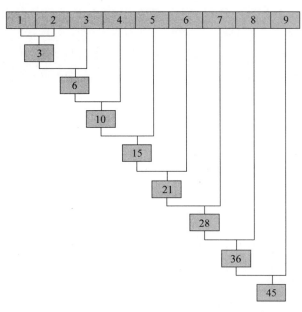

图 2-3　reduce()函数执行过程示意图

```
<filter object at 0x000000000305D898>
>>>list(filter(func, seq))                          #把 filter 对象转换为列表
['foo', 'x41']
>>>seq                                              #不对原列表做任何修改
['foo', 'x41', '?!', '***']
>>>[x for x in seq if x.isalnum()]                  #使用列表推导式实现相同功能
['foo', 'x41']
>>>list(filter(lambda x: x.isalnum(), seq))         #使用 lambda 表达式实现相同功能
['foo', 'x41']
>>>list(filter(None, [1, 2, 3, 0, 0, 4, 0, 5]))     #指定函数为 None
[1, 2, 3, 4, 5]
```

小提示：max()、min()、len()、zip()、enumerate()、sorted()、reversed()、map()、filter()等内置函数以及成员测试运算符 in 也适用于列表对象，以及元组、字符串。

拓展知识：使用列表模拟向量运算。Python 列表支持与整数的乘法运算，表示列表元素进行重复并生成新列表，不对原列表进行任何修改。

```
>>>[1, 2, 3] * 3
[1, 2, 3, 1, 2, 3, 1, 2, 3]
```

Python 列表不支持与整数的加、减、除运算，也不支持列表之间的减、乘、除操作。列表之间的加法运算表示列表元素的合并，生成新列表，而不是向量意义的加法。

```
>>>[1, 2, 3]+[4, 5, 6]
[1, 2, 3, 4, 5, 6]
```

然而,向量运算经常涉及这样的操作,例如,向量所有分量同时加、减、乘、除同一个数,或者向量之间的加、减、乘运算,Python列表对象本身不支持这样的操作,不过可以借助于内置函数和标准库operator中的方法来实现,或者使用扩展库numpy实现更加强大的功能。

```
>>> import random
>>> x=[random.randint(1,100) for i in range(10)]   #生成10个[1,100]区间内的随机数
>>> x
[46, 76, 47, 28, 5, 15, 57, 29, 9, 40]
>>> list(map(lambda i: i+5, x))                    #所有元素同时加5
[51, 81, 52, 33, 10, 20, 62, 34, 14, 45]
>>> x=[random.randint(1,10) for i in range(10)]
>>> x
[2, 2, 9, 6, 7, 9, 2, 1, 2, 7]
>>> y=[random.randint(1,10) for i in range(10)]
>>> y
[8, 1, 9, 7, 1, 5, 8, 4, 1, 9]
>>> import operator
>>> sum(map(operator.mul, x, y))                   #向量内积
278
>>> sum((i*j for i, j in zip(x, y)))               #使用内置函数计算向量内积
278
>>> list(map(operator.add, x, y))                  #两个等长的向量对应元素相加
[10, 3, 18, 13, 8, 14, 10, 5, 3, 16]
>>> list(map(lambda i,j: i+j, x, y))               #使用lambda表达式实现同样效果
[10, 3, 18, 13, 8, 14, 10, 5, 3, 16]
```

列表可以说是Python功能最强大的数据类型之一,提供了大量的方法支持各种操作,同时很多Python内置函数也可以作用于列表。看起来非常复杂的样子,确实不太好记。当暂时还是菜鸟的你看到Python高手运指如飞,行云流水一样完成一个又一个功能,然后一脸崇拜地问高手如何才能记得住那么复杂的东西,高手(按最近网络小说流行的风格,此处高手应该会先打个响指)很可能会淡淡地给你一个令人抓狂的答案"无他,唯手熟尔"。是的,高手没有忽悠你,要想精通,就是要多练,多动手才能真正理解,才能出神入化。不要只羡慕高手编写代码时的潇洒和写意,更要看到高手背后的努力和拼搏。

2.1.3 列表推导式

列表推导式可以说是Python程序开发时应用最多的技术之一,在1.3.2节曾经展示过如何使用列表推导式生成包含随机数的列表。列表推导式可以使用非常简洁的方式来快速生成满足特定需求的列表,代码具有非常强的可读性。另外,Python的内部实现对列表推导式做了大量优化,可以保证很快的运行速度。列表推导式的语法形式为

[表达式 for 变量 in 序列或迭代对象]

列表推导式在逻辑上相当于一个循环,只是形式更加简洁,例如:

```
>>>aList=[x*x for x in range(10)]
```

相当于

```
>>>aList=[]
>>>for x in range(10):
    aList.append(x*x)
```

当然也等价于

```
>>>aList=list(map(lambda x: x*x, range(10)))
```

而

```
>>>freshfruit=[' banana', ' loganberry ', 'passion fruit ']
>>>aList=[w.strip()for w in freshfruit]
```

则等价于下面的代码:

```
>>>freshfruit=[' banana', ' loganberry ', 'passion fruit ']
>>>aList=[]
>>>for item in freshfruit:
    aList.append(item.strip())
```

也等价于

```
>>>freshfruit=[' banana', ' loganberry ', 'passion fruit ']
>>>aList=list(map(lambda x: x.strip(), freshfruit))
```

或

```
>>>freshfruit=[' banana', ' loganberry ', 'passion fruit ']
>>>aList=list(map(str.strip, freshfruit))
```

大家应该看过一个故事,说是阿凡提(也有的说是阿基米德,这不是重点)与国王比赛下棋,国王说要是自己输了的话阿凡提想要什么他都可以拿得出来。阿凡提说那就要点米吧,棋盘一共 64 个小格子,在第一个格子里放 1 粒米,第二个格子里放 2 粒米,第三个格子里放 4 粒米,第四个格子里放 8 粒米,以此类推,后面每个格子里的米都是前一个格子里的 2 倍,一直把 64 个格子都放满。结果可想而知,最后国王没有办法拿出那么多米。那么到底需要多少粒米呢,其实使用列表推导式再结合内置函数 sum()就很容易知道答案。

```
>>>sum([2**i for i in range(64)])
18446744073709551615
```

接下来再通过几个示例来进一步展示列表推导式的强大功能。

1. 使用列表推导式实现嵌套列表的平铺

```
>>>vec=[[1, 2, 3], [4, 5, 6], [7, 8, 9]]
>>>[num for elem in vec for num in elem]
[1, 2, 3, 4, 5, 6, 7, 8, 9]
```

在这个列表推导式中有 2 个循环，其中第一个循环可以看作是外循环，执行得慢；而第二个循环可以看作是内循环，执行得快。上面代码的执行过程等价于下面的写法：

```
>>>vec=[[1, 2, 3], [4, 5, 6], [7, 8, 9]]
>>>result=[]
>>>for elem in vec:
       for num in elem:
           result.append(num)
>>>result
[1, 2, 3, 4, 5, 6, 7, 8, 9]
```

2. 过滤不符合条件的元素

在列表推导式中可以使用 if 子句来进行筛选，只在结果列表中保留符合条件的元素。例如，下面的代码可以列出当前文件夹下所有 Python 源文件：

```
>>>import os
>>>[filename for filename in os.listdir('.')if filename.endswith('.py')]
```

下面的代码用于从列表中选择符合条件的元素组成新的列表：

```
>>>aList=[-1, -4, 6, 7.5, -2.3, 9, -11]
>>>[i for i in aList if i>0]                    #所有大于 0 的数字
[6, 7.5, 9]
```

再例如，已知有一个包含一些同学成绩的字典，计算成绩的最高分、最低分、平均分，并查找所有最高分的同学，代码可以这样编写：

```
>>>scores={"Zhang San": 45, "Li Si": 78, "Wang Wu": 40, "Zhou Liu": 96, "Zhao Qi": 65, "Sun Ba": 90, "Zheng Jiu": 78, "Wu Shi": 99, "Dong Shiyi": 60}
>>>highest=max(scores.values())                 #最高分
>>>lowest=min(scores.values())                  #最低分
>>>highest
99
>>>lowest
40
>>>average=sum(scores.values())/len(scores)     #平均分
>>>average
72.33333333333333
>>>highestPerson=[name for name, score in scores.items()if score ==highest]
```

```
>>>highestPerson
['Wu Shi']
```

下面的代码使用列表推导式查找列表中最大元素的位置：

```
>>>from random import randint
>>>x=[randint(1, 10)for i in range(20)]
>>>m=max(x)
>>>m
10
>>>[index for index, value in enumerate(x)if value ==m]
[0, 5, 6, 10]
>>>x
[10, 2, 3, 4, 5, 10, 10, 9, 2, 4, 10, 8, 2, 2, 9, 7, 6, 2, 5, 6]
```

3. 在列表推导式中使用多个循环，实现多序列元素的任意组合，并且可以结合条件语句过滤特定元素

```
>>>[(x, y)for x in [1, 2, 3] for y in [3, 1, 4] if x !=y]
[(1, 3),(1, 4),(2, 3),(2, 1),(2, 4),(3, 1),(3, 4)]
```

注意：对于包含多个循环的列表推导式，一定要清楚多个循环的执行顺序或"嵌套关系"。例如，上面的代码等价于

```
>>>result=[]
>>>for x in [1, 2, 3]:
      for y in [3, 1, 4]:
          if x !=y:
              result.append((x,y))
>>>result
[(1, 3),(1, 4),(2, 3),(2, 1),(2, 4),(3, 1),(3, 4)]
```

4. 使用列表推导式实现矩阵转置

下面的代码使用列表推导式实现矩阵转置：

```
>>>matrix=[ [1, 2, 3, 4], [5, 6, 7, 8], [9, 10, 11, 12]]
>>>[[row[i] for row in matrix] for i in range(4)]
[[1, 5, 9], [2, 6, 10], [3, 7, 11], [4, 8, 12]]
```

或者，也可以使用内置函数 zip()和 list()来实现矩阵转置：

```
>>>list(map(list,zip(*matrix)))
[[1, 5, 9], [2, 6, 10], [3, 7, 11], [4, 8, 12]]
```

注意：对于嵌套了列表推导式的列表推导式，一定要清楚其执行顺序。例如，上面列表推导式的执行过程等价于下面的代码，可以看出，使用列表推导式更加简洁，代码

可读性更强。

```
>>>matrix=[[1, 2, 3, 4], [5, 6, 7, 8], [9, 10, 11, 12]]
>>>result=[]
>>>for i in range(len(matrix)):
    temp=[]
    for row in matrix:
        temp.append(row[i])
    result.append(temp)
>>>result
[[1, 5, 9], [2, 6, 10], [3, 7, 11], [4, 8, 12]]
```

5. 列表推导式中可以使用函数或复杂表达式

```
>>>def f(v):
    if v%2==0:
        v=v**2
    else:
        v=v+1
    return v
>>>print([f(v) for v in [2, 3, 4, -1] if v>0])
[4, 4, 16]
>>>print([v**2 if v%2==0 else v+1 for v in [2, 3, 4, -1] if v>0])
[4, 4, 16]
```

6. 列表推导式支持文件对象迭代

```
>>>fp=open('C:\install.log', 'r')
>>>print([line for line in fp])        #为节约篇幅,这里没有给出代码的运行结果
>>>fp.close()
```

7. 使用列表推导式生成 100 以内的所有素数

```
>>>[p for p in range(2, 100) if 0 not in [p%d for d in range(2, int(sqrt(p))+1)]]
[2, 3, 5, 7, 11, 13, 17, 19, 23, 29, 31, 37, 41, 43, 47, 53, 59, 61, 67, 71, 73, 79, 83, 89, 97]
```

2.1.4 切片

切片也是 Python 序列的重要操作之一,在形式上,切片使用 2 个冒号分隔的 3 个数字来完成,第一个数字表示切片的开始位置(默认为 0),第二个数字表示切片的截止(但不包含)位置(默认为列表长度),第三个数字表示切片的步长(默认为 1),当步长省略时可以同时省略最后一个冒号。

切片适用于列表、元组、字符串、range 对象等类型,应用于列表时具有最为强大的功

能。可以使用切片来截取列表中的任何部分返回得到一个新列表,也可以通过切片来修改和删除列表中的部分元素,甚至可以通过切片操作为列表对象增加元素。

1. 使用切片获取列表中的部分元素

使用切片可以返回列表原有元素的一个"子集"。与使用下标访问列表元素的方法不同,切片操作不会因为下标越界而抛出异常,而是简单地在列表尾部截断或者返回一个空列表,代码具有更强的健壮性。

```
>>>aList=[3, 4, 5, 6, 7, 9, 11, 13, 15, 17]
>>>aList[::]                        #返回包含原列表中所有元素的新列表
[3, 4, 5, 6, 7, 9, 11, 13, 15, 17]
>>>aList[::-1]                      #返回包含原列表中所有元素的逆序列表
[17, 15, 13, 11, 9, 7, 6, 5, 4, 3]
>>>aList[::2]                       #隔一个元素取一个元素,获取偶数位置的元素
[3, 5, 7, 11, 15]
>>>aList[1::2]                      #隔一个元素取一个元素,获取奇数位置的元素
[4, 6, 9, 13, 17]
>>>aList[3:6]                       #指定切片的开始位置和结束位置
[6, 7, 9]
>>>aList[0:100]                     #切片的结束位置大于列表长度时,从列表尾部截断
[3, 4, 5, 6, 7, 9, 11, 13, 15, 17]
>>>aList[100:]                      #切片的开始位置大于列表长度时,返回空列表
[]
>>>aList[100]                       #抛出异常,不允许越界访问
IndexError: list index out of range
```

2. 使用切片对列表元素进行增、删、改

可以使用切片操作来快速实现很多目的,例如,原地修改列表内容,列表元素的增、删、改以及元素替换等操作都可以通过切片来实现,并且不影响列表对象的内存地址。

```
>>>aList=[3, 5, 7]
>>>aList[len(aList):]
[]
>>>aList[len(aList):]=[9]           #在列表尾部增加元素
>>>aList
[3, 5, 7, 9]
>>>aList[:3]=[1, 2, 3]              #替换列表元素
>>>aList
[1, 2, 3, 9]
>>>aList[:3]=[]                     #删除列表元素
>>>aList
[9]
>>>aList=list(range(10))
```

```
>>>aList
[0, 1, 2, 3, 4, 5, 6, 7, 8, 9]
>>>aList[::2]=[0] * (len(aList)//2)      #替换列表元素
>>>aList
[0, 1, 0, 3, 0, 5, 0, 7, 0, 9]
>>>aList[3:3]=[4, 5, 6]                  #在列表指定位置插入元素
>>>aList
[0, 1, 0, 4, 5, 6, 3, 0, 5, 0, 7, 0, 9]
>>>len(aList)
13
>>>aList[20:30]=[3] * 2                  #这样也可以在尾部追加元素,注意切片的范围
>>>aList
[0, 1, 0, 4, 5, 6, 3, 0, 5, 0, 7, 0, 9, 3, 3]
```

另外,也可以结合使用 del 命令与切片结合来删除列表中的部分元素。

```
>>>aList=[3, 5, 7, 9, 11]
>>>del aList[:3]
>>>aList
[9, 11]
```

❖ **注意**:切片返回的是列表元素的浅复制,与列表对象的直接赋值并不一样。

例如:

```
>>>aList=[3, 5, 7]
>>>bList=aList                  #bList 与 aList 指向同一个内存
>>>bList
[3, 5, 7]
>>>bList[1]=8
>>>aList
[3, 8, 7]
>>>aList ==bList                #两个列表的值是相等的
True
>>>aList is bList               #两个列表是同一个对象
True
>>>id(aList)==id(bList)         #两个列表的内存地址相等
True
>>>
19061816
>>>aList=[3, 5, 7]
>>>bList=aList[::]              #切片,浅复制
>>>aList ==bList                #两个列表的值相等
True
>>>aList is bList               #浅复制,不是同一个对象
False
```

```
>>>id(aList)==id(bList)         #两个列表对象的地址不相等
False
>>>id(x[0])==id(y[0])           #不要怀疑,相同值在内存中仍然只有一份
True
>>>bList[1]=8                   #修改 bList 列表元素的值不会影响 aList
>>>bList                        #bList 的值发生改变
[3, 8, 7]
>>>aList                        #aList 的值没有发生改变
[3, 5, 7]
>>>aList ==bList
False
>>>aList is bList
False
```

顺便再补充一点,虽然直接把一个列表变量赋值给另一个变量时两个变量指向同一个内存地址,但是把一个列表分别赋值给 2 个变量时就不是这样的情况了,例如:

```
>>>x=[1, 2, 3]
>>>y=[1, 2, 3]
>>>id(x)==id(y)
False
>>>y.append(4)                  #修改 y 的值,不影响 x
>>>x
[1, 2, 3]
>>>y
[1, 2, 3, 4]
```

然而,当列表中包含其他可变序列时,情况变得更加复杂,例如:

```
>>>x=[[1], [2], [3]]
>>>y=x[:]
>>>y
[[1], [2], [3]]
>>>y[0]=[4]                     #直接修改 y 中下标为 0 的元素值,不影响 x
>>>y
[[4], [2], [3]]
>>>y[1].append(5)               #通过列表对象的方法原地增加元素,影响 x
>>>y
[[4], [2, 5], [3]]
>>>x
[[1], [2, 5], [3]]
```

估计有的读者已经开始犯晕了,深吸一口气,别紧张,关键是记住两点:①Python 采用的是基于值的内存管理模式;②Python 变量中并不直接存放值,而是存放值的引用。时刻记住这两句话,就不太容易晕了。

注意:使用切片修改列表元素值时,如果左侧切片是连续的,那么等号两侧的列

表长度可以不一样;如果左侧切片不连续,则右侧列表中元素个数必须与左侧相等;使用del命令和切片删除列表中部分元素时,切片可以不连续。例如:

```
>>>x=list(range(10))
>>>x[::2]=[3,5]                    #等号两侧不等长,抛出异常
ValueError: attempt to assign sequence of size 2 to extended slice of size 5
>>>x[::2]=[1, 1, 1, 1, 1]          #等号两侧等长,可以执行
>>>x
[1, 1, 1, 3, 1, 5, 1, 7, 1, 9]
>>>del x[::2]                      #可以删除列表中不连续的元素
>>>x
[1, 3, 5, 7, 9]
```

2.2 元组与生成器推导式

2.2.1 元组

元组也是 Python 的一个重要序列结构。从形式上,元组的所有元素放在一对圆括号中,元素之间使用逗号分隔。下面的代码演示了创建元组的方法:

```
>>>x = (1, 2, 3)                   #直接把元组赋值给一个变量
>>>x
(1, 2, 3)
>>>type(x)                         #使用 type()函数查看变量类型
<class 'tuple'>
>>>x = (3)                         #这和 x=3 是一样的
>>>x
3
>>>x = (3,)                        #如果元组中只有一个元素,必须在后面多写一个逗号
>>>x
(3,)
>>>x = ()                          #空元组
>>>x
()
>>>x=tuple()                       #空元组
>>>x
()
>>>tuple(range(5))                 #将其他迭代对象转换为元组
(0, 1, 2, 3, 4)
```

元组属于不可变(immutable)序列,一旦创建,没有任何方法可以修改元组中元素的值,也无法为元组增加或删除元素。因此,元组没有提供 append()、extend()和 insert()等方法,无法向元组中添加元素;同样,元组也没有 remove()和 pop()方法,也不支持对元组元素进行 del 操作,不能从元组中删除元素,而只能使用 del 命令删除整个元组。元

组也支持切片操作，但是只能通过切片来访问元组中的元素，而不支持使用切片来修改元组中元素的值，也不支持使用切片操作来为元组增加或删除元素。从一定程度上讲，可以认为元组是轻量级的列表，或者"常量列表"。

Python 的内部实现对元组做了大量优化，访问和处理速度比列表更快。如果定义了一系列常量值，主要用途仅是对它们进行遍历或其他类似用途，而不需要对其元素进行任何修改，那么一般建议使用元组而不用列表。元组在内部实现上不允许修改其元素值，从而使得代码更加安全，例如，调用函数时使用元组传递参数可以防止在函数中修改元组，而使用列表则很难做到这一点。

另外，作为不可变序列，与整数、字符串一样，元组可用作字典的键，而列表则永远都不能当做字典键使用，也不能作为集合中的元素，因为列表不是不可变的，或者说不可哈希。

注意：虽然元组属于不可变序列，其元素的值是不可改变的，但是如果元组中包含可变序列，情况就又变得复杂了。例如，下面的代码：

```
>>>x = ([1, 2], 3)              #包含列表的元组
>>>x[0][0]=5                    #修改元组中的列表元素
>>>x
([5, 2], 3)
>>>x[0].append(8)               #为元组中的列表增加元素
>>>x
([5, 2, 8], 3)
>>>x[0]=x[0]+[10]               #试图修改元组的值，失败
Traceback(most recent call last):
  File "<pyshell#83>", line 1, in <module>
    x[0]=x[0]+[10]
TypeError: 'tuple' object does not support item assignment
```

2.2.2 生成器推导式

从形式上看，生成器推导式与列表推导式非常接近，只是生成器推导式使用圆括号而不是列表推导式所使用的方括号。与列表推导式不同的是，生成器推导式的结果是一个生成器对象，而不是列表，也不是元组。使用生成器对象的元素时，可以根据需要将其转化为列表或元组，也可以使用生成器对象的__next__()方法或者内置函数 next()进行遍历，或者直接将其作为迭代器对象来使用。但是不管用哪种方法访问其元素，当所有元素访问结束以后，如果需要重新访问其中的元素，必须重新创建该生成器对象。

```
>>>g = ((i+2)**2 for i in range(10))    #创建生成器对象
>>>g
<generator object <genexpr>at 0x0000000003095200>
>>>tuple(g)                              #将生成器对象转换为元组
(4, 9, 16, 25, 36, 49, 64, 81, 100, 121)
```

```
>>>list(g)                                    #生成器对象已遍历结束,没有元素了
[]
>>>g = ((i+2)**2 for i in range(10))
>>>g.__next__()                               #使用生成器对象的__next__()方法获取元素
4
>>>g.__next__()                               #获取下一个元素
9
>>>next(g)                                    #使用内置函数next()获取生成器对象中的元素
16
>>>next(g)                                    #获取下一个元素
25
>>>g = ((i+2)**2 for i in range(10))
>>>for item in g:                             #使用循环直接遍历生成器对象中的元素
    print(item, end=' ')
4 9 16 25 36 49 64 81 100 121
```

小提示：①在 Python 2.x 中，生成器对象用于获取元素的方法是 next()，而不是 __next__()；②生成器对象具有惰性求值的特点，只在需要时返回元素，比列表推导式具有更高的效率，尤其适合大量数据的遍历。

拓展知识：生成器对象。包含 yield 语句的函数可以用来创建可迭代的生成器对象。下面的代码演示了如何使用生成器来生成斐波那契数列。

```
>>>def f():
    a, b=1, 1                                 #序列解包,同时为多个元素赋值
    while True:
        yield a                               #暂停执行,需要时再产生一个新元素
        a, b=b, a+b                           #序列解包,继续生成新元素
>>>a=f()                                      #创建生成器对象
>>>for i in range(10):                        #斐波那契数列中前 10 个元素
    print(a.__next__(), end=' ')
1 1 2 3 5 8 13 21 34 55
>>>for i in f():                              #斐波那契数列中第一个大于 100 的元素
    if i>100:
        print(i, end=' ')
        break
144
>>>a=f()
>>>next(a)                                    #使用内置函数next()获取生成器对象中的元素
1
>>>next(a)                                    #每次索取新元素时,由 yield 语句生成
1
>>>next(a)
2
>>>next(a)
3
```

2.3 字　　典

字典(dictionary)是包含若干"键:值"元素的无序可变序列,字典中的每个元素包含"键"和"值"两部分,表示一种映射或对应关系,也称为关联数组。定义字典时,每个元素的"键"和"值"用冒号分隔,不同元素之间用逗号分隔,所有的元素放在一对大括号"{"和"}"中。

字典中的"键"可以是 Python 中任意不可变数据,如整数、实数、复数、字符串、元组等,但不能使用列表、集合、字典或其他可变类型作为字典的"键"。另外,字典中的"键"不允许重复,而"值"是可以重复的。

2.3.1　字典创建和元素添加、修改与删除

使用赋值运算符"="将一个字典赋值给一个变量即可创建一个字典变量。

```
>>>a_dict={'server': 'db.diveintopython3.org', 'database': 'mysql'}
>>>a_dict
{'database': 'mysql', 'server': 'db.diveintopython3.org'}
```

也可以使用内置函数 dict()通过已有数据快速创建字典:

```
>>>keys=['a', 'b', 'c', 'd']
>>>values=[1, 2, 3, 4]
>>>dictionary=dict(zip(keys, values))
>>>print(dictionary)
{'a': 1, 'c': 3, 'b': 2, 'd': 4}
>>>x=dict()                          #空字典
>>>x
{}
>>>type(x)                           #查看对象类型
<class 'dict'>
>>>x={}                              #空字典
>>>x
{}
>>>type(x)
<class 'dict'>
```

还可以使用内置函数 dict()根据给定的"键:值"来创建字典:

```
>>>d=dict(name='Dong', age=37)
>>>d
{'age': 37, 'name': 'Dong'}
```

还可以以给定内容为"键",创建"值"为空的字典:

```
>>>adict=dict.fromkeys(['name', 'age', 'sex'])
```

```
>>>adict
{'age': None, 'name': None, 'sex': None}
```

当以指定"键"为下标为字典元素赋值时,有两种含义:①若该"键"存在,则表示修改该"键"对应的值;②若该键不存在,则表示添加一个新的"键:值",也就是添加一个新元素。

```
>>>aDict
{'age': 35, 'name': 'Dong', 'sex': 'male'}
>>>aDict['age']=38                    #修改元素值
>>>aDict
{'age': 38, 'name': 'Dong', 'sex': 'male'}
>>>aDict['address']='SDIBT'           #添加新元素
>>>aDict
{'age': 38, 'address': 'SDIBT', 'name': 'Dong', 'sex': 'male'}
```

使用字典对象的 update() 方法可以将另一个字典的"键:值"一次性全部添加到当前字典对象,如果两个字典中存在相同的"键",则以另一个字典中的"值"为准对当前字典进行更新。

```
>>>aDict
{'age': 37, 'score': [98, 97], 'name': 'Dong', 'sex': 'male'}
>>>aDict.items()                      #返回所有元素
dict_items([('sex', 'male'),('score', [98, 97]),('age', 37),('name', 'Dong')])
>>>aDict.update({'a':97, 'age':39})   #修改'age'键的值,同时添加新元素'a':97
>>>aDict
{'score': [98, 97], 'sex': 'male', 'a': 97, 'age': 39, 'name': 'Dong'}
```

使用 del 命令可以删除整个字典,也可以删除字典中指定的元素。例如,继续上面的代码:

```
>>>del aDict['age']                   #删除字典元素
>>>aDict
{'score': [98, 97], 'sex': 'male', 'a': 97, 'name': 'Dong'}
>>>del aDict                          #删除整个字典
>>>aDict                              #字典对象被删除后不再存在
Traceback(most recent call last):
  File "<pyshell#291>", line 1, in <module>
    aDict
NameError: name 'aDict' is not defined
```

也可以使用字典对象的 pop() 和 popitem() 方法弹出并删除指定的元素,例如:

```
>>>aDict={'age': 37, 'score': [98, 97], 'name': 'Dong', 'sex': 'male'}
>>>aDict.popitem()                    #弹出一个元素,对空字典会抛出异常
('age', 37)
>>>aDict.pop('sex')                   #弹出指定键对应的元素
```

```
'male'
>>>aDict
{'score': [98, 97], 'name': 'Dong'}
```

字典对象的 clear() 方法用于清空字典对象中的所有元素，copy() 方法返回字典对象的浅复制，不再赘述。

2.3.2 访问字典对象的数据

字典中的每个元素表示一种映射关系或对应关系，根据提供的"键"作为下标就可以访问对应的"值"，如果字典中不存在这个"键"会抛出异常，例如：

```
>>>aDict={'age': 37, 'score': [98, 97], 'name': 'Dong', 'sex': 'male'}
>>>aDict['age']                       #指定的"键"存在,返回对应的"值"
37
>>>aDict['address']                   #指定的"键"不存在,抛出异常
Traceback(most recent call last):
  File "<pyshell#26>", line 1, in <module>
    aDict['address']
KeyError: 'address'
>>>assert 'address' in aDict, 'Key "address" not in dict'
Traceback(most recent call last):
  File "<pyshell#44>", line 1, in <module>
    assert 'address' in aDict, 'Key "address" not in dict'
AssertionError: Key "address" not in dict
```

为了避免程序运行时引发异常而导致崩溃，在使用下标的方式访问字典元素时，最好能配合条件判断或者异常处理结构，例如：

```
>>>aDict={'age': 37, 'score': [98, 97], 'name': 'Dong', 'sex': 'male'}
>>>if 'Age' in aDict:                 #首先判断字典中是否存在指定的"键"
    print(aDict['Age'])
else:
    print('Not Exists.')

Not Exists.
>>>try:                               #使用异常处理结构
    print(aDict['address'])
except:
    print('Not Exists.')

Not Exists.
```

上面的方法虽然能够满足要求，但是代码显得非常啰唆，有没有更好的办法呢？答案是肯定的，强大的 Python 怎么可能会这么不人性化呢。字典对象提供了一个 get() 方法用来返回指定"键"对应的"值"，更妙的是这个方法允许指定该键不存在时返回特定的

"值"。例如：

```
>>>aDict.get('age')                              #如果字典中存在该"键",则返回对应的"值"
37
>>>aDict.get('address', 'Not Exists.')           #指定的"键"不存在时返回指定的默认值
'Not Exists.'
>>>aDict
{'age': 37, 'score': [98, 97], 'name': 'Dong', 'sex': 'male'}
```

字典对象的setdefault()方法用于返回指定"键"对应的"值"，如果字典中不存在该"键"，就添加一个新元素并设置该"键"对应的"值"，例如：

```
>>>aDict.setdefault('address', 'SDIBT')  #增加新元素
'SDIBT'
>>>aDict
{'age': 37, 'score': [98, 97], 'name': 'Dong', 'address': 'SDIBT', 'sex': 'male'}
```

最后，当对字典对象进行迭代时，默认是遍历字典的"键"，这一点必须清醒地记在脑子里。当然，可以使用字典对象的items()方法返回字典中的元素，即所有"键:值"对，字典对象的keys()方法返回所有"键"，values()方法返回所有"值"。例如：

```
>>>aDict={'age': 37, 'score': [98, 97], 'name': 'Dong', 'sex': 'male'}
>>>for item in aDict:                            #默认遍历字典的"键"
    print(item)

score
age
sex
name
>>>for item in aDict.items():                    #明确指定遍历字典的元素
    print(item)

('score', [98, 97])
('age', 37)
('sex', 'male')
('name', 'Dong')
>>>aDict.items()
dict_items([('age', 37), ('score', [98, 97]), ('name', 'Dong'), ('sex', 'male')])
>>>aDict.keys()
dict_keys(['age', 'score', 'name', 'sex'])
>>>aDict.values()
dict_values([37, [98, 97], 'Dong', 'male'])
```

小提示：内置函数len()、max()、min()、sum()、sorted()以及成员测试运算符in也适用于字典对象，但默认是作用于字典的"键"。若想作用于元素，即"键:值"对，则

需要使用字典对象的items()方法明确指定;若想作用于"值",则需要使用values()明确指定。

2.3.3 案例精选

统计分析在很多领域都有重要用途,如密码破解、图像直方图等。下面的代码首先生成包含1000个随机字符的字符串,然后统计每个字符的出现次数。

```
>>>import string
>>>import random
>>>x=string.ascii_letters+string.digits+string.punctuation
>>>x
'abcdefghijklmnopqrstuvwxyzABCDEFGHIJKLMNOPQRSTUVWXYZ0123456789!"#$%&\'()*+,-./:;<=>?@[\\]^_`{|}~'
>>>y=[random.choice(x)for i in range(1000)]    #生成包含1000个随机字符的列表
>>>z=''.join(y)                                 #把列表中的字符连接成为字符串
>>>d=dict()                                     #空字典
>>>for ch in z:
       d[ch]=d.get(ch,0)+1                      #修改每个字符的频次
```

也可以使用collections模块的defaultdict类来实现该功能。

```
>>>import string
>>>import random
>>>x=string.ascii_letters+string.digits+string.punctuation
>>>y=[random.choice(x)for i in range(1000)]
>>>z=''.join(y)
>>>from collections import defaultdict
>>>frequences=defaultdict(int)                  #所有值默认为0
>>>frequences
defaultdict(<class 'int'>, {})
>>>for item in z:
       frequences[item] +=1                     #修改每个字符的频次
>>>frequences.items()
```

使用collections模块的Counter类可以快速实现这个功能,并且能够满足其他需要,例如,查找出现次数最多的元素。下面的代码演示了Counter类的用法:

```
>>>from collections import Counter
>>>frequences=Counter(z)                        #这里的z还是前面代码中的字符串对象
>>>frequences.items()
>>>frequences.most_common(1)                    #返回出现次数最多的1个字符及其频率
>>>frequences.most_common(3)                    #返回出现次数最多的前3个字符及其频率
```

拓展知识:内置函数globals()和locals()分别返回包含当前作用域内所有全局变量和局部变量的名称及值的字典。例如:

```
>>>a = (1, 2, 3, 4, 5)                          #全局变量
>>>b='Hello world.'                             #全局变量
>>>def demo():
    a=3                                         #局部变量
    b=[1, 2, 3]                                 #局部变量
    print('locals:', locals())
    print('globals:', globals())
>>>demo()
locals: {'a': 3, 'b': [1, 2, 3]}
globals: {'a':(1, 2, 3, 4, 5), 'b': 'Hello world.', '__builtins__': <module '__builtin__'(built-in)>, 'demo': <function demo at 0x013907F0>, '__package__': None, '__name__': '__main__', '__doc__': None}
```

拓展知识：有序字典。Python 内置字典是无序的,如果需要一个可以记住元素插入顺序的字典,可以使用 collections.OrderedDict。例如：

```
>>>import collections
>>>x=collections.OrderedDict()                  #有序字典
>>>x['a']=3
>>>x['b']=5
>>>x['c']=8
>>>x
OrderedDict([('a', 3),('b', 5),('c', 8)])
```

拓展知识：内置函数 sorted()可以对字典元素进行排序并返回新列表,充分利用 key 参数可以实现丰富的排序功能。例如：

```
>>>phonebook={'Linda':'7750', 'Bob':'9345', 'Carol':'5834'}
>>>from operator import itemgetter
>>>sorted(phonebook.items(), key=itemgetter(1))         #按字典的"值"进行排序
[('Carol', '5834'),('Linda', '7750'),('Bob', '9345')]
>>>sorted(phonebook.items(), key=itemgetter(0))         #按字典的"键"进行排序
[('Bob', '9345'),('Carol', '5834'),('Linda', '7750')]
>>>sorted(phonebook.items(), key=lambda item:item[0])   #按字典的"键"进行排序
[('Bob', '9345'),('Carol', '5834'),('Linda', '7750')]
>>>persons=[ {'name':'Dong', 'age':37},
             {'name':'Zhang', 'age':40},
             {'name':'Li', 'age':50},
             {'name':'Dong', 'age':43}]
>>>print(persons)
[{'age': 37, 'name': 'Dong'}, {'age': 40, 'name': 'Zhang'}, {'age': 50, 'name': 'Li'}, {'age': 43, 'name': 'Dong'}]
#使用 key 来指定排序依据,先按姓名升序排序,姓名相同的按年龄降序排序
#注意,在某一项前面加负号表示降序排序,这一点只适用于数字类型,不通用
>>>print(sorted(persons, key=lambda x:(x['name'], -x['age'])))
```

```
[{'age': 43, 'name': 'Dong'}, {'age': 37, 'name': 'Dong'}, {'age': 50, 'name': 'Li
'}, {'age': 40, 'name': 'Zhang'}]
```

拓展知识：Python 支持字典推导式快速生成符合特定条件的字典。

```
>>>{i:str(i)for i in range(1, 5)}
{1: '1', 2: '2', 3: '3', 4: '4'}
>>>x=['A', 'B', 'C', 'D']
>>>y=['a', 'b', 'b', 'd']
>>>{i:j for i,j in zip(x,y)}
{'A': 'a', 'C': 'b', 'B': 'b', 'D': 'd'}
```

2.4 集　　合

2.4.1 集合基础知识

集合是无序可变序列，使用一对大括号(作者温馨提示：这一点和字典很类似，千万不要搞混啊)作为界定符，元素之间使用逗号分隔，同一个集合内的每个元素都是唯一的，元素之间不允许重复。

在 Python 中，直接将集合赋值给变量即可创建一个集合对象。

```
>>>a={3, 5}                              #创建集合对象
>>>a
{3, 5}
>>>type(a)                               #查看对象类型
<class 'set'>
```

也可以使用 set()函数将列表、元组等其他可迭代对象转换为集合，如果原来的数据中存在重复元素，则在转换为集合的时候只保留一个。

```
>>>a_set=set(range(8, 14))               #把 range 对象转换为集合
>>>a_set
{8, 9, 10, 11, 12, 13}
>>>b_set=set([0, 1, 2, 3, 0, 1, 2, 3, 7, 8])   #转换时自动去掉重复元素
>>>b_set
{0, 1, 2, 3, 7, 8}
>>>x=set()                               #空集合
>>>x
{}
```

当不再使用某个集合时，可以使用 del 命令删除整个集合。

注意：集合中只能包含数字、字符串、元组等不可变类型(或者说可哈希)的数据，而不能包含列表、字典、集合等可变类型的数据。Python 提供了一个内置函数 hash()来计算对象的哈希值，凡是无法计算哈希值(调用 hash()函数时抛出异常)的对象都不能作

为集合的元素,也不能作为字典对象的"键"。

拓展知识:字典和集合的 in 操作比列表快很多。相信各位读者也能有这样的体验,一个功能可以使用多种方法实现,也可以采用不同的数据类型实现。如果仔细分析比较一下会发现,不同数据类型之间某些操作的效率相差还是很大的,在选用时应多加注意,因为不同的选择意味着不同的速度和效率。例如,由于 Python 字典和集合都使用 hash 表来存储元素,因此元素查找操作的速度非常快,这就直接决定了关键字 in 作用于字典和集合时比作用于列表要快得多。

```python
import random
import time

x=list(range(10000))                        #生成列表
y=set(range(10000))                         #生成集合
z=dict(zip(range(1000),range(10000)))       #生成字典
r=random.randint(0, 9999)                   #生成随机数

start=time.time()
for i in range(9999999):
    r in x                                  #测试列表中是否包含某个元素
print('list,time used:', time.time()-start)

start=time.time()
for i in range(9999999):
    r in y                                  #测试集合中是否包含某个元素
print('set,time used:', time.time()-start)

start=time.time()
for i in range(9999999):
    r in z                                  #测试字典中是否包含某个元素
print('dict,time used:', time.time()-start)
```

上面的代码运行结果如下,对于成员测试运算符 in,列表的效率远远不如字典和集合,差距简直太惊人了。大家修改一下上面代码中列表、字典和集合的长度会发现,随着序列的变长,列表的速度越来越慢,而字典和集合基本上不受影响。

```
list,time used: 889.4648745059967
set,time used: 1.5110864639282227
dict,time used: 1.0640606880187988
```

2.4.2 集合操作与运算

1. 集合元素增加与删除

使用集合对象的 add()方法可以为其增加新元素,如果该元素已存在于集合则忽略

该操作;update()方法用于合并另外一个集合中的元素到当前集合中。例如：

```
>>>s={1, 2, 3}
>>>s.add(3)                              #添加元素,重复元素自动忽略
>>>s
{1, 2, 3}
>>>s.update({3,4})                       #更新当前集合,自动忽略重复的元素
>>>s
{1, 2, 3, 4}
```

集合对象的 pop()方法用于随机删除并返回集合中的一个元素,如果集合为空则抛出异常;remove()方法用于删除集合中的元素,如果指定元素不存在则抛出异常;discard()用于从集合中删除一个特定元素,如果元素不在集合中则忽略该操作;clear()方法清空集合删除所有元素。例如：

```
>>>s.discard(5)                          #删除元素,不存在则忽略该操作
>>>s
{1, 2, 3, 4}
>>>s.remove(5)                           #删除元素,不存在就抛出异常
Traceback (most recent call last):
  File "<pyshell#425>", line 1, in <module>
    s.remove(5)
KeyError: 5
>>>s.pop()                               #删除并返回一个元素
1
```

2. 集合运算

Python 集合支持交集、并集、差集等运算,例如：

```
>>>a_set=set([8, 9, 10, 11, 12, 13])
>>>b_set={0, 1, 2, 3, 7, 8}
>>>a_set | b_set                         #并集
{0, 1, 2, 3, 7, 8, 9, 10, 11, 12, 13}
>>>a_set.union(b_set)                    #并集
{0, 1, 2, 3, 7, 8, 9, 10, 11, 12, 13}
>>>a_set & b_set                         #交集
{8}
>>>a_set.intersection(b_set)             #交集
{8}
>>>a_set.difference(b_set)               #差集
{9, 10, 11, 12, 13}
>>>a_set-b_set
{9, 10, 11, 12, 13}
>>>a_set.symmetric_difference(b_set)     #对称差集
```

```
{0, 1, 2, 3, 7, 9, 10, 11, 12, 13}
>>>a_set ^ b_set
{0, 1, 2, 3, 7, 9, 10, 11, 12, 13}
>>>x={1, 2, 3}
>>>y={1, 2, 5}
>>>z={1, 2, 3, 4}
>>>x<y                                    #比较集合大小
False
>>>x<z
True
>>>y<z
False
>>>x.issubset(y)                          #测试是否为子集
False
>>>x.issubset(z)
True
```

小提示：内置函数 len()、max()、min()、sum()、sorted()以及成员测试运算符 in 也适用于集合。

注意：关系运算符＞、＞＝、＜、＜＝作用于集合时表示集合之间的包含关系，而不是集合中元素的大小关系。例如，两个集合 A 和 B，如果 A＜B 不成立，不代表 A＞B 就一定成立。

拓展知识：自定义枚举类型。除了本章介绍的常用数据类型，Python 还通过 collections、enum、array、heapq、fractions 等标准库提供了其他丰富的类型，这里简单介绍如何使用 enum 模块提供的 Enum 类来创建枚举类型，其他标准库将在后续章节中根据内容的组织逐步进行介绍。

```
>>>from enum import Enum                  #导入模块中的类
>>>class Color(Enum):                     #创建自定义枚举类
    red=1
    blue=2
    green=3
>>>Color.red                              #访问枚举类的成员
<Color.red: 1>
>>>type(Color.green)                      #查看枚举类成员的类型
<enum 'Color'>
>>>isinstance(Color.red, Color)
True
>>>x=dict()
>>>x[Color.red]='red'                     #枚举类成员可以哈希,可以作为字典的"键"
>>>x
{<Color.red: 1>: 'red'}
>>>Color(2)                               #返回指定值对应的枚举类成员
```

```
<Color.blue: 2>
>>>Color['red']
<Color.red: 1>
>>>r=Color.red
>>>r.name
'red'
>>>r.value
1
>>>list(Color)                                    #枚举类是可以迭代的
[<Color.red: 1>, <Color.blue: 2>, <Color.green: 3>]
```

2.4.3 案例精选

作为集合的具体应用，可以使用集合快速提取序列中单一元素，即提取出序列中所有不重复元素。如果使用传统方式，需要编写下面的代码：

```
>>>import random
#生成 100 个介于 0~9999 之间的随机数
>>>listRandom=[random.choice(range(10000)) for i in range(100)]
>>>noRepeat=[]
>>>for i in listRandom :
    if i not in noRepeat :
        noRepeat.append(i)
>>>len(listRandom)
>>>len(noRepeat)
```

而如果使用集合，只需要下面这么一行代码就可以了，可以参考上面的代码对结果进行验证。

```
>>>newSet=set(listRandom)
```

拓展知识：集合中的元素不允许重复，Python 集合的内部实现为此做了大量相应的优化，判断集合中是否包含某元素时比列表速度快很多。前面已经介绍了这个内容，这里再给一个例子，下面的代码用于返回指定范围内一定数量的不重复数字，使用集合的效率明显优于使用列表。

```
import random
import time

def RandomNumbers1(number, start, end):
    '''使用列表来生成 number 个介于 start 和 end 之间的不重复随机数'''
    data=[]
    while True:
        element=random.randint(start, end)
        if element not in data:
```

```
            data.append(element)
        if len(data)==number:
            break
    return data

def RandomNumbers2(number, start, end):
    '''使用集合来生成 number 个介于 start 和 end 之间的不重复随机数'''
    data=set()
    while True:
        element=random.randint(start, end)
        data.add(element)
        if len(data)==number:
            return data

start=time.time()
for i in range(10000):
    d1=RandomNumbers1(500, 1, 10000)
print('Time used:', time.time()-start)

start=time.time()
for i in range(10000):
    d2=RandomNumbers2(500, 1, 10000)
print('Time used:', time.time()-start)
```

运行结果为

```
Time used: 41.77738952636719
Time used: 13.330762386322021
```

上面的代码只是为了展示 Python 获取不重复元素的原理，如果在项目中需要这样一个功能时，还是直接使用下面的方法更好一些，random 模块的 sample() 方法可以直接从指定序列中选取指定数量个不重复的元素。

```
>>>import random
>>>random.sample(range(1000), 20)
[61, 538, 873, 815, 708, 609, 995, 64, 7, 719, 922, 859, 807, 464, 789, 651, 31, 702, 504, 25]
```

思考题：我给学生讲这段代码的时候，有同学修改参数进行调用，例如 RandomNumbers2(500, 1, 100)，结果导致死循环。你能想到原因吗？

拓展知识：Python 也支持集合推导式。集合推导式是什么？请看下面的代码：

```
>>>{x.strip() for x in ('  he  ', 'she   ', '   I')}
{'I', 'she', 'he'}
>>>import random
```

```
>>>x={random.randint(1,500)for i in range(100)}
                                    #生成随机数,自动去除重复元素
>>>len(x)                           #一般而言输出结果会小于100
>>>{str(x)for x in range(10)}
{'3', '0', '1', '8', '4', '7', '5', '6', '9', '2'}
```

2.5 序列解包

在实际开发中,序列解包是非常重要和常用的一个功能,可以使用非常简洁的形式完成复杂的功能,大幅度提高了代码的可读性,并且减少了程序员的代码输入量。例如,可以使用序列解包功能对多个变量同时进行赋值,下面都是合法的 Python 赋值方法。

```
>>>x, y, z=1, 2, 3                  #多个变量同时赋值
>>>v_tuple = (False, 3.5, 'exp')
>>> (x, y, z)=v_tuple
>>>x, y, z=v_tuple
>>>x, y, z=range(3)                 #可以使用 range 对象进行序列解包
>>>x, y, z=map(str, range(3))       #使用迭代对象进行序列解包
```

序列解包也可以用于列表和字典,但是对字典使用时,默认是对字典"键"进行操作,如果需要对"键:值"对进行操作,需要使用字典的 items()方法说明,如果需要对字典"值"进行操作,则需要使用字典的 values()方法明确指定。下面的代码演示了列表与字典的序列解包操作:

```
>>>a=[1, 2, 3]
>>>b, c, d=a                        #列表也支持序列解包的用法
>>>b
1
>>>x, y, z=sorted([1, 3, 2])        #sorted()函数返回排序后的列表
>>>s={'a':1, 'b':2, 'c':3}
>>>b, c, d=s.items()
>>>b                                #这里的结果如果和你的不一样是正常的
('c', 3)
>>>b, c, d=s                        #使用字典时不用太多考虑元素的顺序
>>>b                                #多执行几次试一试,或许结果会有变化
'c'
>>>b, c, d=s.values()
>>>print(b, c, d)
1 3 2
```

使用序列解包可以很方便地同时遍历多个序列。

```
>>>keys=['a', 'b', 'c', 'd']
>>>values=[1, 2, 3, 4]
>>>for k, v in zip(keys, values):
```

```
        print(k, v)
a 1
b 2
c 3
d 4
```

下面代码演示了对内置函数 enumerate() 返回的迭代对象进行遍历时序列解包的用法：

```
>>>x=['a', 'b', 'c']
>>>for i, v in enumerate(x):
    print('The value on position {0} is {1}'.format(i,v))
The value on position 0 is a
The value on position 1 is b
The value on position 2 is c
```

下面对字典的操作也使用到序列解包：

```
>>>s={'a':1, 'b':2, 'c':3}
>>>for k, v in s.items():                   #字典中的每个元素包含"键"和"值"两部分
    print(k, v)
a 1
c 3
b 2
```

另外，序列解包还支持下面的用法：

```
>>>print(*[1, 2, 3], 4, *(5, 6))
1 2 3 4 5 6
>>> *range(4),4
(0, 1, 2, 3, 4)
>>>{*range(4), 4, *(5, 6, 7)}
{0, 1, 2, 3, 4, 5, 6, 7}
>>>{'x': 1, **{'y': 2}}
{'y': 2, 'x': 1}
```

小提示：①序列解包的有些用法在低版本的 Python 中不支持；②在调用函数时，在实参前面加上一个星号(*)也可以进行序列解包，从而实现将序列中的元素值依次传递给相同数量的形参，详见3.3节。

第 3 章 程序控制结构与函数设计

3.1 选择结构

生活中处处充满了选择：如果周末不下雨我就约朋友去爬山，否则就去教学楼大厅里打太极拳；如果某个同学平时学习很认真但是期末考试前确实因为临时有事耽误了复习而没有考好，只要差得不太多我也一样给打及格；去市场买菜的时候比较一下，哪家的菜又好又便宜就买哪家的；诸如此类，我们时刻都在根据实际条件做出这样或那样的选择。编写程序也是如此，当某个条件得到满足时就去做特定的事情，否则就做另一件事情，这就是选择结构。

3.1.1 条件表达式

在选择结构和循环结构中，都要根据条件表达式的值来确定下一步的执行流程。条件表达式的值只要不是 False、0(或 0.0、0j 等)、空值 None、空列表、空元组、空集合、空字典、空字符串、空 range 对象或其他空迭代对象，Python 解释器均认为与 True 等价。从这个意义上来讲，所有的 Python 合法表达式都可以作为条件表达式，包括含有函数调用的表达式。例如：

```
>>>if 3:                        #使用整数作为条件表达式
    print(5)
5
>>>a=[1, 2, 3]
>>>if a:                        #使用列表作为条件表达式
    print(a)
[1, 2, 3]
>>>a=[]
>>>if a:
    print(a)
else:
    print('empty')
empty
>>>i=s=0
>>>while i <=10:                #使用关系表达式作为条件表达式
```

```
        s +=i
        i +=1
>>>print(s)
55
>>>i=s=0
>>>while True:                          #使用常量 True 作为条件表达式
        s +=i
        i +=1
        if i>10:                        #符合特定条件时使用 break 语句退出循环
            break
>>>print(s)
55
>>>s=0
>>>for i in range(0, 11, 1):            #遍历序列元素
        s +=i
>>>print(s)
55
```

关于表达式和运算符的详细内容在 1.3.3 节中已有介绍,这里不再赘述,只重点介绍一下比较特殊的几个运算符。首先是关系运算符,与很多语言不同的是,在 Python 中的关系运算符可以连续使用,例如:

```
>>>print(1<2<3)
True
>>>print(1<2>3)
False
>>>print(1<3>2)
True
```

在 Python 中,条件表达式中不允许使用赋值运算符"=",避免了其他语言中误将关系运算符"=="写作赋值运算符"="带来的麻烦,例如,下面的代码在条件表达式中使用赋值运算符"="将抛出异常,提示语法错误。

```
>>>if a=3:                              #条件表达式中不允许使用赋值运算符
SyntaxError: invalid syntax
>>>if(a=3)and(b=4):
SyntaxError: invalid syntax
```

比较特殊的运算符还有逻辑运算符 and 和 or,这两个运算符具有短路求值或惰性求值的特点,简单地说,就是只计算必须计算的表达式的值。以 and 为例,对于表达式"表达式 1 and 表达式 2"而言,如果"表达式 1"的值为 False 或其他等价值时,不论"表达式 2"的值是什么,整个表达式的值都是 False,此时"表达式 2"的值无论是什么都不影响整个表达式的值,因此将不会被计算,从而减少不必要的计算和判断。另外,Python 中的逻辑运算符在某些方面和其他语言也有所不同,例如:

```
>>>3 and 5
5
>>>3 or 5
3
>>>0 and 5
0
>>>0 or 5
5
>>>not 3
False
>>>not 0
True
```

下面的函数使用指定的分隔符把多个字符串连接成一个字符串,如果用户没有指定分隔符则使用逗号。

```
>>>def Join(chList, sep=None):
    return(sep or ',').join(chList)      #注意:参数 sep 不是字符串时会抛出异常
>>>chTest=['1', '2', '3', '4', '5']
>>>Join(chTest)
'1,2,3,4,5'
>>>Join(chTest, ':')
'1:2:3:4:5'
>>>Join(chTest, ' ')
'1 2 3 4 5'
```

当然,也可以把上面的函数直接定义为下面带有默认值参数的形式:

```
>>>def Join(chList, sep=','):
    return sep.join(chList)
```

小技巧:在设计包含多个条件的条件表达式时,如果能够大概预测不同条件失败的概率,并将多个条件根据 and 和 or 运算符的短路求值特性来组织顺序,可以大幅度提高程序运行效率,减少不必要的计算,这也属于代码优化的内容。

拓展知识:逻辑运算符与常见电路连接方式的相似之处。大家应该都学过高中物理,还记得并联电路、串联电路、短路这样的概念吗? 可以做个简单类比,or 运算符类似于并联电路,只要有一个开关是通的那么灯就是亮的;and 运算符类似于串联电路,必须所有开关是通的那么灯才是亮的;not 运算符类似于短路电路,如果开关通了那么灯就灭了,如图 3-1 所示。

3.1.2 选择结构的几种形式

选择结构通过判断某些特定条件是否满足来决定下一步的执行流程,是非常重要的控制结构。常见的有单分支选择结构、双分支选择结构、多分支选择结构以及嵌套的分支

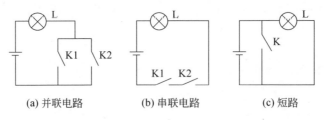

图 3-1　逻辑运算符与几种电路的类比关系

结构,形式比较灵活多变,具体使用哪一种最终还是取决于要实现的业务逻辑。循环结构和异常处理结构中也可以带有 else 子句,也可以看作是特殊形式的选择结构。

1. 单分支选择结构

单分支选择结构是最简单的一种形式,其语法如下所示,其中表达式后面的冒号":"是不可缺少的,表示一个语句块的开始,后面几种其他形式的选择结构和循环结构中的冒号也是必须要有的。

```
if 表达式:
    语句块
```

当表达式值为 True 或其他等价值时,表示条件满足,语句块将被执行,否则该语句块将不被执行,继续执行后面的代码(如果有),如图 3-2 所示。

下面的代码简单演示了单分支选择结构的用法:

```
x=input('Input two numbers:')
a,b=map(int,x.split())
if a>b:
    a,b=b,a                    #序列解包,交换两个变量的值
print(a,b)
```

图 3-2　单分支选择结构

注意:在 Python 中,代码的缩进非常重要,缩进是体现代码逻辑关系的重要方式,同一个代码块必须保证相同的缩进量。有的老师让学生学习 Python 的原因之一就是 Python 对代码排版或布局的严格要求可以培养学生严谨的习惯。而实际上,只要遵循一定的约定,Python 代码的排版是可以降低要求的,例如下面的代码:

```
>>>if 3>2: print('ok')              #如果语句较短,可以直接写在分支语句后面
ok
>>>if True:print(3);print(5)        #在一行写多个语句,使用分号分隔
3
5
```

小提示:在上面代码中,"a,b=b,a"是 Python 序列解包的用法,用来交换两个变量的值,等价于 C 语言的如下 3 条代码(假设变量已经声明并且类型正确),关于序列

解包更多内容请参考 2.5 节。

```
c=a;
a=b;
b=c;
```

2．双分支选择结构

双分支选择结构的语法为

```
if 表达式：
    语句块 1
else：
    语句块 2
```

当表达式值为 True 或其他等价值时，执行语句块 1，否则执行语句块 2，如图 3-3 所示。

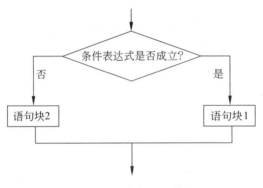

图 3-3　双分支选择结构

下面的代码演示了双分支选择结构的用法：

```
>>>chTest=['1','2','3','4','5']
>>>if chTest:
    print(chTest)
else:
    print('Empty')
['1','2','3','4','5']
```

🔖 **拓展知识**：Python 还提供了一个三元运算符，可以实现与选择结构相似的效果。语法为

```
value1 if condition else value2
```

当条件表达式 condition 的值与 True 等价时，表达式的值为 value1，否则表达式的值为 value2。另外，value1 和 value2 本身也可以是复杂表达式，也可以包含函数调用。下面的代码演示了上面的表达式的用法，可以看出，这个结构的表达式也具有惰性求值的

特点。

```
>>>a=5
>>>print(6) if a>3 else print(5)
6
>>>print(6 if a>3 else 5)              #注意,虽然结果与上一行代码一样,但代码含义不同
6
>>>b=6 if a>13 else 9                  #赋值运算符优先级低
>>>b
9
>>>x=math.sqrt(9) if 5>3 else random.randint(1, 100)
                                       #此时还没有导入math模块
Traceback(most recent call last):
  File "<pyshell#23>", line 1, in <module>
    x=math.sqrt(9) if 5>3 else random.randint(1,100)
NameError: name 'math' is not defined
>>>import math
#此时没有导入random模块,由于条件表达式 5>3 的值为 True,所以可以正常运行
>>>x=math.sqrt(9) if 5>3 else random.randint(1,100)
#第一个条件表达式 2>3 的值为 False,需要计算第二个表达式的值,然而此时还没有导入
random模块,从而出错
>>>x=math.sqrt(9) if 2>3 else random.randint(1, 100)
Traceback(most recent call last):
  File "<pyshell#26>", line 1, in <module>
    x=math.sqrt(9) if 2>3 else random.randint(1,100)
NameError: name 'random' is not defined
>>>import random
>>>x=math.sqrt(9) if 2>3 else random.randint(1, 100)
```

最后,三元运算符是可以嵌套使用的,可以实现多分支选择的效果,但这样的代码可读性非常不好,不建议使用。

```
>>>x=3
>>>(1 if x>2 else 0) if f(x)>5 else ('a' if x<5 else 'b')
                                       #可以嵌套使用,但不提倡这样用
1
>>>x=0
>>>(1 if x>2 else 0) if f(x)>5 else ('a' if x<5 else 'b')
'a'
```

3. 多分支选择结构

多分支选择结构为用户提供了更多的选择,可以实现复杂的业务逻辑,多分支选择结构的语法为

```
if 表达式1:
```

```
    语句块 1
elif 表达式 2:
    语句块 2
elif 表达式 3:
    语句块 3
  ⋮
else:
    语句块 n
```

其中,关键字 elif 是 else if 的缩写。下面的代码演示了如何利用多分支选择结构将成绩从百分制变换到等级制。

```
>>>def func(score):
    if score>100:
        return 'wrong score.must <=100.'
    elif score >=90:
        return 'A'
    elif score >=80:
        return 'B'
    elif score >=70:
        return 'C'
    elif score >=60:
        return 'D'
    elif score >=0:
        return 'E'
    else:
        return 'wrong score.must >0'
>>>func(120)
'wrong score.must <=100.'
>>>func(99)
'A'
>>>func(87)
'B'
>>>func(62)
'D'
>>>func(3)
'E'
>>>func(-10)
'wrong score.must >0'
```

4. 选择结构的嵌套

选择结构可以进行嵌套来表达复杂的业务逻辑,语法如下:

```
if 表达式 1:
```

```
        语句块 1
        if 表达式 2:
            语句块 2
        else:
            语句块 3
    else:
        if 表达式 4:
            语句块 4
```

上面语法示意中的代码层次和隶属关系如图 3-4 所示,注意相同层次的代码必须具有相同的缩进量。

使用嵌套选择结构时,一定要严格控制好不同级别代码块的缩进量,因为这决定了不同代码块的从属关系和业务逻辑是否被正确地实现,以及代码是否能够被 Python 正确理解和执行。例如,前面百分制转等级制的代码,作为一种编程技巧,还可以尝试下面的写法:

```
if 表达式1:
    语句块1
    if 表达式2:
    3 │ 语句块2
    else:
    3 │ 语句块3
else:
    if 表达式4:
    3 │ 语句块4
```

图 3-4 代码层次与隶属关系

```
>>>def func(score):
    degree='DCBAAE'
    if score>100 or score<0:
        return 'wrong score.must between 0 and 100.'
    else:
        index=(score-60)//10
        if index>=0:
            return degree[index]
        else:
            return degree[-1]
>>>func(-10)
'wrong score.must between 0 and 100.'
>>>func(30)
'E'
>>>func(50)
'E'
>>>func(60)
'D'
>>>func(93)
'A'
>>>func(100)
'A'
```

小提示:①在 IDLE 交互式环境中每次只能执行一条语句,如果需要编写多条语句实现复杂的业务逻辑,需要创建一个 Python 程序文件;②嵌套选择结构、嵌套循环结构以及选择结构与循环结构的互相嵌套代码中,缩进一定要控制好,保证代码的隶属关系正确。

3.1.3 案例精选

例 3-1 面试资格确认。

```
age=24
subject="计算机"
college="非重点"
if(age>25 and subject=="电子信息工程")or(college=="重点" and subject=="电子信息工程")or(age<=28 and subject=="计算机"):
    print("恭喜,您已获得我公司的面试机会!")
else:
    print("抱歉,您未达到面试要求")
```

建议:在编写条件表达式时,建议适当使用括号,这样可以更准确地表达业务逻辑,同时提高代码可读性。

例 3-2 用户输入若干个成绩,求所有成绩的平均分。每输入一个成绩后询问是否继续输入下一个成绩,回答 yes 就继续输入下一个成绩,回答 no 就停止输入成绩。

```
numbers=[]
while True:
    x=input('请输入一个整数:')
    try:                                    #异常处理结构有关知识见第 7 章
        numbers.append(int(x))
    except:
        print('不是整数')
    while True:
        flag=input('继续输入吗? (yes/no)')
        if flag.lower()not in('yes', 'no'):  #限定用户输入内容必须为 yes 或 no
            print('只能输入 yes 或 no')
        else:
            break
    if flag.lower()=='no':
        break

print(sum(numbers)/len(numbers))
```

例 3-3 编写程序,判断今天是今年的第几天。

```
import time
date=time.localtime()                        #获取当前日期时间
year,month,day=date[:3]
day_month=[31, 28, 31, 30, 31, 30, 31, 31, 30, 31, 30, 31]
if year%400==0 or(year%4==0 and year%100!=0):  #判断是否为闰年
    day_month[1]=29
```

```
if month==1:
    print(day)
else:
    print(sum(day_month[:month-1])+day)
```

拓展知识：Python 标准库 datetime。这个标准库提供了 timedelta 对象可以很方便地计算指定年、月、日、时、分、秒之前或之后的日期时间，还提供了返回结果中包含"今天是今年第几天"、"今天是本周第几天"等答案的 timetuple()函数，等等。

```
>>>import datetime
>>>Today=datetime.date.today()
>>>Today
datetime.date(2015, 12, 6)
>>>Today-datetime.date(Today.year, 1, 1)+datetime.timedelta(days=1)
datetime.timedelta(340)
>>>Today.timetuple().tm_yday                #今天是今年的第几天
340
>>>Today.replace(year=2013)                 #替换日期中的年
datetime.date(2013, 12, 6)
>>>Today.replace(month=1)                   #替换日期中的月
datetime.date(2015, 1, 6)
>>>now=datetime.datetime.now()
>>>now
datetime.datetime(2015, 12, 6, 16, 1, 6, 313898)
>>>now.replace(second=30)                   #替换日期时间中的秒
datetime.datetime(2015, 12, 6, 16, 1, 30, 313898)
>>>now+datetime.timedelta(days=5)           #计算 5 天后的日期时间
datetime.datetime(2015, 12, 11, 16, 1, 6, 313898)
>>>now+datetime.timedelta(weeks=-5)         #计算 5 周前的日期时间
datetime.datetime(2015, 11, 1, 16, 1, 6, 313898)
```

拓展知识：标准库 calendar 也提供了一些与日期操作有关的方法。例如：

```
>>>import calendar                          #导入模块
>>>print(calendar.calendar(2016))           #查看 2016 年日历表,结果略
>>>print(calendar.month(2016, 4))           #查看 2016 年 4 月份的日历表
>>>calendar.isleap(2016)                    #判断是否为闰年
True
>>>calendar.weekday(2016, 4, 26)            #查看指定日期是周几
1
```

拓展知识：也可以自己编写代码模拟 Python 标准库 calendar 中查看日历的方法。

```
from datetime import date
```

```python
daysOfMonth=[31, 28, 31, 30, 31, 30, 31, 31, 30, 31, 30, 31]

def myCalendar(year, month):
    #获取 year 年 month 月 1 日是周几
    start=date(year, month, 1).timetuple().tm_wday
    #打印头部信息
    print('{0}年{1}月日历'.format(year,month).center(56))
    print('\t'.join('日 一 二 三 四 五 六'.split()))
    #获取该月有多少天,如果是 2 月并且是闰年,适当调整一下
    day=daysOfMonth[month-1]
    if month==2:
        if year%400==0 or(year%4==0 and year%100!=0):
            day +=1
    #生成数据,根据需要在前面填充空白
    result=[' '*8 for i in range(start+1)]
    result +=list(map(lambda d: str(d).ljust(8), range(1, day+1)))
    #打印数据
    for i, day in enumerate(result):
        if i!=0 and i%7==0:
            print()
        print(day, end='')
    print()
def main(year, month=-1):
    if type(year)!=int or year<1000 or year>10000:
        print('Year error')
        return
    if type(month)==int:
        #如果没有指定月份,就打印全年的日历
        if month==-1:
            for m in range(1, 13):
                myCalendar(year, m)
        #如果指定了月份,就只打印这一个月的日历
        elif month in range(1,13):
            myCalendar(year, month)
        else:
            print('Month error')
            return
    else:
        print('Month error')
        return
main(2017)
```

3.2 循环结构

《道德经》云:"反者,道之动",认为循环是道的运动方式,这充分说明了循环的重要性。

3.2.1 for 循环与 while 循环的基本语法

一只羊、两只羊、三只羊、四只羊、五只羊……有过失眠经历的朋友应该都数过羊(为什么是羊而不是其他动物呢?我研究的非权威结果是羊的发音口型变化少并且动作幅度小,而很多人在默念的时候嘴会不由自主地跟着发音动。于是结论来了,试着数一数马、牛、猪、鸡、鸭、鱼、老虎、桌子或者其他东西,口型变化较大而对注意力要求较高,所以不容易入睡),不停地重复一件事情,时间久了会非常无聊,然后大脑就会由于疲劳而容易入睡。当然了,前提是不要数着数着羊却想起了烤羊肉串和扎啤。

重复性的劳动会使人疲劳,而计算机不会,只要代码写得正确,计算机就会孜孜不倦、不知疲劳地重复工作。在 Python 中主要有两种形式的循环结构:for 循环和 while 循环。while 循环一般用于循环次数难以提前确定的情况,当然也可以用于循环次数确定的情况;for 循环一般用于循环次数可以提前确定的情况,尤其适用于枚举或遍历序列或迭代对象中元素的场合。当循环带有 else 子句时,如果循环因为条件表达式不成立或序列遍历结束而自然结束时则执行 else 结构中的语句,如果循环是因为执行了 break 语句而导致循环提前结束则不执行 else 中的语句。其完整语法形式为

```
while 条件表达式:
    循环体
[else:
    else 子句代码块]
```

和

```
for 取值 in 序列或迭代对象:
    循环体
[else:
    else 子句代码块]
```

其中,方括号内的 else 子句可以没有,也可以有。下面的代码演示了带有 else 子句的循环结构,该代码用来计算 1+2+3+…+99+100 的结果。

```
>>>s=0
>>>for i in range(1, 101):          #不包括 101
    s +=i
else:
    print(s)
5050
```

下面的代码使用 while 循环实现了同样的功能：

```
>>>s=i=0
>>>while i<=100:
    s+=i
    i+=1
else:
    print(s)
5050
```

下面的代码巧妙运用 range()函数来控制循环次数,输出由星号(*)组成的菱形图案：

```
def main(n):
    for i in range(n):
        print((' * '*i).center(n*3))
    for i in range(n, 0, -1):
        print((' * '*i).center(n*3))

main(6)
```

建议：编程时一般优先考虑使用 for 循环。

3.2.2　break 与 continue 语句

break 语句和 continue 语句在 while 循环及 for 循环中都可以使用,并且一般常与选择结构结合使用。一旦 break 语句被执行,将使得 break 语句所属层次的循环提前结束。continue 语句的作用是提前结束本次循环,并忽略 continue 之后的所有语句,直接回到循环的顶端,提前进入下一次循环。

下面的代码用来计算小于 100 的最大素数,可以看出 break 语句在循环中的作用。

```
>>>for n in range(100, 1, -1):
    for i in range(2, n):
        if n%i==0:
            break                    #结束内循环
    else:
        print(n)
        break                        #结束外循环
97
```

删除上面代码中最后一个 break 语句,并对输出语句略加修改,则可以用来输出 100 以内的所有素数,例如：

```
>>>for n in range(100, 1, -1):
    for i in range(2, n):
        if n%i==0:
```

```
            break
    else:
        print(n, end=' ')
97 89 83 79 73 71 67 61 59 53 47 43 41 37 31 29 23 19 17 13 11 7 5 3 2
```

> **注意**：过多的 break 和 continue 语句会降低程序的可读性。除非 break 或 continue 语句可以让代码更简单或更清晰，否则不要轻易使用。

3.2.3 循环代码优化技巧

学过太极拳的朋友应该听说过一句话：要把拳练好，必把圈练小。一般而言，功夫越深的人外形动作越小，编写代码也是同样的道理。在满足功能要求的前提下应该追求代码的短小精悍，而这也是对 Python 内功的考验。虽然不至于"两句三年得，一吟双泪流"，但确实也是需要经过反复推敲的。有不少人编写代码之前没有经过系统地规划，也没有经过深思熟虑，完全是凭着感觉走，想到哪写到哪，写的时候还反复地修改前面的代码。对于一些小的程序这样做是可以的，但是对于大型软件开发，这种"不管黑猫白猫，抓到老鼠就是好猫"的态度是可怕的。编写代码时，不仅要考虑功能，还要考虑性能以及可维护性、可扩展性、可移植性等。

虽然现在的计算机配置越来越高，内存越来越大，速度越来越快，计算能力越来越强，各种云平台更是提供了惊人的存储空间和计算能力，但作为一种美德，还是要合理利用资源，能少用一点资源就省一点，能让代码运行更快一点就尽量优化一下，同时也能给代码进行适当"减肥"来增加可读性。在编写循环语句时，应尽量减少循环内部不必要或无关的计算，将与循环变量无关的代码尽可能地提取到循环之外，这样可以提高代码的执行效率。对于使用多重循环嵌套的情况，应尽量减少内层循环中不必要的计算，尽可能地向外提。例如，下面的代码，第二段明显比第一段的运行效率要高。

```
import time
digits = (1, 2, 3, 4)

start=time.time()
for i in range(1000):
    result=[]
    for i in digits:
        for j in digits:
            for k in digits:
                result.append(i*100+j*10+k)
print(time.time()-start)

start=time.time()
for i in range(1000):
    result=[]
    for i in digits:
```

```
            i=i*100
            for j in digits:
                j=j*10
                for k in digits:
                    result.append(i+j+k)
    print(time.time()-start)
```

运行结果如下:

```
0.043002367019653322
0.0210011005401611333
```

另外,在循环中应尽量引用局部变量,局部变量的查询和访问速度比全局变量略快,在使用模块中的方法时,可以通过将其转换为局部变量来提高运行速度。例如下面的代码:

```
import time
import math

start=time.time()                                #获取当前时间
for i in range(10000000):
    math.sin(i)
print('Time Used:', time.time()-start)           #输出所用时间

loc_sin=math.sin
start=time.time()
for i in range(10000000):
    loc_sin(i)
print('Time Used:', time.time()-start)
```

运行结果如下:

```
Time Used: 3.600205898284912
Time Used: 3.0221729278564453
```

虽然速度提高并不是非常多,但是对于某些对实时性要求特别高的应用场景一点点的提高也是有意义的。而实际上,上面这段代码还有优化的空间,大家能想到吗?本书其他地方介绍过,尝试着找一下。

《太极尺寸分毫解》曰"功夫先练开展,后练紧凑"。编写代码也是同样的道理,首先要把代码写对,保证完全符合功能要求,然后再进行必要的优化来提高性能。过早地追求性能优化有时可能会带来灾难而浪费大量精力。代码优化涉及的面非常广,对程序员的功底要求很高。除了上面介绍的循环代码优化,第2章和第5章中介绍的内容中也涉及一些优化的内容。例如,如果经常需要测试一个序列是否包含一个元素就应该尽量使用字典或集合而不使用列表,连接多个字符串时尽量使用join()方法而不要使用运算符+,对列表进行元素的插入和删除操作时应尽量从列表尾部进行,等等。

3.2.4 案例精选

例 3-4 输出序列中的元素。

对于类似元素遍历的问题,一般也优先考虑使用 for 循环,参考代码如下:

```
a_list=['a', 'b', 'mpilgrim', 'z', 'example']
for i, v in enumerate(a_list):
    print('列表的第', i+1, '个元素是:', v)
```

例 3-5 求 1~100 之间能被 7 整除,但不能同时被 5 整除的所有整数。

该例主要介绍条件表达式的写法,参考代码如下:

```
for i in range(1, 101):
    if i %7 ==0 and i %5 !=0:
        print(i)
```

例 3-6 输出"水仙花数"。

所谓水仙花数是指一个 3 位的十进制数,其各位数字的立方和恰好等于该数本身。例如,153 是水仙花数,因为 $153=1^3+5^3+3^3$。

```
for i in range(100, 1000):
    ge=i %10
    shi=i // 10 %10
    bai=i // 100
    if ge**3+shi**3+bai**3 ==i:
        print(i)
```

例 3-7 求平均分。

```
score=[70, 90, 78, 85, 97, 94, 65, 80]
s=0
for i in score:
    s +=i
print(s / len(score))
```

当然也可以使用下面的内置函数来计算平均分:

```
print(sum(score) / len(score))
```

注意:在 Python 2.x 中,/运算符与 Python 3.x 的解释不一样,对于上面的代码,在 Python 2.x 中需要写成下面的样子:

```
print(sum(score) * 1.0 / len(score))
```

例 3-8 打印九九乘法表。

该例主要介绍循环结构的嵌套用法和循环条件的控制,参考代码如下:

```
for i in range(1, 10):
```

```
        for j in range(1, i+1):
            print('{0}*{1}={2}'.format(i,j,i*j), end=' ')
    print()                                           #打印空行
```

例 3-9 求 200 以内能被 17 整除的最大正整数。

熟练掌握 range()函数的用法,对于很多循环来说可能起到事半功倍的效果,参考代码如下:

```
for i in range(200, 0, -1):
    if i%17 == 0:
        print(i)
        break
```

例 3-10 判断一个数是否为素数。

在该例中,重点演示循环结构中 else 子句的用法。

```
import math

n=input("Input an integer:")
n=int(n)
m=int(math.sqrt(n)+2)
for i in range(2, m):
    if n%i == 0:
        print('No')
        break
else:
    print('Yes')
```

拓展知识:math 是用于数学计算的标准库。除了用于平方根函数 sqrt()和取整函数 ceil(),Python 标准库 math 还提供了最大公约数函数 gcd()、sin()、asin()等三角函数与反三角函数、弧度与角度转换函数 degrees()和 radians()、误差函数 erf()、剩余误差函数 erfc()、伽马函数 gamma()、对数函数 log()、log2()、log10()、阶乘函数 factorial()、常数 pi 和 e,等等。

例 3-11 鸡兔同笼问题。假设共有鸡、兔 30 只,脚 90 只,求鸡、兔各有多少只?

```
for ji in range(0, 31):
    if 2*ji+(30-ji)*4 == 90:
        print('ji:', ji, ' tu:', 30-ji)
```

趣味拓展:所有鸡、兔听口令,抬起一条腿,再抬起一条腿,好吧,现在所有的鸡都目瞪口呆地坐地上了(难道这就是传说中的呆若木鸡),站着的都是还有两条腿站立的兔子(兔子表示压力也很大),这时站立着的腿的数量的一半是兔子,当然如果得到的数字不是整数则表示无解。代码如下:

```
def demo(jitu, tui):
```

```
            tu=(tui-jitu*2)/2
            if int(tu)==tu:
                return(int(jitu-tu), int(tu))

print(demo(30, 90))
```

例 3-12 编写程序,输出由 1、2、3、4 这 4 个数字组成的每位数都不相同的所有三位数。

```
digits = (1, 2, 3, 4)
for i in digits:
    ii=i*100
    for j in digits:
        if j==i:
            continue
        jj=j*10
        for k in digits:
            if k==i or k==j:
                continue
            print(ii+jj+k)
```

例 3-13 编写程序,计算组合数 $C(n,i)$,即从 n 个元素中任选 i 个,有多少种选法?

根据组合数的定义,需要计算 3 个数的阶乘,在很多编程语言中都很难直接使用整型变量表示大数的阶乘结果,虽然 Python 并不存在这个问题,但是计算大数的阶乘仍需要相当多的时间。例如:

```
>>>def Cni1(n, i):
    import math
    return int(math.factorial(n)/math.factorial(i)/math.factorial(n-i))
>>>Cni2(6,2)
15
```

现在我们换个角度来看这个问题。容易知道,Cni(8,3)= 8!/3!/(8-3)!=(8×7×6×5×4×3×2×1)/(3×2×1)/(5×4×3×2×1),简单分析可以发现,对于(5,8]区间的数,分子上出现一次而分母上没出现;(3,5]区间的数在分子、分母上各出现一次;[1,3]区间的数分子上出现一次而分母上出现两次。根据这一规律,可以编写如下非常高效的组合数计算程序。

```
def Cni2(n,i):
    if not(isinstance(n,int)and isinstance(i,int)and n>=i):
        print('n and i must be integers and n must be larger than or equal to i.')
        return
    result=1
    Min, Max=sorted((i,n-i))
    for i in range(n,0,-1):
```

```
            if i>Max:
                result *=i
            elif i<=Min:
                result /=i
    return result
print(Cni1(6,2))
```

拓展知识：有时候换个角度来思考和解决问题，或许会更加有效和快捷。例如，在信号处理领域，时域或空域的卷积对应于变换域中的乘法。再例如，现有 100 名乒乓球运动员采用淘汰赛的方式进行比赛，问至少需要多少场比赛才能决出冠军呢？肯定有读者立刻就提笔计算了，50+25+12+…，计算的时候遇到剩余人数为奇数的时候还要纠结一会儿。实际上换个角度可以这样来想，所谓淘汰赛就是每比一场就下去一个人，而所谓决出冠军就是最后只剩下一个人，所以需要淘汰 99 个人，至少需要比赛 99 场。

拓展知识：也可以直接使用 Python 标准库 itertools 提供的函数来解决组合数计算的问题。

```
>>>import itertools
>>>len(tuple(itertools.combinations(range(60),2)))
1770
```

计算组合数时如果数值 n 和 i 较大，建议使用前面定义的 Cni2()函数，不建议使用 combinations()和 Cni1()函数，因为这会增加大量的额外操作甚至导致死机。

combinations()更多的时候是用来返回迭代对象进行惰性求值，而不是像上面代码所演示的用法。除了 combinations()函数，itertools 还提供了排列函数 permutations()、用于循环遍历可迭代对象元素的函数 cycle()、根据一个序列的值对另一个序列进行过滤的函数 compress()、根据函数返回值对序列进行分组的函数 groupby()。

```
>>>import itertools
>>>x='Private Key'
>>>y=itertools.cycle(x)                #循环遍历序列中的元素
>>>for i in range(20):
    print(next(y), end=',')
P,r,i,v,a,t,e, ,K,e,y,P,r,i,v,a,t,e, ,K,
>>>for i in range(5):
    print(next(y), end=',')
e,y,P,r,i,
>>>x=range(1, 20)
>>>y = (1,0) * 9+ (1,)
>>>y
(1, 0, 1, 0, 1, 0, 1, 0, 1, 0, 1, 0, 1, 0, 1, 0, 1, 0, 1)
>>>list(itertools.compress(x, y))      #根据一个序列的值对另一个序列进行过滤
[1, 3, 5, 7, 9, 11, 13, 15, 17, 19]
>>>def group(v):
```

```
        if v>10:
            return 'greater than 10'
        elif v<5:
            return 'less than 5'
        else:
            return 'between 5 and 10'
>>>x=range(20)
>>>y=itertools.groupby(x, group)             #根据函数返回值对序列元素进行分组
>>>for k, v in y:
    print(k, ':', list(v))
less than 5 : [0, 1, 2, 3, 4]
between 5 and 10 : [5, 6, 7, 8, 9, 10]
greater than 10 : [11, 12, 13, 14, 15, 16, 17, 18, 19]
>>>list(itertools.permutations([1, 2, 3, 4], 3))
                                             #从 4 个元素中任选 3 个的所有排列
>>>x=itertools.permutations([1,2,3,4], 4)    #4 个元素全排列
>>>next(x)
(1, 2, 3, 4)
>>>next(x)
(1, 2, 4, 3)
>>>next(x)
(1, 3, 2, 4)
```

例 3-14 编写程序，计算理财产品收益，假设利息和本金一起滚动。

```
def licai(base, rate, days):
    result=base                              #初始投资金额
    times=365//days                          #整除,用来计算一年可以滚动多少期
    for i in range(times):
        result=result+result*rate/365*days
    return result
print(licai(100000, 0.0385, 14))             #14 天理财,利率为 0.0385,投资 10 万元
```

3.3 函数设计与使用

大家在初中数学课程中就学习过函数的概念，函数表示从自变量到因变量之间的一种映射或对应关系。软件开发中的函数也具有相似的含义，也是把输入经过一定的变换和处理最后得到预定的输出，如图 3-5 所示。从外部来看，函数就像一个黑盒子，不需要了解内部原理，只需要了解其接口或使用方法即可。

输入 ——→ 函数 ——→ 输出

图 3-5 函数示意图

在软件开发过程中，经常有很多操作是完全相同或者是非常相似的，仅仅是要处理的数据不同而已，因此经常会在不同的代码位置多次执行相似或完全相同的代码块。很显然，从软件设计和代码复用的角度来讲，直接将该代码块复制

到多个相应的位置然后进行简单修改绝对不是一个好主意。虽然这样可以使得多份复制的代码可以彼此独立地进行修改,但这样不仅增加了代码量,也增加了代码阅读、理解和维护的难度,更重要的是为代码测试和纠错带来了很大的困难。一旦被复制的代码块将来某天被发现存在问题而需要修改,则必须对所有的复制都做同样正确的修改,这在实际中是很难完成的一项任务。由于代码量的大幅度增加,导致代码之间的关系更加复杂,很可能在修补旧漏洞的同时又引入了新漏洞。因此,应尽量减少使用直接复制代码块的方式来实现复用。解决这个问题的有效方法是设计函数(function)和类(class)。本章介绍函数的设计与使用,第 4 章介绍面向对象程序设计。

将可能需要反复执行的代码封装为函数,并在需要执行该段代码功能的地方进行调用,这不仅可以实现代码的复用,更重要的是可以保证代码的一致性,只需要修改该函数代码则所有调用位置均得到体现。同时,把大任务拆分成多个函数也是分治法的经典应用,复杂问题简单化,使得软件开发像搭积木一样简单。当然,在实际开发中,需要对函数进行良好的设计和优化才能充分发挥其优势。在编写函数时,有很多原则需要参考和遵守,例如,不要在同一个函数中执行太多的功能,尽量只让其完成一个高度相关且大小合适的功能,以提高模块的内聚性。另外,尽量减少不同函数之间的隐式耦合,例如减少全局变量的使用,使得函数之间仅通过调用和参数传递来显式体现其相互关系。

3.3.1 基本语法

在 Python 中,定义函数的语法如下:

```
def 函数名([参数列表]):
    '''注释'''
    函数体
```

在 Python 中使用 def 关键字来定义函数,然后是一个空格和函数名称,接下来是一对圆括号,在圆括号内是形式参数列表,如果有多个参数则使用逗号分隔开,圆括号之后是一个冒号和换行,最后是必要的注释和函数体代码。定义函数时需要注意的问题:①函数形参不需要声明其类型,也不需要指定函数返回值类型;②即使该函数不需要接收任何参数,也必须保留一对空的圆括号;③括号后面的冒号必不可少;④函数体相对于 def 关键字必须保持一定的空格缩进。

小提示:注释可以说是软件开发人员的笔记,对代码测试人员和维护人员来说也非常重要。在 Python 中有两种注释的方式:符号#后面的内容表示注释,不属于任何语句的一对三引号中的内容也表示注释。

小技巧:不少程序员是编写完代码之后再添加适当的注释,我恰恰相反。我一般都是先写注释,以注释的形式用自然语言把程序思路描述出来,然后再把这些注释"翻译"成程序语言,正所谓"代码未动,注释先行"。

下面的函数用来计算斐波那契数列中小于参数 n 的所有值:

```
def fib(n):                    #定义函数,括号里的 n 是形参
```

```
    a, b=1, 1
    while a<n:
        print(a, end=' ')
        a, b=b, a+b
    print()
```

该函数的调用方式为

```
fib(1000)                                        #调用函数,括号里的 1000 是实参
```

在定义函数时,开头部分的注释并不是必需的,但是如果为函数的定义加上一段注释的话,可以为用户提供友好的提示和使用帮助。例如,把上面生成斐波那契数列的函数定义修改为下面的形式,在函数开头加上一段注释。

```
>>>def fib(n):
    '''accept an integer n.
     return the numbers less than n in Fibonacci sequence.'''
    a, b=1, 1
    while a<n:
        print(a, end=' ')
        a, b=b, a+b
    print()
```

如此一来,可以使用内置函数 help()来查看函数的使用帮助,并且在调用该函数时输入左侧圆括号之后,立刻就会得到该函数的使用说明,如图 3-6 所示。

```
>>> def fib(n):
      '''accept an integer n.
      return the numbers less than n in Fibonacci sequence.'''
      a, b = 1, 1
      while a < n:
          print(a, end='')
          a, b = b, a+b
      print()


>>> print(fib.__doc__)
accept an integer n.
      return the numbers less than n in Fibonacci sequence.
>>> help(fib)
Help on function fib in module __main__:

fib(n)
    accept an integer n.
    return the numbers less than n in Fibonacci sequence.

>>> fib(
(n)
accept an integer n.
return the numbers less than n in Fibonacci sequence.
```

图 3-6 使用注释来为用户提示函数使用说明

建议:如果代码本身不能提供非常好的可读性,那么最好加上适当的注释来说明,要不然,自己写的代码自己都看不懂了。很多程序员都有过这样的经历。

在 Python 中,定义函数时不需要声明函数的返回值类型,而是使用 return 语句结束函数的执行的同时返回任意类型的值,函数返回值类型与 return 语句返回表达式的类型一致。无论 return 语句出现在函数的什么位置,一旦得到执行将直接结束函数的执行。如果函数没有 return 语句或者执行了不返回任何值的 return 语句,Python 将认为该函

数以return None结束,即返回空值。

> **小提示**:作为使用者,在调用函数时,一定要注意函数有没有返回值,以及是否会对函数实参的值进行修改。例如,前面第2章介绍过的列表对象方法sort()属于原地操作,没有返回值,而内置函数sorted()则返回排序后的列表,并不对原列表做任何修改。

```
>>>a_list=[1, 2, 3, 4, 9, 5, 7]
>>>print(sorted(a_list))
[1, 2, 3, 4, 5, 7, 9]
>>>print(a_list)                          #原列表内容没变
[1, 2, 3, 4, 9, 5, 7]
>>>print(a_list.sort())                   #列表对象的sort()方法没有返回值
None
>>>print(a_list)
[1, 2, 3, 4, 5, 7, 9]
```

> **拓展知识**:函数属于可调用对象。由于构造函数的存在,类也是可调用的。另外,任何包含__call__()方法的类的对象都是可调用的。例如,下面的代码演示了函数嵌套定义的情况:

```
def linear(a, b):
    def result(x):                        #在Python中,函数是可以嵌套定义的
        return a * x+b
    return result
```

下面的代码演示了可调用对象类的定义:

```
class linear:
    def __init__(self, a, b):
        self.a, self.b=a, b
    def __call__(self, x):
        return self.a * x+self.b
```

使用上面的嵌套函数和类这两种方式中任何一个,都可以通过以下的方式来定义一个可调用对象:

```
taxes=linear(0.3, 2)
```

然后通过下面的方式来调用该对象:

```
taxes(5)
```

下面的代码完整地演示了嵌套函数定义与使用的方法,有效利用了用户名检查功能的代码,关于面向对象编程的知识请参考第4章。

```
def check_permission(func):
    def wrapper(* args, **kwargs):
        if kwargs.get('username')!='admin':
```

```
        raise Exception('Sorry. You are not allowed.')
    return func(*args, **kwargs)
return wrapper

class ReadWriteFile(object):
    #把函数 check_permission 作为装饰器使用
    @check_permission
    def read(self, username, filename):
        return open(filename,'r').read()

    def write(self, username, filename, content):
        open(filename,'a+').write(content)
    #把函数 check_permission 作为普通函数使用
    write=check_permission(write)

t=ReadWriteFile()
print('Originally…')
print(t.read(username='admin', filename=r'd:\sample.txt'))
print('Now, try to write to a file…')
t.write(username='admin', filename=r'd:\sample.txt', content='\nhello world')
print('After calling to write…')
print(t.read(username='admin', filename=r'd:\sample.txt'))
```

3.3.2 函数参数不得不说的几件事

函数定义时圆括号内是使用逗号分隔开的形参列表（parameters），一个函数可以没有参数，但是定义和调用时一对圆括号必须要有，表示这是一个函数并且不接收参数。函数调用时向其传递实参（arguments），根据不同的参数类型，将实参的值或引用传递给形参。

在定义函数时，对参数个数并没有限制，如果有多个形参，则需要使用逗号进行分隔。例如，下面的函数用来接收 2 个参数，并输出其中的最大值。

```
def printMax(a, b):
    if a>b:
        pirnt(a, 'is the max')
    else:
        print(b, 'is the max')
```

🍀注意：这里只是为了演示，忽略了一些细节，如果输入的参数不支持比较运算，则会出错，可以参考后面第 7 章中介绍的异常处理结构来解决这个问题。

对于绝大多数情况下，在函数内部直接修改形参的值不会影响实参。例如：

```
>>>def addOne(a):
    print(a)                              #输出原变量 a 的值
```

```
        a +=1                            #这条语句会得到一个新的变量 a
        print(a)
>>>a=3
>>>addOne(a)
3
4
>>>a
3
```

从运行结果可以看出,在函数内部修改了形参 a 的值,但是当函数运行结束以后,实参 a 的值并没有被修改。然而,在有些情况下,可以通过特殊的方式在函数内部修改实参的值,例如:

```
>>>def modify(v):                        #修改列表元素值
        v[0]=v[0]+1
>>>a=[2]
>>>modify(a)
>>>a
[3]
>>>def modify(v, item):                  #为列表增加元素
        v.append(item)
>>>a=[2]
>>>modify(a, 3)
>>>a
[2, 3]
>>>def modify(d):                        #修改字典元素值或为字典增加元素
        d['age']=38
>>>a={'name':'Dong', 'age':37, 'sex':'Male'}
>>>a
{'age': 37, 'name': 'Dong', 'sex': 'Male'}
>>>modify(a)
>>>a
{'age': 38, 'name': 'Dong', 'sex': 'Male'}
```

也就是说,如果传递给函数的是 Python 可变序列,并且在函数内部使用下标或序列自身支持的方式为可变序列增加、删除元素或修改元素值时,修改后的结果是可以反映到函数之外的,即实参也得到了相应的修改。

1. 默认值参数

在定义函数时,Python 支持默认值参数,即在定义函数时为形参设置默认值。在调用带有默认值参数的函数时,可以不用为设置了默认值的形参进行传值,此时函数将会直接使用函数定义时设置的默认值,也可以通过显式赋值来替换其默认值。也就是说,在调用函数时是否为默认值参数传递实参是可选的,具有较大的灵活性。带有默认值参数的函数定义语法如下:

```
def 函数名(…,形参名=默认值):
    函数体
```

可以使用"函数名.__defaults__"随时查看函数所有默认值参数的当前值,其返回值为一个元组,其中的元素依次表示每个默认值参数的当前值。例如下面的函数定义:

```
>>>def say(message, times =1):
    print((message+' ') * times)
>>>say.func_defaults
(1,)
```

调用该函数时,如果只为第一个参数传递实参,则第二个参数使用默认值1;如果为第二个参数传递实参,则不再使用默认值1,而是使用调用者显式传递的值。

```
>>>say('hello')
hello
>>>say('hello', 3)
hello hello hello
>>>say('hi', 7)
hi hi hi hi hi hi hi
```

注意:在定义带有默认值参数的函数时,默认值参数必须出现在函数形参列表的最右端,任何一个默认值参数右边都不能再出现非默认值参数。

注意:一般情况下,都是调用函数时为其传递参数,这时形参的值由调用函数时实参的值确定。但如果函数的默认值参数不是调用时传递的,而是通过其他方式对其赋值,那么默认值参数的值可能会在函数定义时确定,而不是函数调用时。例如:

```
>>>i=5
>>>def demo(v):
    print(v)
>>>i=6
>>>demo(i)                               #调用时明确传递参数值
6
>>>i=5
>>>def demo(v=i):
    print(v)
>>>i=6
>>>demo()                                #调用时没有传递参数值
5
```

注意:多次调用函数并且不为默认值参数传递值时,默认值参数只在第一次调用时进行解释,对于列表、字典这样可变类型的默认值参数,这一点可能会导致很严重的逻辑错误,而这种错误或许会耗费大量精力来定位和纠正。例如下面的代码:

```
def demo(newitem, old_list=[]):
```

```
        old_list.append(newitem)
        return old_list

print(demo('5', [1, 2, 3, 4]))
print(demo('aaa', ['a', 'b']))
print(demo('a'))
print(demo('b'))
```

上面的函数使用列表作为默认参数,由于其可记忆性,连续多次调用该函数而不给该参数传值时,再次调用时将保留上一次调用的结果,从而导致很难发现的错误。下面的代码就不存在这个问题:

```
def demo(newitem, old_list=None):
    if old_list is None:
        old_list=[]
    old_list.append(newitem)
    return old_list
```

2. 关键参数

关键参数主要指调用函数时的参数传递方式,与函数定义无关。通过关键参数可以按参数名字传递值,实参顺序可以和形参顺序不一致,但不影响参数值的传递结果,避免了用户需要牢记参数位置和顺序的麻烦,使得函数的调用和参数传递更加灵活方便。

```
>>>def demo(a, b, c=5):
    print(a, b, c)
>>>demo(3, 7)
3 7 5
>>>demo(a=7, b=3, c=6)
7 3 6
>>>demo(c=8, a=9, b=0)
9 0 8
```

3. 可变长度参数

可变长度参数在定义函数时主要有两种形式:*parameter 和**parameter,前者用来接收任意多个实参并将其放在一个元组中,后者接收类似于关键参数一样显式赋值形式的多个实参并将其放入字典中。

下面的代码演示了第一种形式可变长度参数的用法,即无论调用该函数时传递了多少实参,一律将其放入元组中:

```
>>>def demo(*p):
    print(p)
>>>demo(1, 2, 3)
(1, 2, 3)
```

```
>>>demo(1, 2, 3, 4, 5, 6, 7)
(1, 2, 3, 4, 5, 6, 7)
```

下面的代码演示了第二种形式可变长度参数的用法,即在调用该函数时自动将接收的参数转换为字典:

```
>>>def demo(**p):
    for item in p.items():
        print(item)
>>>demo(x=1, y=2, z=3)
('y', 2)
('x', 1)
('z', 3)
```

注意：Python 定义函数时可以同时使用位置参数、关键参数、默认值参数和可变长度参数,但是除非真的很必要,否则请不要这样用,因为这会使得代码非常混乱而严重降低可读性,并导致程序查错非常困难。另外,一般而言,一个函数如果可以接收很多不同类型参数的话,很可能是函数设计得不好,例如函数功能过多,需要进行必要的拆分和重新设计,以满足模块高内聚的要求。

4. 传递参数时的序列解包

调用含有多个参数的函数时,可以使用 Python 列表、元组、集合、字典以及其他可迭代对象作为实参,并在实参名称前加一个星号,Python 解释器将自动进行解包,然后传递给多个单变量形参。

```
>>>def demo(a, b, c):
    print(a+b+c)
>>>seq=[1, 2, 3]
>>>demo(*seq)
6
>>>tup = (1, 2, 3)
>>>demo(*tup)
6
>>>dic={1:'a', 2:'b', 3:'c'}
>>>demo(*dic)
6
>>>Set={1, 2, 3}
>>>demo(*Set)
6
>>>demo(*dic.values())
abc
```

小提示：①字典对象作为实参时默认使用字典的"键",如果需要将字典中"键:值"元素作为参数则需要使用 items()方法明确说明,如果需要将字典的"值"作为参数则需要调用字典的 values() 方法明确说明；②实参中元素个数与形参个数必须相等，否则将出现错误。

注意：调用函数时如果对实参使用一个星号（*）进行序列解包，这么这些解包后的实参将会被当作普通位置参数对待，并且会在关键参数和使用两个星号（**）进行序列解包的参数之前进行处理。

```
>>>def demo(a, b, c):                          #定义函数
    print(a, b, c)
>>>demo(*(1, 2, 3))                            #调用,序列解包
1 2 3
>>>demo(1, *(2, 3))                            #位置参数和序列解包同时使用
1 2 3
>>>demo(1, *(2,), 3)
1 2 3
>>>demo(a=1, *(2, 3))                          #序列解包相当于位置参数,优先处理
Traceback(most recent call last):
  File "<pyshell#26>", line 1, in <module>
    demo(a=1, *(2, 3))
TypeError: demo()got multiple values for argument 'a'
>>>demo(b=1, *(2, 3))
Traceback(most recent call last):
  File "<pyshell#27>", line 1, in <module>
    demo(b=1, *(2, 3))
TypeError: demo()got multiple values for argument 'b'
>>>demo(c=1, *(2, 3))
2 3 1
>>>demo(**{'a':1, 'b':2}, *(3,))               #序列解包不能在关键参数解包之后
SyntaxError: iterable argument unpacking follows keyword argument unpacking
>>>demo(*(3,), **{'a':1, 'b':2})
Traceback(most recent call last):
  File "<pyshell#30>", line 1, in <module>
    demo(*(3,), **{'a':1, 'b':2})
TypeError: demo()got multiple values for argument 'a'
>>>demo(*(3,), **{'c':1, 'b':2})
3 2 1
```

3.3.3 变量作用域

变量起作用的代码范围称为变量的作用域，不同作用域内同名变量之间互不影响，就像不同文件夹的同名文件之间互不影响一样。一个变量在函数外部定义和在函数内部定义，其作用域是不同的，函数内部定义的变量一般为局部变量，在函数外部定义的变量为全局变量。

在函数内定义的普通变量只在该函数内起作用，当函数运行结束后，在其内部定义的

局部变量将被自动删除而不可访问。在函数内部定义的全局变量当函数结束以后仍然存在并且可以访问。

如果想要在函数内部修改一个定义在函数外的变量值,那么这个变量就不能是局部的,其作用域必须为全局的。可以在函数内部通过 global 关键字来声明或定义全局变量,这分两种情况:

(1) 一个变量已在函数外定义,如果在函数内需要修改这个变量的值,并将这个赋值结果反映到函数之外,可以在函数内用 global 明确声明要使用已定义的同名全局变量。

(2) 在函数内部直接使用 global 关键字将一个变量声明为全局变量,如果在函数外没有定义该全局变量,在调用这个函数之后,会自动增加新的全局变量。

或者说,也可以这么理解:①在函数内如果只引用某个变量的值而没有为其赋新值,该变量为(隐式的)全局变量;②如果在函数内任意位置有为变量赋值的操作,该变量即被认为是(隐式的)局部变量,除非在函数内显式地用关键字 global 进行声明。

下面的代码演示了局部变量和全局变量的用法。

```
>>>def demo():
    global x                              #声明或创建全局变量
    x=3                                   #修改全局变量的值
    y=4                                   #局部变量
    print(x, y)
>>>x=5                                    #在函数外部定义了全局变量 x
>>>demo()                                 #本次调用修改了全局变量 x 的值
3 4
>>>x
3
>>>y                                      #局部变量在函数运行结束之后自动删除
Traceback(most recent call last):
  File "<pyshell#11>", line 1, in <module>
    y
NameError: name 'y' is not defined
>>>del x                                  #删除了全局变量 x
>>>x
Traceback(most recent call last):
  File "<pyshell#13>", line 1, in <module>
    x
NameError: name 'x' is not defined
>>>demo()                                 #本次调用创建了全局变量
3 4
>>>x
3
>>>y                                      #局部变量在函数调用结束后自动删除
Traceback(most recent call last):
```

```
    File "<pyshell#11>", line 1, in <module>
        y
NameError: name 'y' is not defined
```

如果局部变量与全局变量具有相同的名字,那么该局部变量会在自己的作用域内隐藏同名的全局变量,例如下面的代码所演示。

```
>>>def demo():
    x=3                             #创建了局部变量,并自动隐藏了同名的全局变量
    print(x)
>>>x=5                              #创建全局变量
>>>x
5
>>>demo()
3
>>>x                                #函数调用结束后,不影响全局变量 x 的值
5
```

最后,如果需要在同一个程序的不同模块之间共享全局变量,可以编写一个专门的模块来实现这一目的。例如,假设在模块 A.py 中有如下变量定义:

```
global_variable=0
```

而在模块 B.py 中使用以下语句修改该全局变量的值:

```
import A
A.global_variable=1
```

在模块 C.py 中使用以下语句来访问全局变量的值:

```
import A
print(A.global_variable)
```

从而实现了在不同模块之间共享全局变量的目的。

小提示:①一般而言,局部变量的引用比全局变量速度快,应优先考虑使用;②应尽量避免过多使用全局变量,因为全局变量会增加不同函数之间的隐式耦合度,降低代码可读性,并使得代码测试和纠错变得很困难。

拓展知识:局部变量的空间是在栈上分配的,而栈空间是由操作系统维护的,每当调用一个函数时,操作系统会为其分配一个栈帧,函数调用结束后立刻释放这个栈帧。因此,函数调用结束后,该函数内部所有的局部变量都不再存在。

拓展知识:除了局部变量和全局变量,Python 还支持使用 nonlocal 关键字定义一种介于两者之间的变量。例如下面的代码:

```
def scope_test():
    def do_local():
        spam="我是局部变量"
```

```
    def do_nonlocal():
        nonlocal spam                          #这时要求 spam 必须是已存在的变量
        spam="我不是局部变量,也不是全局变量"

    def do_global():
        global spam                            #如果全局作用域内没有 spam,就自动新建一个
        spam="我是全局变量"

    spam="原来的值"
    do_local()
    print("局部变量赋值后:", spam)
    do_nonlocal()
    print("nonlocal 变量赋值后:", spam)
    do_global()
    print("全局变量赋值后", spam)

scope_test()
print("全局变量:", spam)
```

上面的代码运行结果为

```
局部变量赋值后: 原来的值
nonlocal 变量赋值后: 我不是局部变量,也不是全局变量
全局变量赋值后 我不是局部变量,也不是全局变量
全局变量: 我是全局变量
```

3.3.4 lambda 表达式

  lambda 表达式常用来声明匿名函数,即没有函数名字的临时使用的小函数,例如第 2 章中列表对象的 sort()方法以及内置函数 sorted()中的 key 参数。lambda 表达式只可以包含一个表达式,不允许包含其他复杂的语句,但在表达式中可以调用其他函数,并支持默认值参数和关键参数,该表达式的计算结果相当于函数的返回值。下面的代码演示了不同情况下 lambda 表达式的应用。

```
>>>f=lambda x, y, z: x+y+z
>>>print(f(1, 2, 3))                    #把 lambda 表达式当作函数使用
6
>>>g=lambda x, y=2, z=3: x+y+z          #含有默认值参数
>>>print(g(1))
6
>>>print(g(2, z=4, y=5))                #调用时使用关键参数
11
>>>L=[(lambda x: x**2),(lambda x: x**3),(lambda x: x**4)]
>>>print(L[0](2), L[1](2), L[2](2))
```

```
4 8 16
>>>D={'f1':(lambda: 2+3), 'f2':(lambda: 2*3), 'f3':(lambda: 2**3)}
>>>print(D['f1'](), D['f2'](), D['f3']())
5 6 8
>>>L=[1, 2, 3, 4, 5]
>>>print(map((lambda x: x+10), L))        #没有名字的lambda表达式,作为函数参数
[11, 12, 13, 14, 15]
>>>L
[1, 2, 3, 4, 5]
>>>def demo(n):
    return n*n
>>>demo(5)
25
>>>a_list=[1, 2, 3, 4, 5]
>>>map(lambda x: demo(x), a_list)         #在lambda表达式中调用函数
[1, 4, 9, 16, 25]
>>>data=list(range(20))
>>>print(data)
[0, 1, 2, 3, 4, 5, 6, 7, 8, 9, 10, 11, 12, 13, 14, 15, 16, 17, 18, 19]
>>>import random
>>>random.shuffle(data)
>>>data
[4, 3, 11, 13, 12, 15, 9, 2, 10, 6, 19, 18, 14, 8, 0, 7, 5, 17, 1, 16]
>>>data.sort(key=lambda x: x)             #用在列表的sort()方法中,作为函数参数
>>>data
[0, 1, 2, 3, 4, 5, 6, 7, 8, 9, 10, 11, 12, 13, 14, 15, 16, 17, 18, 19]
>>>data.sort(key=lambda x: len(str(x)))   #使用lambda表达式指定排序规则
>>>data
[0, 1, 2, 3, 4, 5, 6, 7, 8, 9, 10, 11, 12, 13, 14, 15, 16, 17, 18, 19]
>>>data.sort(key=lambda x: len(str(x)), reverse=True)
>>>data
[10, 11, 12, 13, 14, 15, 16, 17, 18, 19, 0, 1, 2, 3, 4, 5, 6, 7, 8, 9]
```

注意:在使用lambda表达式时,要注意变量作用域可能会带来的问题,例如,下面的代码中变量x是在外部作用域中定义的,对lambda表达式而言不是局部变量,从而导致出现了错误。

```
>>>r=[]
>>>for x in range(10):
    r.append(lambda: x**2)
>>>r[0]()
81
>>>r[1]()
81
```

```
>>>r[2]()
81
```

而修改为下面的代码,则可以得到正确的结果。

```
>>>r=[]
>>>for x in range(10):
    r.append(lambda n=x: n**2)
>>>r[0]()
0
>>>r[1]()
1
>>>r[5]()
25
>>>r[8]()
64
```

3.3.5 案例精选

例 3-15 编写函数计算圆的面积。

```
from math import pi as PI

def CircleArea(r):
    if isinstance(r,(int, float)):        #确保接收的参数为数值
        return PI * r * r
    else:
        print('You must give me an integer or float as radius.')

print(CircleArea(3))
```

例 3-16 编写函数,接收任意多个实数,返回一个元组,其中第一个元素为所有参数的平均值,其他元素为所有参数中大于平均值的实数。

```
def demo(*para):
    avg=sum(para)/len(para)               #平均值
    g=[i for i in para if i>avg]          #列表推导式
    return (avg,)+tuple(g)

print(demo(1, 2, 3, 4))
```

例 3-17 编写函数,接收字符串参数,返回一个元组,其中第一个元素为大写字母的个数,第二个元素为小写字母的个数。

```
def demo(s):
    result=[0, 0]
    for ch in s:
```

```
        if 'a'<=ch<='z':
            result[1] +=1
        elif 'A'<=ch<='Z':
            result[0] +=1
    return result

print(demo('aaaabbbbC'))
```

小提示：上面代码中使用关系运算符判断一个字符是否为大写字母或小写字母，只是为了演示一种用法，在实际开发中还是建议使用字符串对象自身提供的 isupper()和 islower()方法，这样速度会更快一些。

例 3-18 编写函数，接收包含 20 个整数的列表 lst 和一个整数 k 作为参数，返回新列表。处理规则：将列表 lst 中下标 k 之前的元素逆序，下标 k 之后的元素逆序，然后将整个列表 lst 中的所有元素逆序。

```
def demo(lst, k):
    x=lst[:k]
    x.reverse()
    y=lst[k:]
    y.reverse()
    r=x+y
    return list(reversed(r))

lst=list(range(1, 21))
print(lst)
print(demo(lst, 5))
```

拓展知识：例 3-18 描述的实际上是将列表循环左移 k 位的算法，下面的代码使用了更加直接的方法，但对于长列表来说效率远不如上面的代码高，比下面小技巧中提到的方法更是相差很多。

```
def demo(lst, k):
    temp=lst[:]
    for i in range(k):
        temp.append(temp.pop(0))
    return temp
```

小技巧：对于本例中描述的问题，使用切片可以直接实现，可以达到最快的速度。

```
def demo(lst, k):
    return lst[k:]+lst[:k]
```

例 3-19 编写函数，接收整数参数 t，返回斐波那契数列中大于 t 的第一个数。

```
def demo(t):
    a, b=1, 1
    while b<t:
        a, b=b, a+b
    else:
        return b

print(demo(50))
```

例 3-20　编写函数,接收一个包含若干整数的列表参数 lst,返回一个元组,其中第一个元素为列表 lst 中的最小值,其余元素为最小值在列表 lst 中的下标。

```
import random

def demo(lst):
    m=min(lst)
    result = (m,)
    positions=[index for index, value in enumerate(lst) if value==m]
    result=result+tuple(positions)
    return result

x=[random.randint(1, 20) for i in range(50)]
print(x)
print(demo(x))
```

例 3-21　编写函数,接收一个整数 t 为参数,打印杨辉三角前 t 行。

```
def demo(t):
    result=[[1], [1, 1]]
    line=[1, 1]
    for i in range(2, t):
        r=[]
        for j in range(0, len(line)-1):
            r.append(line[j]+line[j+1])
        line=[1]+r+[1]
        result.append(line)
    return result

def output(result):
    for item in result:
        print(item)

output(demo(10))
```

上面的代码运行结果为

[1]
[1, 1]
[1, 2, 1]
[1, 3, 3, 1]
[1, 4, 6, 4, 1]
[1, 5, 10, 10, 5, 1]
[1, 6, 15, 20, 15, 6, 1]
[1, 7, 21, 35, 35, 21, 7, 1]
[1, 8, 28, 56, 70, 56, 28, 8, 1]
[1, 9, 36, 84, 126, 126, 84, 36, 9, 1]

例 3-22　编写函数，接收一个正偶数为参数，输出两个素数，并且这两个素数之和等于原来的正偶数。如果存在多组符合条件的素数，则全部输出。

```
import math

def IsPrime(n):
    m=int(math.sqrt(n))+1
    for i in range(2, m):
        if n%i==0:
            return False
    return True

def demo(n):
    if isinstance(n, int) and n>0 and n%2==0:
        for i in range(3, int(n/2)+1):
            if i%2==1 and IsPrime(i) and IsPrime(n-i):
                print(i, '+', n-i, '=', n)

demo(60)
```

例 3-23　编写函数，接收两个正整数作为参数，返回一个元组，其中第一个元素为最大公约数，第二个元素为最小公倍数。

```
def demo(m, n):
    if m>n:
        m, n=n, m
    p=m*n
    while m!=0:
        r=n%m
        n=m
        m=r
    return(n, int(p/n))

print(demo(20, 30))
```

拓展知识：Python 标准库 fractions 中提供了 gcd() 函数用来计算最大公约数，在 Python 3.5 版本中，标准库 math 也提供了计算最大公约数的函数 gcd()。利用 gcd() 函数，上面的代码也可以写作：

```python
def demo(m, n):
    import math
    r=math.gcd(m,n)
    return(r, int(m*n/r))
```

例 3-24　编写函数，接收一个所有元素值都不相等的整数列表 x 和一个整数 n，要求将值为 n 的元素作为支点，将列表中所有值小于 n 的元素全部放到 n 的前面，所有值大于 n 的元素放到 n 的后面。

```python
import random

def demo(x, n):
    if n not in x:
        print(n, ' is not an element of ', x)
        return

    i=x.index(n)                        #获取指定元素在列表中的索引
    x[0], x[i]=x[i], x[0]               #将指定元素与第 0 个元素交换
    key=x[0]

    i=0
    j=len(x)-1
    while i<j:
        while i<j and x[j]>=key:        #从后向前寻找第一个比指定元素小的元素
            j-=1
        x[i]=x[j]

        while i<j and x[i]<=key:        #从前向后寻找第一个比指定元素大的元素
            i+=1
        x[j]=x[i]

    x[i]=key

x =list(range(1, 10))
random.shuffle(x)                       #将元素打乱顺序
print(x)
demo(x, 4)
print(x)
```

拓展知识：例 3-24 给出的算法是快速排序算法中非常重要的一个步骤，当然也可以使用下面更加简洁的代码来实现。

```
>>>import random
>>>def demo(x, n):
    t1=[i for i in x if i<n]
    t2=[i for i in x if i>n]
    return t1+[n]+t2
>>>x=list(range(1,10))
>>>random.shuffle(x)
>>>x
[1, 9, 3, 6, 5, 2, 4, 7, 8]
>>>demo(x, 4)
[1, 3, 2, 4, 9, 6, 5, 7, 8]
```

例 3-25 编写函数,计算字符串匹配的准确率。

以打字练习程序为例,假设 origin 为原始内容,userInput 为用户输入的内容,下面的代码用来测试用户输入的准确率。

```
def Rate(origin, userInput):
    if not(isinstance(origin, str)and isinstance(userInput, str)):
        print('The two parameters must be strings.')
        return
    if len(origin)<len(userInput):
        print('Sorry. I suppose the second parameter string is shorter.')
        return
    right=0                                    #精确匹配的字符个数
    for origin_char, user_char in zip(origin, userInput):
        if origin_char==user_char:
            right +=1
    return right/len(origin)

origin='Shandong Institute of Business and Technology'
userInput='ShanDong institute of business and technolog'
print(Rate(origin, userInput))                 #输出测试结果
```

例 3-26 编写函数,对整数进行因数分解。

```
from random import randint
from math import sqrt

def factoring(n):
    '''对大数进行因数分解'''
    if not isinstance(n, int):
        print('You must give me an integer')
        return
    #开始分解,把所有因数都添加到 result 列表中
    result=[]
```

```
            for p in primes:
                while n!=1:
                    if n%p ==0:
                        n=n/p
                        result.append(p)
                    else:
                        break
                else:
                    result=map(str, result)
                    result='*'.join(result)
                    return result
        #考虑参数本身就是素数的情况
        if not result:
            return n

testData=[randint(10, 100000) for i in range(50)]
#随机数中的最大数
maxData=max(testData)
#小于 maxData 的所有素数
primes=[p for p in range(2, maxData) if 0 not in [ p%d for d in range(2, int(sqrt(p))
        +1)] ]

for data in testData:
    r=factoring(data)
    print(data, '=', r)
    #测试分解结果是否正确
    print(data==eval(r))
```

例 3-27 韩信点兵。

韩信为了不让敌人知道自己的兵力有多少,让士兵报数时先从 1 至 3 报数,再从 1 至 5 重新报数,然后再从 1 至 7 重新报数,只需要记下最后一名士兵每次报数是几,即可快速计算出自己有多少士兵。

```
from functools import reduce
from math import gcd

def isCoPrime(p):
    '''判断 p 中每个元组的第 1 个数 (即 mi) 之间是否互素'''
    for index, item1 in enumerate(p):
        for item2 in p[index+1:]:
            if gcd(item1[0], item2[0])!=1:
                return False
    return True
```

```
def extEuclid(Mi, mi):
    '''暴力穷举,求 Mi 对 mi 的乘法逆元,也可以使用扩展欧几里得算法快速求解'''
    for i in range(1, mi):
        if i*Mi %mi ==1:
            return i

def chineseRemainder(p):
    '''p 为[(3, 2),(7, 1),(13, 5),(mi, ai)…]形式的参数,其中 3/7/13 为商,2/1/5 为余数'''
    #先判断所给数据中的 mi 是否互素,如果不是则提示数据错误并退出
    if not isCoPrime(p):
        return 'Data error.'
    #切片浅复制,临时变量,防止修改实参中的数据
    pp=p[:]
    #求 M=m1*m2*m3*…*mn
    ppp=[item[0] for item in pp]
    M=reduce(lambda x,y: x*y, ppp)
    for index, item in enumerate(pp):
        Mi=int(M/item[0])
        bi=extEuclid(Mi, item[0])
        pp[index]=item+ (Mi, bi)
    #求解最终结果,sum(ai*bi*Mi)mod M
    result=sum([item[1]*item[2]*item[3] for item in pp])
    result=result %M
    #考虑特殊情况,不允许结果为 1
    if result==1:
        result=result+M
    return result

data=[[(3,2),(5,3),(7,2)],
      [(5,1),(3,2)],
      [(5,1),(3,1)],
      [(5,4),(3,2)],
      [(7,2),(8,4),(9,3)],
      [(5,2),(6,4),(7,4)],
      [(3,2),(5,3),(7,4)]]
for p in data:
    print(p)
    print(chineseRemainder(p))
```

作为一种练习,也可以用暴力枚举法求解韩信点兵的问题。例如:

```
def chineseRemainder(p):
    '''p 为[(3, 2),(7, 1),(13, 5),…]形式的参数,其中 3/7/13 为商,2/1/5 为余数'''
    #检查数据是否合法,若有相同商对应不同余数则认为给的数据不合法
    for index1, pair1 in enumerate(p):
```

```
                for pair2 in p[index1+1:]:
                    if pair1[0]==pair2[0] and pair1[1]!=pair2[1]:
                        print('Data Error.')
                        return
    #对给定数据按商从大到小排序
    p=sorted(p, key=lambda x:x[0], reverse=True)
    #生成嵌套列表
    possibleValues=list(map(lambda x: list((i * x[0]+x[1] for i in range(1,
10000))), p))
    #寻找第一个共同包含的数,该数即符合条件的最小数
    for value in possibleValues[0]:
        flag=True
        for rest in possibleValues[1:]:
            if value not in rest:
                flag=False
        if flag:
            print(value)
            return
    else:
        print('Can not find a number')

p=[[(5,3),(9,3),(13,3),(17,3)],
   [(9,7),(5,2),(4,3)],
   [(3,2),(5,3),(7,2)],
   [(3,2),(4,1)],
   [(2,1),(4,3),(5,2),(7,3),(9,4)],
   [(2,1),(3,2),(5,4),(6,5),(7,0)]]
for pp in p:
    chineseRemainder(pp)
```

例 3-28 模拟发红包算法。

据说这世间没啥问题是一个红包解决不了的,如果有,那就两个红包。微信红包不仅是好朋友之间沟通感情的方式,也是情侣之间在结婚纪念日、生日等重要日期表达爱意的形式,还是一种比较流行的交易手段。随着微信红包的流行,还出现了很多好玩的表情(见图 3-7)和段子,亲朋好友之间偶尔玩玩挺好的。当然,也有人因为玩红包游戏而倾家荡产,下面的代码将为大家揭秘如何控制红包金额的分配。

```
import random

def hongbao(total, num):
    #total 表示拟发红包总金额
    #num 表示拟发红包数量
    each=[]
    #已发红包总金额
```

```
        already=0
        for i in range(1, num):
            #为当前抢红包的人随机分配金额
            #至少给剩下的人每人留一分钱
            t=random.randint(1,(total-already)-(num-i))
            each.append(t)
            already=already+t
        #剩余所有钱发给最后一个人
        each.append(total-already)
        return each

if __name__=='__main__':
    total=5
    num=5
    #模拟30次
    for i in range(30):
        each=hongbao(total, num)
        print(each)
```

💡**小提示**：通过修改代码可以控制红包分配的规律，然后就可以大致控制每个人领取的钱数。

图3-7　几个好玩的表情

例3-29　编写函数，将YYYY-MM-DD的日期形式转换为YYYYQ的形式，其中Q表示季度。

```
def convert(YearMonthDay):
    if not isinstance(YearMonthDay, str):
        return 'Type Error. Must be str'
    if YearMonthDay.count('-')!=2:
        return 'Parameter Error. Must contains 2 -'
    data=YearMonthDay.split('-')
```

```
        if(len(data[0])!=4)or(len(data[1])not in(1,2))or(len(data[2])not in(1,2)):
            return 'Parameter Error. Must be YYYY-MM-DD'
    try:
        year, month, day=map(int, data)
        quarter=[[3, 4, 5], [6, 7, 8], [9, 10, 11], [12, 1, 2]]
        for q, m in enumerate(quarter):
            if month in m:
                return str(year)+str(q+1)
    except:
        return 'Parameter Error. Must be YYYY-MM-DD, and all be digits'

print(convert('2016-a9-27'))
```

例 3-30 模拟一维信号卷积,并模拟整数乘法。

在数字信号处理中经常会用到卷积计算,例如各种滤波器的设计。两个序列的卷积计算大体需要 3 步:①翻转其中一个序列;②移动翻转后的序列,并计算每次移动后两个序列的重叠面积;③重复第②步,直至两个序列没有重叠部分。假设一个序列为[1,2,3],另一个序列为[4,5],这两个序列的卷积计算步骤如图 3-8 所示。

除了滤波器设计,一维序列卷积还可以用来计算大整数乘法和多项式乘法,下面的代码以大整数乘法来演示其用法,当然在 Python 中大整数的乘法直接计算即可。首先需要把大整数使用列表来表示,列表中的每个元素用来表示大整数中的一位数字,例如,数字 123 表示为[1,2,3],数字 45 表示为[4,5],使用卷积计算得到结果为[4,13,22,15],把卷积结果转换为数字 5535 的步骤和原理如图 3-9 所示,图中中间一排数字表示进位,最下面一排数字表示第一排数字加上进位以后对 10 的余数。

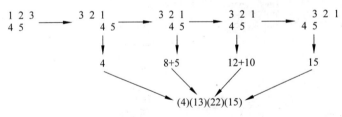
图 3-8 一维序列卷积计算原理示意图

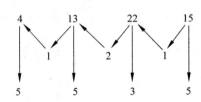
图 3-9 把卷积结果转换为数字

```
def conv(lst1, lst2):
    '''用来计算两个列表所表示的信号的卷积,返回一个列表'''
    result=[]
    #翻转第一个列表
    lst1.reverse()
    length1=len(lst1)
    length2=len(lst2)
    #移动翻转后的第一个列表,直到"完全移入"
    for i in range(1, length1+1):
        t=lst1[length1-i:]
```

```
        #计算重叠"面积"
        v=sum((item1 * item2 for item1, item2 in zip(t,lst2)))
        result.append(v)
    #继续移动翻转后的第一个列表,直到"完全移出"
    for i in range(1, length2):
        t=lst2[i:]
        v=sum((item1 * item2 for item1, item2 in zip(lst1,t)))
        result.append(v)
    return result

def mul(lst):
    '''把列表中的数字转换为普通整数的形式'''
    result=''
    c=0
    for item in lst[::-1]:
        item=item+c
        #计算当前位的余数和向前一位进位的数字
        n, c=str(item%10), item //10
        #使用字符串记录临时结果
        result +=n
    if c:
        result +=str(c)
    return eval(result[::-1])

def main(num1, num2):
    lst1=list(map(int, str(num1)))
    lst2=list(map(int, str(num2)))
    result=conv(lst1, lst2)
    print(mul(result)==num1 * num2)

from random import randint
for i in range(100):
    num1=randint(1, 99999999)
    num2=randint(1, 99999999999)
    main(num1, num2)
```

例 3-31 猜数游戏。系统随机产生一个数,玩家最多可以猜 5 次,系统会根据玩家的猜测进行提示,玩家则可以根据系统的提示对下一次的猜测进行适当调整。

```
from random import randint

def guess():
    #随机生成一个整数
    value=randint(1,1000)
    #最多允许猜 5 次
```

```
    maxTimes=5
    for i in range(maxTimes):
        prompt='Start to GUESS:' if i==0 else 'Guess again:'
        #使用异常处理结构,防止输入不是数字的情况
        try:
            x=int(input(prompt))
            #猜对了
            if x ==value:
                print('Congratulations!')
                break
            elif x>value:
                print('Too big')
            else:
                print('Too little')
        except:
            print('Must input an integer between 1 and 999')
    else:
    #次数用完还没猜对,游戏结束,提示正确答案
        print('Game over. FAIL.')
        print('The value is ', value)

guess()
```

例 3-32 计算形式如 a+aa+aaa+aaaa+…+aaa…aaa 的表达式的值,其中 a 为小于 10 的自然数。

```
def demo(v, n):
    assert 0<v<10, 'v must between 1 and 9'
    assert type(n)==int, 'n must be integer'
    result, t=0, 0
    for i in range(n):
        t=t*10+v
        result +=t
    return result

print(demo(3, 4))
```

例 3-33 有 n 个人围成一圈,顺序排号。从第一个人开始从 1 到 k(假设 k=3)报数,报到 k 的人退出圈子,然后圈子缩小,从下一个人继续游戏,问最后留下的是原来的第几号?

```
from itertools import cycle

def demo(lst, k):
    #切片,以免影响原来的数据
```

```
        t_lst=lst[:]
        #游戏一直进行到只剩下最后一个人
        while len(t_lst)>1:
            #创建cycle对象
            c=cycle(t_lst)
            #从1到k报数
            for i in range(k):
                t=next(c)
            #一个人出局,圈子缩小
            index=t_lst.index(t)
            t_lst=t_lst[index+1:]+t_lst[:index]
            #测试用,查看每次一个人出局之后剩余人的编号
            print(t_lst)
        #游戏结束
        return t_lst[0]

lst=list(range(1,11))
print(demo(lst, 3))
```

例 3-34 汉诺塔问题。

据说古代有一个梵塔,塔内有 3 个底座 A、B、C,A 座上有 64 个盘子,盘子大小不等,大的在下,小的在上。有一个和尚想把这 64 个盘子从 A 座移到 C 座,但每次只能允许移动一个盘子,在移动盘子的过程中可以利用 B 座,但任何时刻 3 个座上的盘子都必须始终保持大盘在下、小盘在上的顺序。如果只有一个盘子,则不需要利用 B 座,直接将盘子从 A 移到 C 即可。和尚想知道这项任务的详细移动步骤和顺序。这实际上是一个非常巨大的工程,是一个不可能完成的任务。根据数学知识我们可以知道,移动 n 个盘子需要 2^{n-1} 步,64 个盘子需要 18 446 744 073 709 551 615 步。如果每步需要一秒钟,那么就需要 584 942 417 355.072 年。

```
def hannuo(num, src, dst, temp=None):
    #声明用来记录移动次数的变量为全局变量
    global times
    #确认参数类型和范围
    assert type(num)==int, 'num must be integer'
    assert num>0, 'num must>0'
    #只剩最后或只有一个盘子需要移动,这也是函数递归调用的结束条件
    if num ==1:
        print('The {0} Times move:{1}==>{2}'.format(times, src, dst))
        times +=1
    else:
        #递归调用函数自身,
        #先把除最后一个盘子之外的所有盘子移到临时柱子上
        hannuo(num-1, src, temp, dst)
        #把最后一个盘子直接移到目标柱子上
```

```
        hannuo(1, src, dst)
        #把除最后一个盘子之外的其他盘子从临时柱子上移到目标柱子上
        hannuo(num-1, temp, dst, src)
#用来记录移动次数的变量
times=1
#A 表示最初放置盘子的柱子,C 是目标柱子,B 是临时柱子
hannuo(3, 'A', 'C', 'B')
```

拓展知识:函数递归调用。函数的递归调用是函数调用的一种特殊情况,函数调用自己,自己再调用自己……当某个条件得到满足时就不再调用了,最后再一层一层地返回直到该函数的第一次调用,如图 3-10 所示。从图中可以看出,每次调用函数时必须要记住离开的位置才能保证函数运行结束以后回到正确的位置,这个过程称为保存现场,这需要一定的栈空间。因此,递归深度如果太深的话,可能会使栈空间不足进而导致程序崩溃。

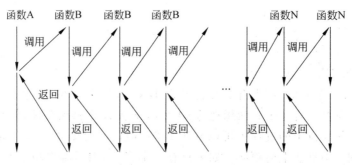

图 3-10 函数递归调用示意图

例 3-35 编写函数计算任意位数的黑洞数。黑洞数是指这样的整数:由这个数字每位数字组成的最大数减去每位数字组成的最小数仍然得到这个数自身。例如,3 位黑洞数是 495,因为 954－459＝495,4 位数字是 6174,因为 7641－1467＝6174。

```
def main(n):
    '''参数 n 表示数字的位数,例如 n=3 时,返回 495'''
    #待测试数范围的起点和结束值
    start=10**(n-1)+2
    end=start * 10-20
    #依次测试每个数
    for i in range(start, end):
        i=str(i)
        #由这几个数字组成的最大的数
        big=''.join(sorted(i,reverse=True))
        big=int(big)
        #由这几个数字组成的最小的数
        little=''.join(sorted(i))
        little=int(little)
```

```
            if big-little==int(i):
                print(i)
n=4
main(n)
```

这个问题还有另外一个计算方法,以计算三位黑洞数为例,任意找一个三位数789,依次做如下计算:987-879=198 ==> 981-189=792 ==> 972-279=693 ==> 963-369=594 ==> 954-459=495。按这个思路可以编写出下面的代码,但是如果初始数字选取得不合适,会需要大量递归调用,可能会导致无法求解或者栈溢出而使程序崩溃。

```
def blackHole(n):
    data=str(n)
    big=sorted(data, reverse=True)
    big=int(''.join(big))
    little=sorted(data)
    little=int(''.join(little))
    data=int(data)
    if big-little ==data:
        return data
    else:
        return blackHole(big-little)

print(blackHole(126))
```

例 3-36 24 点游戏是指随机选取 4 张扑克牌(不包括大小王),然后通过四则运算来构造表达式,如果表达式的值恰好等于 24 就赢一次。下面的代码定义了一个函数用来测试随机给定的 4 个数是否符合 24 点游戏规则,如果符合就输出所有可能的表达式。

```
from random import randint
from itertools import permutations

#4 个数字和 2 个运算符可能组成的表达式形式
exps = ('((%s %s %s)%s %s)%s %s',
        '(%s %s %s)%s(%s %s %s)',
        '(%s %s(%s %s %s))%s %s',
        '%s %s((%s %s %s)%s %s)',
        '%s %s(%s %s(%s %s %s))')
ops=r'+-*/'

def test24(v):
    result=[]
    #Python 允许函数的嵌套定义
    #这个函数对字符串表达式求值并验证是否等于 24
    def check(exp):
```

```
            try:
                #有可能会出现除0异常,所以放到异常处理结构中
                return int(eval(exp))==24
            except:
                return False
        #全排列,枚举4个数的所有可能顺序
        for a in permutations(v):
            #查找4个数的当前排列能实现24的表达式
            t=[exp %(a[0], op1, a[1], op2, a[2], op3, a[3])for op1 in ops for op2 in ops
              for op3 in ops for exp in exps if check(exp %(a[0], op1, a[1], op2, a[2],
              op3, a[3]))]
            if t:
                result.append(t)
    return result

for i in range(20):
    print('=' * 20)
    #生成随机数字进行测试
    lst=[randint(1, 14)for j in range(4)]
    r=test24(lst)
    if r:
        print(r)
    else:
        print('No answer for ', lst)
```

例 3-37 双色球是一种比较常见的彩票玩法,每一注彩票由 6 个介于 1 到 33 之间的不重复数字和 1 个介于 1 到 16 之间的数字组成。下面的代码用来随机生成一注双色球彩票,结果是完全随机的。

```
import random

def doubleColor():
    red=random.sample(range(1,34), 6)
    blue=random.choice(range(1, 17))
    return str(red)+'-'+str(blue)

print(doubleColor())
```

例 3-38 八皇后问题。八皇后问题是高斯先生(就是小时候就把 $1+2+3+\cdots+100$ 转换成 $(1+100)\times 50$ 的那个数学家)在 60 多年以前提出来的,是一个经典的回溯算法问题,其核心为:在国际象棋棋盘(8 行 8 列)上摆放 8 个皇后,要求 8 个皇后中任意两个都不能位于同一行、同一列或同一斜线上。

```
def isValid(s, col):
    '''这个函数用来检查最后一个皇后的位置是否合法'''
```

```python
    #当前皇后的行号
    row=len(s)
    #检查当前的皇后们是否有冲突
    for r, c in enumerate(s):
        #如果这一列已有皇后,或者某个皇后与当前皇后的水平与垂直距离相等
        #就表示当前皇后位置不合法,不允许放置
        if c ==col or abs(row-r)==abs(col-c):
            return False

    return True


def queen(n, s=()):
    '''这个函数返回的结果是每个皇后所在列号'''
    #已是最后一个皇后,保存本次结果
    if len(s)==n:
        return [s]

    res=[]
    for col in range(n):
        if not isValid(s, col): continue
        for r in queen(n, s + (col,)):
            res.append(r)

    return res

#形式转换,最终结果中包含每个皇后所在的行号和列号
result=[[(r, c)for r, c in enumerate(s)] for s in queen(8)]
#输出合法结果的数量
print(len(result))
#输出所有可能的结果,也就是所有皇后的摆放位置
#结果中每个皇后的位置是一个元组,里面两个数分别是行号和列号
for r in result:
    print(r)
```

第 4 章 面向对象程序设计

面向对象程序设计(Object Oriented Programming,OOP)的思想主要针对大型软件设计而提出,使得软件设计更加灵活,能够很好地支持代码复用和设计复用,代码具有更好的可读性和可扩展性,大幅度降低了软件开发的难度。面向对象程序设计的一个关键性观念是将数据以及对数据的操作封装在一起,组成一个相互依存、不可分割的整体,即对象,不同对象之间通过消息机制来通信或者同步。对于相同类型的对象(instance)进行分类、抽象后,得出共同的特征而形成了类(class),面向对象程序设计的关键就是如何合理地定义这些类并且合理组织多个类之间的关系。

Python 是真正面向对象的高级动态编程语言,完全支持面向对象的基本功能,如封装、继承、多态以及对基类方法的覆盖或重写。Python 中对象的概念很广泛,Python 中的一切内容都可以称为对象,函数也是对象。创建类时用变量形式表示对象特征的成员称为数据成员(attribute),用函数形式表示对象行为的成员称为成员方法(method),数据成员和成员方法统称为类的成员。

4.1 基 础 知 识

4.1.1 类的定义与使用

Python 使用 class 关键字来定义类,class 关键字之后是一个空格,接下来是类的名字,如果派生自其他基类的话则需要把所有基类放到一对圆括号中并使用逗号分隔,然后是一个冒号,最后换行并定义类的内部实现。类名的首字母一般要大写,当然也可以按照自己的习惯定义类名,但是一般推荐参考惯例来命名,并在整个系统的设计和实现中保持风格一致,这一点对于团队合作非常重要。例如:

```
class Car(object):              #定义一个类,派生自 object 类
    def infor(self):            #定义成员方法
        print(" This is a car ")
```

定义了类之后,就可以用来实例化对象,并通过"对象名.成员"的方式来访问其中的数据成员或成员方法,例如:

```
>>>car=Car()                    #实例化对象
```

```
>>>car.infor()                          #调用对象的方法
This is a car
```

在 Python 中,可以使用内置方法 isinstance()来测试一个对象是否为某个类的实例,例如:

```
>>>isinstance(car, Car)
True
>>>isinstance(car, str)
False
```

最后,Python 提供了一个关键字 pass,执行的时候什么也不会发生,可以用在类和函数的定义中或者选择结构中,表示空语句。如果暂时没有确定如何实现某个功能,或者为以后的软件升级预留空间,可以使用关键字 pass 来"占位"。例如,下面的代码都是合法的:

```
>>>class A:
    pass

>>>def demo():
    pass

>>>if 5>3:
    Pass
```

💡 **小提示**:可以使用三引号为类进行必要的注释,例如:

```
>>>class Test:
    '''This is only a test.'''
    pass
>>>Test.__doc__
'This is only a test.'
```

4.1.2 私有成员与公有成员

从形式上看,在定义类的成员时,如果成员名以两个下划线(__)开头则表示是私有成员,但是Python 并没有对私有成员提供严格的访问保护机制。私有成员在类的外部不能直接访问,一般是在类的内部进行访问和操作,或者在类外部通过调用对象的公有成员方法来访问。另外,Python 提供了一种特殊方式"对象名._类名__xxx"可以访问私有成员,但这会破坏类的封装性,不推荐这样做(不过真的很难阻止别人这么做)。公有属性是可以公开使用的,既可以在类的内部进行访问,也可以在外部程序中使用。

```
>>>class A:
    def __init__(self, value1=0, value2=0):        #构造函数
        self._value1=value1
```

```
        self.__value2=value2              #私有成员
    def setValue(self, value1, value2):    #成员方法
        self._value1=value1
        self.__value2=value2               #在类内部可以直接访问私有成员
    def show(self):                        #成员方法
        print(self._value1)
        print(self.__value2)
>>>a=A()
>>>a._value1                               #在类外部可以直接访问非私有成员
0
>>>a._A__value2                            #在外部访问对象的私有数据成员
0
```

在 IDLE 环境中，在对象或类名后面加上一个圆点"."，稍等一秒钟则会自动列出其所有公开成员，如图 4-1 所示，模块也具有同样的特点。

而如果在圆点"."后面再加一个下画线，则会列出该对象或类的所有成员，包括私有成员，如图 4-2 所示。

图 4-1 列出对象公开成员　　图 4-2 列出对象所有成员

在 Python 中，以下画线开头和结束的成员名有特殊的含义，类定义中用下画线作为变量名和方法名前缀和后缀来表示类的特殊成员。

（1）_xxx：保护成员，不能用'from module import *'导入，只有类对象和子类对象可以访问这些成员。

（2）__xxx__：系统定义的特殊成员，详见 4.1.7 节。

（3）__xxx：类中的私有成员，一般只有类对象自己能访问，子类对象也不能访问到这个成员，但在对象外部可以通过"对象名._类名__xxx"这样的特殊方式来访问。

注意：Python 中不存在严格意义上的私有成员。

小提示：在 IDLE 交互模式下，下画线(_)表示解释器中最后一次语句正确执行的输出结果。例如：

```
>>>3+5
8
>>>_+2
10
```

```
>>>_ * 3
30
>>>_ / 5
6.0
>>>3
3
>>>1/0
Traceback(most recent call last):
  File "<pyshell#2>", line 1, in <module>
    1/0
ZeroDivisionError: integer division or modulo by zero
>>>_
3
```

4.1.3 数据成员

数据成员用来说明对象特有的一些属性,如人的身份证号、姓名、年龄、性别、身高、学历,汽车的品牌、颜色、最高时速,蛋糕的名称、尺寸、配料,书的名字、作者、ISBN、出版社、出版日期,等等。

数据成员可以大致分为两类:属于对象的数据成员和属于类的数据成员。属于对象的数据成员主要指在构造函数__init__()中定义的(当然也可以在其他成员方法中定义),定义和使用时必须以 self 作为前缀(这一点是必需的),同一个类的不同对象(实例)之间的数据成员之间互不影响;属于类的数据成员是该类所有对象共享的,不属于任何一个对象,在定义类时这类数据成员不在任何一个成员方法的定义中。在主程序中或类的外部,对象数据成员属于实例(对象),只能通过对象名访问;而类数据成员属于类,可以通过类名或对象名访问。另外,在 Python 中可以动态地为类和对象增加成员,这也是 Python 动态类型的一种重要体现。

```
class Car(object):
    price=100000                              #属于类的数据成员
    def __init__(self, c):
        self.color=c                          #属于对象的数据成员

car1=Car("Red")                               #实例化对象
car2=Car("Blue")
print(car1.color, Car.price)                  #访问对象和类的数据成员
Car.price=110000                              #修改类的属性
Car.name='QQ'                                 #动态增加类的属性
car1.color="Yellow"                           #修改实例的属性
print(car2.color, Car.price, Car.name)
print(car1.color, Car.price, Car.name)
def setSpeed(self, s):
    self.speed=s
```

```
import types
car1.setSpeed=types.MethodType(setSpeed, car1)    #动态为对象增加成员方法
car1.setSpeed(50)                                  #调用对象的成员方法
print(car1.speed)
```

拓展知识：利用类数据成员的共享性，可以实时获得该类的对象数量，并且可以控制该类可以创建的对象最大数量。例如：

```
>>>class Demo(object):
    total=0
    def __new__(cls, * args, **kwargs):          #该方法在__init__()之前被调用
        if cls.total>=3:                          #最多允许创建3个对象
            raise Exception('最多只能创建3个对象')
        else:
            return object.__new__(cls)
    def __init__(self):
        Demo.total=Demo.total+1
>>>t1=Demo()
>>>t1
<__main__.Demo object at 0x00000000034A0278>
>>>t2=Demo()
>>>t3=Demo()
>>>t4=Demo()
Traceback(most recent call last):
  File "<pyshell#8>", line 1, in <module>
    t4=Demo()
  File "<pyshell#3>", line 5, in __new__
    raise Exception('最多只能创建3个对象')
Exception: 最多只能创建3个对象
>>>t4
Traceback(most recent call last):
  File "<pyshell#9>", line 1, in <module>
    t4
NameError: name 't4' is not defined
```

4.1.4 方法

方法用来描述对象所具有的行为，例如，列表对象的追加元素、插入元素、删除元素、排序，字符串对象的分隔、连接、排版、替换，烤箱的温度设置、烘烤，等等。

在类中定义的方法可以粗略分为四大类：公有方法、私有方法、静态方法和类方法。公有方法、私有方法一般是指属于对象的实例方法，其中私有方法的名字以两个下画线（__）开始。每个对象都有自己的公有方法和私有方法，在这两类方法中都可以访问属于类和对象的成员；公有方法通过对象名直接调用，私有方法不能通过对象名直接调用，只能在实例方法中通过self调用或在外部通过Python支持的特殊方式来调用。

类的所有实例方法都必须至少有一个名为 self 的参数,并且必须是方法的第一个形参(如果有多个形参的话),self 参数代表对象自身。在类的实例方法中访问实例属性时需要以 self 为前缀,但在外部通过对象名调用对象方法时并不需要传递这个参数,如果在外部通过类名调用属于对象的公有方法,需要显式为该方法的 self 参数传递一个对象名,用来明确指定访问哪个对象的数据成员。

静态方法和类方法都可以通过类名和对象名调用,但不能直接访问属于对象的成员,只能访问属于类的成员。一般将 cls 作为类方法的第一个参数,表示该类自身,在调用类方法时不需要为该参数传递值。例如下面的代码所演示:

```
>>>class Root:
    __total=0
    def __init__(self, v):                  #构造函数
        self.__value=v
        Root.__total += 1

    def show(self):                         #普通实例方法
        print('self.__value:', self.__value)
        print('Root.__total:', Root.__total)

    @classmethod                            #修饰器,声明类方法
    def classShowTotal(cls):                #类方法
        print(cls.__total)

    @staticmethod                           #修饰器,声明静态方法
    def staticShowTotal():                  #静态方法
        print(Root.__total)
>>>r=Root(3)
>>>r.classShowTotal()                       #通过对象来调用类方法
1
>>>r.staticShowTotal()                      #通过对象来调用静态方法
1
>>>r.show()
self.__value: 3
Root.__total: 1
>>>rr=Root(5)
>>>Root.classShowTotal()                    #通过类名调用类方法
2
>>>Root.staticShowTotal()                   #通过类名调用静态方法
2
>>>Root.show()                              #试图通过类名直接调用实例方法,失败
Traceback(most recent call last):
  File "<pyshell#9>", line 1, in <module>
    Root.show()
```

```
TypeError: unbound method show () must be called with Root instance as first
argument(got nothing instead)
>>>Root.show(r)                    #可以通过这种方法来调用方法并访问实例成员
self.__value: 3
Root.__total: 2
>>>r.show()
self.__value: 3
Root.__total: 2
>>>Root.show(rr)                   #通过类名调用实例方法时为self参数显式传递对象名
self.__value: 5
Root.__total: 2
>>>rr.show()
self.__value: 5
Root.__total: 2
```

小提示：在 Python 中，在类中定义实例方法时将第一个参数定义为 self 只是一个习惯，并不必须使用 self 这个名字，但是一般也不建议使用别的名字。同样，属于类的方法中使用 cls 作为第一个参数也是一种习惯，也可以使用其他的名字作为第一个参数，虽然不建议这样做。

注意：不同对象实例的数据成员之间互不影响，是不共享的。但同一个类的所有实例方法是在不同对象之间共享的，所有对象都执行相同的代码，通过 self 参数来判断要处理哪个对象的数据。

拓展知识：在 Python 中，函数和方法是有区别的。方法一般指与特定实例绑定的函数，通过对象调用方法时，对象本身将被作为第一个参数传递过去，普通函数并不具备这个特点。

```
>>>class Demo:
    pass
>>>t=Demo()
>>>def test(self, v):
    self.value=v
>>>t.test=test                            #动态增加普通函数
>>>t.test
<function test at 0x00000000034B7EA0>
>>>t.test(t, 3)
>>>print(t.value)
3
>>>import types
>>>t.test=types.MethodType(test, t)       #动态增加绑定的方法
>>>t.test
<bound method test of <__main__.Demo object at 0x000000000074F9E8>>
>>>t.test(5)
```

```
>>>print(t.value)
5
```

4.1.5 属性

公开的数据成员可以在外部随意访问和修改,很难控制用户修改时新数据的合法性。解决这一问题的常用方法是定义私有数据成员,然后设计公开的成员方法来提供对私有数据成员的读取和修改操作,修改私有数据成员时可以对值进行合法性检查,提高了程序的健壮性,保证了数据的完整性。属性结合了公开数据成员和成员方法的优点,既可以像成员方法那样对值进行必要的检查,又可以像数据成员一样灵活地访问。

Python 2.x 中属性的实现有很多不如人意的地方。在 Python 3.x 中,属性得到了较为完整的实现,支持更加全面的保护机制。如果设置属性为只读,则无法修改其值,也无法为对象增加与属性同名的新成员,同时,也无法删除对象属性。例如:

```
>>>class Test:
    def __init__(self, value):
        self.__value=value                  #私有数据成员

    @property                               #修饰器,定义属性,提供对私有数据成员的访问
    def value(self):                        #只读属性,无法修改和删除
        return self.__value
>>>t=Test(3)
>>>t.value
3
>>>t.value=5                                #只读属性不允许修改值
Traceback(most recent call last):
  File "<pyshell#151>", line 1, in <module>
    t.value=5
AttributeError: can't set attribute
>>>t.v=5                                    #动态增加新成员
>>>t.v
5
>>>del t.v                                  #动态删除成员
>>>del t.value                              #试图删除对象属性,失败
Traceback(most recent call last):
  File "<pyshell#152>", line 1, in <module>
    del t.value
AttributeError: can't delete attribute
>>>t.value
3
```

下面的代码则把属性设置为可读、可修改,而不允许删除。

```
>>>class Test:
```

```python
    def __init__(self, value):
        self.__value=value

    def __get(self):                              #读取私有数据成员的值
        return self.__value

    def __set(self, v):                           #修改私有数据成员的值
        self.__value=v

    value=property(__get, __set)                  #可读可写属性,指定相应的读写方法

    def show(self):
        print(self.__value)
>>>t=Test(3)
>>>t.value                                        #允许读取属性值
3
>>>t.value=5                                      #允许修改属性值
>>>t.value
5
>>>t.show()                                       #属性对应的私有变量也得到了相应的修改
5
>>>del t.value                                    #试图删除属性,失败
Traceback(most recent call last):
  File "<pyshell#152>", line 1, in <module>
    del t.value
AttributeError: can't delete attribute
```

当然,也可以将属性设置为可读、可修改、可删除。

```python
>>>class Test:
    def __init__(self, value):
        self.__value=value

    def __get(self):
        return self.__value

    def __set(self, v):
        self.__value=v

    def __del(self):                              #删除对象的私有数据成员
        del self.__value

    value=property(__get, __set, __del)           #可读、可写、可删除的属性

    def show(self):
```

```
        print(self.__value)
>>>t=Test(3)
>>>t.show()
3
>>>t.value
3
>>>t.value=5
>>>t.show()
5
>>>t.value
5
>>>del t.value
>>>t.value                                          #相应的私有数据成员已删除,访问失败
Traceback(most recent call last):
  File "<pyshell#165>", line 1, in <module>
    t.value
  File "<pyshell#157>", line 6, in __get
    return self.__value
AttributeError: 'Test' object has no attribute '_Test__value'
>>>t.show()
Traceback(most recent call last):
  File "<pyshell#166>", line 1, in <module>
    t.show()
  File "<pyshell#157>", line 17, in show
    print(self.__value)
AttributeError: 'Test' object has no attribute '_Test__value'
>>>t.value =1                                       #为对象动态增加属性和对应的私有数据成员
>>>t.show()
1
>>>t.value
1
```

4.1.6 继 承

俗话说得好,"虎父无犬子"、"龙生龙,凤生凤,老鼠的儿子会打洞",这在一定程度上说明了继承的重要性。在面向对象编程中,继承是代码复用和设计复用的重要途径,是面向对象程序设计的重要特性之一,继承也是实现多态的必要条件之一。

设计一个新类时,如果可以继承一个已有的设计良好的类然后进行二次开发,无疑会大幅度减少开发工作量,并且可以很大程度地保证质量。在继承关系中,已有的、设计好的类称为父类或基类,新设计的类称为子类或派生类。派生类可以继承父类的公有成员,但是不能继承其私有成员。如果需要在派生类中调用基类的方法,可以使用内置函数super()或者通过"基类名.方法名()"的方式来实现这一目的。

例 4-1 设计 Person 类,并根据 Person 派生 Teacher 类,分别创建 Person 类与

Teacher类的对象。

```python
#基类必须继承于object,否则在派生类中将无法使用super()函数
class Person(object):
    def __init__(self, name='', age=20, sex='man'):
        self.setName(name)                    #通过调用方法进行初始化
        self.setAge(age)                      #这样可以对参数进行更好地控制
        self.setSex(sex)

    def setName(self, name):
        if not isinstance(name, str):
            print('name must be string.')
            #如果数据不合法,就使用默认值
            self.__name=''
            return
        self.__name=name

    def setAge(self, age):
        if type(age)!=int:
            print('age must be integer.')
            self.__age=20
            return
        self.__age=age

    def setSex(self, sex):
        if sex not in ('man', 'woman'):
            print('sex must be "man" or "woman"')
            self.__sex='man'
            return
        self.__sex=sex

    def show(self):
        print(self.__name, self.__age, self.__sex, sep='\n')

#派生类
class Teacher(Person):
    def __init__(self, name='', age=30, sex='man', department='Computer'):
        #调用基类构造方法初始化基类的私有数据成员
        super(Teacher, self).__init__(name, age, sex)
        #也可以这样初始化基类的私有数据成员
        #Person.__init__(self, name, age, sex)
        #初始化派生类的数据成员
        self.setDepartment(department)
```

```python
    def setDepartment(self, department):
        if type(department)!=str:
            print('department must be a string.')
            self.__department='Computer'
            return
        self.__department=department

    def show(self):
        super(Teacher, self).show()
        print(self.__department)

if __name__=='__main__':
    #创建基类对象
    zhangsan=Person('Zhang San', 19, 'man')
    zhangsan.show()
    print('='*30)

    #创建派生类对象
    lisi=Teacher('Li si', 32, 'man', 'Math')
    lisi.show()
    #调用继承的方法修改年龄
    lisi.setAge(40)
    lisi.show()
```

小提示：Python支持多继承，如果父类中有相同的方法名，而在子类中使用时没有指定父类名，则Python解释器将从左向右按顺序进行搜索。

下面的代码完整地描述了类的继承机制，请认真体会构造函数、私有方法以及普通公开方法的继承原理。

```python
>>>class A():
    def __init__(self):
        self.__private()
        self.public()

    def __private(self):
        print('__private()method of A')

    def public(self):
        print('public()method of A')

    def __test(self):
        print('__test()method of A')
>>>class B(A):                              #注意,B类没有构造函数
    def __private(self):
```

```
            print('__private()method of B')

    def public(self):
        print('public()method of B')
>>>b=B()                                            #创建派生类对象
__private()method of A
public()method of B
>>>b.__test()                                       #派生类没有继承基类中的私有成员方法
Traceback (most recent call last):
  File "<pyshell#215>", line 1, in <module>
    b.__test()
AttributeError: 'B' object has no attribute '__test'
>>>class C(A):                                      #注意,C类有构造函数
    def __init__(self):
        self.__private()
        self.public()

    def __private(self):
        print('__private()method of C')

    def public(self):
        print('public()method of C')
>>>c=C()
__private()method of C
public()method of C
```

> **拓展知识**：所谓多态,是指基类的同一个方法在不同派生类对象中具有不同的表现和行为。龙生九子,子子皆不同。禅宗说"一花开五叶",也是这个道理。派生类继承了基类的行为和属性之后,还会增加某些特定的行为和属性,同时还可能会对继承来的某些行为进行一定的改变,这恰恰是多态的表现形式。在Python中主要通过重写基类的方法来实现多态。

4.1.7 特殊方法与运算符重载

Python类有大量的特殊方法,其中比较常见的是构造函数和析构函数。Python中类的构造函数是__init__(),一般用来为数据成员设置初始值或进行其他必要的初始化工作,在创建对象时被自动调用和执行。如果用户没有设计构造函数,Python将提供一个默认的构造函数用来进行必要的初始化工作。Python中类的析构函数是__del__(),一般用来释放对象占用的资源,在Python删除对象和收回对象空间时被自动调用和执行。如果用户没有编写析构函数,Python将提供一个默认的析构函数进行必要的清理工作。

在Python中,除了构造函数和析构函数之外,还有大量的特殊方法支持更多的功能,例如,运算符重载就是通过在类中重写特殊函数来实现的。在自定义类时如果重写了

某个特殊方法即可支持对应的运算符,具体实现什么工作则完全可以根据需要来定义(有读者说,好像有点多态的意思)。表 4-1 列出了其中一部分特殊方法,完整列表请参考网址 https://docs.python.org/3/reference/datamodel.html#special-method-names。

表 4-1　Python 类特殊方法

方　　法	功　能　说　明
__new__()	类的静态方法,用于确定是否要创建对象
__init__()	构造函数,生成对象时调用
__del__()	析构函数,释放对象时调用
__add__()	＋
__sub__()	－
__mul__()	＊
__truediv__()	/
__floordiv__()	//
__mod__()	％
__pow__()	＊＊
__repr__()	打印、转换
__setitem__()	按照索引赋值
__getitem__()	按照索引获取值
__len__()	计算长度
__call__()	函数调用
__contains__()	in
__eq__()、__ne__()、__lt__()、__le__()、__gt__()、__ge__()	＝＝、!＝、＜、＜＝、＞、＞＝
__str__()	转化为字符串
__lshift__()、__rshift__()	＜＜、＞＞
__and__()、__or__()、__invert__()、__xor__()	＆、｜、～、^
__iadd__()、__isub__()	＋＝、－＝

4.2　案例精选

4.2.1　自定义数组

例 4-2　自定义一个数组类,支持数组与数字之间的四则运算,数组之间的加法运算、内积运算和大小比较,数组元素访问和修改,以及成员测试等功能。

```python
class MyArray:
    '''All the elements in this array must be numbers'''
    def __IsNumber(self, n):
        if not isinstance(n,(int, float, complex)):
            return False
        return True

    #构造函数,进行必要的初始化
    def __init__(self, *args):
        if not args:
            self.__value=[]
        else:
            for arg in args:
                if not self.__IsNumber(arg):
                    print('All elements must be numbers')
                    return
            self.__value=list(args)

    #析构函数,释放内部封装的列表
    def __del__(self):
        del self.__value

    #重载运算符+
    #数组中每个元素都与数字 n 相加,或两个数组相加,返回新数组
    def __add__(self, n):
        if self.__IsNumber(n):
            #数组中所有元素都与数字 n 相加
            b=MyArray()
            b.__value=[item+n for item in self.__value]
            return b
        elif isinstance(n, MyArray):
            #两个等长的数组对应元素相加
            if len(n.__value)==len(self.__value):
                c=MyArray()
                c.__value=[i+j for i, j in zip(self.__value, n.__value)]
                #for i, j in zip(self.__value, n.__value):
                #    c.__value.append(i+j)
                return c
            else:
                print('Lenght not equal')
        else:
            print('Not supported')

    #重载运算符-
```

```python
##数组中每个元素都与数字n相减,返回新数组
def __sub__(self, n):
    if not self.__IsNumber(n):
        print('-operating with ', type(n), ' and number type is not supported.')
        return
    b=MyArray()
    b.__value=[item-n for item in self.__value]
    return b

#重载运算符*
#数组中每个元素都与数字n相乘,返回新数组
def __mul__(self, n):
    if not self.__IsNumber(n):
        print('* operating with ', type(n), ' and number type is not supported.')
        return
    b=MyArray()
    b.__value=[item*n for item in self.__value]
    return b

#重载运算符/
#数组中每个元素都与数字n相除,返回新数组
def __truediv__(self, n):
    if not self.__IsNumber(n):
        print(r'/ operating with ', type(n), ' and number type is not supported.')
        return
    b=MyArray()
    b.__value=[item/n for item in self.__value]
    return b

#重载运算符//
#数组中每个元素都与数字n整除,返回新数组
def __floordiv__(self, n):
    if not isinstance(n, int):
        print(n, ' is not an integer')
        return
    b=MyArray()
    b.__value=[item//n for item in self.__value]
    return b

#重载运算符%
#数组中每个元素都与数字n求余数,返回新数组
def __mod__(self, n):
    if not self.__IsNumber(n):
        print(r'%operating with ', type(n), ' and number type is not supported.')
```

```python
            return
        b=MyArray()
        b.__value=[item%n for item in self.__value]
        return b

    #重载运算符**
    #数组中每个元素都与数字n进行幂计算,返回新数组
    def __pow__(self, n):
        if not self.__IsNumber(n):
            print('** operating with ', type(n), ' and number type is not supported.')
            return
        b=MyArray()
        b.__value=[item**n for item in self.__value]
        return b


    def __len__(self):
        return len(self.__value)

    #直接使用该类对象作为表达式来查看对象的值
    def __repr__(self):
    #equivalent to return 'self.__value'
        return repr(self.__value)

    #支持使用print()函数查看对象的值
    def __str__(self):
        return str(self.__value)

    #追加元素
    def append(self, v):
        if not self.__IsNumber(v):
            print('Only number can be appended.')
            return
        self.__value.append(v)

    #获取指定下标的元素值,支持使用列表或元组指定多个下标
    def __getitem__(self, index):
        length=len(self.__value)
        #如果指定单个整数作为下标,则直接返回元素值
        if isinstance(index, int) and 0<=index<length:
            return self.__value[index]
        #使用列表或元组指定多个整数下标
        elif isinstance(index,(list,tuple)):
            for i in index:
                if not(isinstance(i,int) and 0<=i<length):
```

```python
                return 'index error'
            result=[]
            for item in index:
                result.append(self.__value[item])
            return result
        else:
            return 'index error'

    #修改元素值,支持使用列表或元组指定多个下标,同时修改多个元素值
    def __setitem__(self, index, value):
        length=len(self.__value)
        #如果下标合法,则直接修改元素值
        if isinstance(index, int) and 0<=index<length:
            self.__value[index]=value
        #支持使用列表或元组指定多个下标
        elif isinstance(index,(list,tuple)):
            for i in index:
                if not(isinstance(i,int) and 0<=i<length):
                    raise Exception('index error')
            #如果下标和给的值都是列表或元组,并且个数一样
            #则分别为多个下标的元素修改值
            if isinstance(value,(list,tuple)):
                if len(index)==len(value):
                    for i, v in enumerate(index):
                        self.__value[v]=value[i]
                else:
                    raise Exception('values and index must be of the same length')
            #如果指定多个下标和一个普通值,则把多个元素修改为相同的值
            elif isinstance(value,(int,float,complex)):
                for i in index:
                    self.__value[i]=value
            else:
                raise Exception('value error')
        else:
            raise Exception('index error')

    #支持成员测试运算符 in,测试数组中是否包含某个元素
    def __contains__(self, v):
        if v in self.__value:
            return True
        return False

    #模拟向量内积
    def dot(self, v):
```

```python
            if not isinstance(v, MyArray):
                print(v, ' must be an instance of MyArray.')
                return
            if len(v)!=len(self.__value):
                print('The size must be equal.')
                return
            return sum([i * j for i,j in zip(self.__value, v.__value)])
            #b=MyArray()
            #for m, n in zip(v.__value, self.__value):
            #    b.__value.append(m * n)
            #return sum(b.__value)

        #重载运算符==,测试两个数组是否相等
        def __eq__(self, v):
            if not isinstance(v, MyArray):
                print(v, ' must be an instance of MyArray.')
                return False
            if self.__value ==v.__value:
                return True
            return False

        #重载运算符<,比较两个数组大小
        def __lt__(self, v):
            if not isinstance(v, MyArray):
                print(v, ' must be an instance of MyArray.')
                return False
            if self.__value<v.__value:
                return True
            return False

if __name__ =='__main__':
    print('Please use me as a module.')
```

将上面的程序保存为 MyArray.py 文件,可以作为 Python 模块导入并使用其中的数组类。

```
>>>from MyArray import MyArray           #导入模块中的自定义类
>>>x=MyArray(1, 2, 3, 4, 5, 6)           #实例化对象
>>>y=MyArray(6, 5, 4, 3, 2, 1)
>>>len(x)                                #返回数组长度,即数组中的元素个数
6
>>>x+5                                   #每个元素加5,返回新数组
[6, 7, 8, 9, 10, 11]
>>>x * 3                                 #每个元素乘以3,返回新数组
[3, 6, 9, 12, 15, 18]
```

```
>>>x.dot(y)                          #计算两个数组(一维向量)的内积
56
>>>x.append(7)                       #在数组尾部追加新元素
>>>x
[1, 2, 3, 4, 5, 6, 7]
>>>x.dot(y)
The size must be equal.
>>>x[9]=8                            #试图修改元素值
Index type error or out of range
>>>x / 2
[0.5, 1.0, 1.5, 2.0, 2.5, 3.0, 3.5]
>>>x // 2
[0, 1, 1, 2, 2, 3, 3]
>>>x %3
[1, 2, 0, 1, 2, 0, 1]
>>>x[2]                              #返回指定位置的元素值
3
>>>'a' in x                          #测试数组中是否包含某个元素
False
>>>3 in x
True
>>>x< y                              #比较数组大小
True
>>>x=MyArray(1, 2, 3, 4, 5, 6)
>>>x+y                               #两个数组中对应元素相加,返回新数组
[7, 7, 7, 7, 7, 7]
>>>x[[2,3,4]]                        #查看多个位置上的元素值
[3, 4, 5]
>>>x[[2, 3]]=[8, 9]                  #同时修改多个元素的值
>>>x
[1, 2, 8, 9, 5, 6]
>>>x[[1,3,5]]=0                      #为多个元素赋值为相同的值
>>>x
[1, 0, 8, 0, 5, 0]
```

4.2.2 自定义矩阵

例 4-3 模拟矩阵运算,支持矩阵转置,修改矩阵大小,矩阵与数字的加、减、乘运算,以及矩阵与矩阵的加、减、乘运算。

```
class simNumpyArray(object):
    def __init__(self, p):
        '''可以接收列表、元组、range对象等类型的数据,并且每个元素都必须为数字'''
        if type(p)not in(list, tuple, range):
```

```python
            print('data type error')
            return
    for item in p:
        #下面这行用来判断参数类型,可以这样写
        #if isinstance(item,(int, float, complex)):
        if type(item) not in (int, float, complex):
            print('data type error')
            return
    self.__data=[list(p)]
    self.__row=1
    self.__col=len(p)

#析构函数
def __del__(self):
    del self.__data

#修改大小,首先检查给定的大小参数是否合适
def reshape(self, size):
    '''参数必须为元组或列表,如(row,col)或[row,col]
       row或col其中一个可以为-1,表示自动计算
    '''
    if not(isinstance(size, list) or isinstance(size, tuple)):
        print('size parameter error')
        return
    if len(size)!=2:
        print('size parameter error')
        return
    if(not isinstance(size[0],int))or(not isinstance(size[1],int)):
        print('size parameter error')
        return
    if size[0] !=-1 and size[1] !=-1 and size[0] * size[1] !=self.__row *
self.__col:
        print('size parameter error')
        return
    #行数或列数为-1表示该值自动计算
    if size[0] ==-1:
        if size[1] ==-1 or(self.__row * self.__col)%size[1] !=0:
            print('size parameter error')
            return
    if size[1] ==-1:
        if size[0] ==-1 or(self.__row * self.__col)%size[0] !=0:
            print('size parameter error')
            return
```

```python
    #重新合并数据
    data=[t for i in self.__data for t in i]
    #修改大小
    if size[0] ==-1:
        self.__row=int(self.__row*self.__col/size[1])
        self.__col=size[1]
    elif size[1] ==-1:
        self.__col=int(self.__row*self.__col/size[0])
        self.__row=size[0]
    else:
        self.__row=size[0]
        self.__col=size[1]
    self.__data=[[data[row*self.__col+col] for col in range(self.__col)]
                 for row in range(self.__row)]

#在交互模式直接使用变量名作为表达式查看值时调用该函数
def __repr__(self):
    #return repr('\n'.join(map(str, self.__data)))
    for i in self.__data:
        print(i)
    return ''

#使用print()函数输出值时调用该函数
def __str__(self):
    return '\n'.join(map(str, self.__data))

#属性,矩阵转置
@property
def T(self):
    b=simNumpyArray([t for i in self.__data for t in i])
    b.reshape((self.__row, self.__col))
    b.__data=list(map(list,zip(*b.__data)))
    b.__row, b.__col=b.__col, b.__row
    return b

#通用代码,适用于矩阵与整数、实数、复数的加、减、乘、除、整除、幂
def __operate(self, n, op):
    b=simNumpyArray([t for i in self.__data for t in i])
    b.reshape((self.__row, self.__col))
    b.__data=[[eval(str(j)+op+str(n)) for j in item] for item in b.__data]
    return b

#通用代码,适用于矩阵之间的加、减
def __matrixAddSub(self, n, op):
```

```python
            c=simNumpyArray([1])
            c.__row=self.__row
            c.__col=self.__col
            c.__data=[[eval(str(x[i])+op+str(y[i]))for i in range(len(x))] for x,y
                    in zip(self.__data, n.__data)]
            return c

    #所有元素统一加一个数字,或者两个矩阵相加
    def __add__(self, n):
        #参数是整数或实数,则返回矩阵
        #其中的每个元素为原矩阵中元素与该整数或实数的加法结果
        if type(n)in(int, float, complex):
            return self.__operate(n, '+')
        elif isinstance(n, simNumpyArray):
            #如果参数为同类型矩阵,且大小一致,则为两个矩阵中对应元素相加
            if n.__row==self.__row and n.__col==self.__col:
                return self.__matrixAddSub(n, '+')
            else:
                print('two matrix must be the same size')
                return
        else:
            print('data type error')
            return

    #所有元素统一减一个数字,或者两个矩阵相减
    def __sub__(self, n):
        #参数是整数或实数,则返回矩阵
        #其中的每个元素为原矩阵中元素与该整数或实数的加法结果
        if type(n)in(int, float, complex):
            return self.__operate(n, '-')
        elif isinstance(n, simNumpyArray):
            #如果参数为同类型矩阵,且大小一致,则为两个矩阵中对应元素相减
            if n.__row==self.__row and n.__col==self.__col:
                #先实例化一个临时对象,其值临时为[1]
                return self.__matrixAddSub(n, '-')
            else:
                print('two matrix must be the same size')
                return
        else:
            print('data type error')
            return

    #所有元素统一乘一个数字,或者两个矩阵相乘
    def __mul__(self, n):
```

```python
            #参数是整数或实数,则返回矩阵
            #其中的每个元素为原矩阵中元素与该整数或实数的加法结果
            if type(n) in(int, float, complex):
                return self.__operate(n, '*')
            elif isinstance(n, simNumpyArray):
                #如果参数为同类型矩阵,且第一个矩阵的列数等于第二个矩阵的行数
                if n.__row==self.__col:
                    data=[]
                    for row in self.__data:
                        t=[]
                        for ii in range(n.__col):
                            col=[c[ii] for c in n.__data]
                            tt=sum([i*j for i,j in zip(row,col)])
                            t.append(tt)
                        data.append(t)
                    c=simNumpyArray([t for i in data for t in i])
                    c.reshape((self.__row, n.__col))
                    return c
                else:
                    print('size error.')
                    return
            else:
                print('data type error')
                return

    #所有元素统一除以一个数字,本程序使用 Python 3.5.1 编写,真除法
    def __truediv__(self, n):
        if type(n) in(int, float, complex):
            return self.__operate(n, '/')
        else:
            print('data type error')
            return

    #矩阵元素与数字计算整商
    def __floordiv__(self, n):
        if type(n) in(int, float, complex):
            return self.__operate(n, '//')
        else:
            print('data type error')
            return

    #矩阵与数字的幂运算
    def __pow__(self, n):
        if type(n) in(int, float, complex):
```

```
            return self.__operate(n, '**')
        else:
            print('data type error')
            return

    #测试两个矩阵是否相等
    def __eq__(self, n):
        if isinstance(n, simNumpyArray):
            if self.__data==n.__data:
                return True
            else:
                return False
        else:
            print('data type error')
            return

    #测试矩阵自身是否小于另一个矩阵
    def __lt__(self, n):
        if isinstance(n, simNumpyArray):
            if self.__data<n.__data:
                return True
            else:
                return False
        else:
            print('data type error')
            return

    #成员测试运算符
    def __contains__(self, v):
        if v in self.__data:
            return True
        else:
            return False

    #支持迭代
    def __iter__(self):
        return iter(self.__data)

    #通用方法,计算三角函数
    def __triangle(self, method):
        try:
            b=simNumpyArray([t for i in self.__data for t in i])
            b.reshape((self.__row, self.__col))
            b.__data=[[eval("__import__('math')."+method+"("+str(j)+")")for j
```

```
                in item] for item in b.__data]
            return b
        except:
            return 'method error'

    #属性,对所有元素求正弦
    @property
    def Sin(self):
        return self.__triangle('sin')

    #属性,对所有元素求余弦
    @property
    def Cos(self):
        return self.__triangle('cos')
```

4.2.3 自定义队列

队列是一种特殊的线性表,只允许在队列尾部进行元组插入操作和在队列头部进行元素删除操作,具有"先入先出(FIFO)"或"后入后出(LILO)"的特点,在多线程编程、作业管理等方面具有重要的应用。

Python 列表对象的 append()方法用于在列表尾部追加元素,pop(0)可以删除并返回列表头部的元素。

```
>>>x=[]
>>>x.append(1)                          #在尾部追加元素,模拟入队操作
>>>x.append(2)
>>>x.append(3)
>>>x
[1, 2, 3]
>>>x.pop(0)                             #在头部弹出元素,模拟出队操作
1
>>>x
[2, 3]
>>>x.pop(0)
2
>>>x.pop(0)
3
>>>x
[]
>>>x.pop(0)                             #空队列弹出头部元素失败,抛出异常
Traceback (most recent call last):
  File "<pyshell#233>", line 1, in <module>
    x.pop(0)
IndexError: pop from empty list
```

从上面的代码可以看出，使用 Python 列表直接模拟队列结构，无法限制队列的大小，并且当列表为空时进行弹出元素的操作会抛出异常。可以对列表进行封装，自定义队列类来避免这些问题。

例 4-4 设计自定义队列类，模拟入队、出队等基本操作。

```
class myQueue:
    #构造函数,默认队列大小为 10
    def __init__(self, size=10):
        self._content=[]
        self._size=size
        self._current=0

    #析构函数
    def __del__(self):
        del self._content

    def setSize(self, size):
        if size<self._current:
            #如果缩小队列,应删除后面的元素
            for i in range(size, self._current)[::-1]:
                del self._content[i]
            self._current=size
        self._size=size

    #入队
    def put(self, v):
        if self._current<self._size:
            self._content.append(v)
            self._current=self._current+1
        else:
            print('The queue is full')

    #出队
    def get(self):
        if self._content:
            self._current=self._current-1
            return self._content.pop(0)
        else:
            print('The queue is empty')

    def show(self):
        if self._content:
            print(self._content)
        else:
```

```
            print('The queue is empty')

    #把队列置空
    def empty(self):
        self._content=[]
        self._current=0

    def isEmpty(self):
        if not self._content:
            return True
        else:
            return False

    def isFull(self):
        if self._current ==self._size:
            return True
        else:
            return False

if __name__ == '__main__':
    print('Please use me as a module.')
```

将上面的代码保存为 myQueue.py 文件,并保存在当前文件夹、Python 3.5 安装文件夹或 sys.path 列表指定的其他文件夹中,当然也可以使用 append()方法把该文件所在文件夹添加到 sys.path 列表中。下面的代码演示了自定义队列类的用法。

```
>>>import myQueue                    #导入包含自定义队列类的模块
>>>q=myQueue.myQueue()               #创建自定义队列对象
>>>q.get()                           #获取元素
The queue is empty
>>>q.put(5)                          #在队列尾部插入元素
>>>q.put(7)
>>>q.isFull()                        #判断队列是否已满
False
>>>q.put('a')
>>>q.put(3)
>>>q.show()                          #查看队列中的所有元素
[5, 7, 'a', 3]
>>>q.setSize(3)                      #修改队列的大小
>>>q.show()
[5, 7, 'a']
>>>q.put(10)
The queue is full
>>>q.setSize(5)
>>>q.put(10)
```

```
>>>q.show()
[5, 7, 'a', 10]
```

拓展知识：Python 标准库 queue 提供了 LILO 队列类 Queue、LIFO 队列类 LifoQueue、优先级队列类 PriorityQueue，标准库 collections 提供了双端队列。例如：

```
>>>from queue import Queue              #LILO 队列
>>>q=Queue()                            #创建队列对象
>>>q.put(0)                             #在队列尾部插入元素
>>>q.put(1)
>>>q.put(2)
>>>print(q.queue)                       #查看队列中的所有元素
deque([0, 1, 2])
>>>q.get()                              #返回并删除队列头部元素
0
>>>q.get()
1
>>>q.queue
deque([2])
>>>q.get()
2

>>>from queue import LifoQueue          #LIFO 队列
>>>q=LifoQueue()                        #创建 LIFO 队列对象
>>>q.put(1)                             #在队列尾部插入元素
>>>q.put(2)
>>>q.put(3)
>>>q.queue                              #查看队列中的所有元素
[1, 2, 3]
>>>q.get()                              #返回并删除队列尾部元素
3
>>>q.queue
[1, 2]
>>>q.get()
2
>>>q.queue
[1]

>>>from queue import PriorityQueue      #优先级队列
>>>q=PriorityQueue()                    #创建优先级队列对象
>>>q.put(3)                             #插入元素
>>>q.put(8)                             #插入元素
>>>q.put(100)
>>>q.queue                              #查看优先级队列中的所有元素
```

```
[3, 8, 100]
>>>q.put(1)                                    #插入元素,自动调整优先级队列
>>>q.put(2)
>>>q.queue
[1, 2, 100, 8, 3]
>>>q.get()                                     #返回并删除优先级最低的元素
1
>>>q.get()
2
>>>q.get()
3
>>>q.get()
8
>>>q.get()
100

>>>from collections import deque               #双端队列
>>>q=deque(['Eric', 'John', 'Smith'])
>>>q
deque(['Eric', 'John', 'Smith'])
>>>q.append('Tom')                             #在右侧插入新元素
>>>q.appendleft('Terry')                       #在左侧插入新元素
>>>q
deque(['Terry', 'Eric', 'John', 'Smith', 'Tom'])
>>>q.rotate(2)                                 #循环右移 2 次
>>>q
deque(['Smith', 'Tom', 'Terry', 'Eric', 'John'])
>>>q.pop()                                     #返回并删除队列最右端元素
'John'
>>>q
deque(['Smith', 'Tom', 'Terry', 'Eric'])
>>>q.popleft()                                 #返回并删除队列最左端元素
'Smith'
>>>q
deque(['Tom', 'Terry', 'Eric'])
```

4.2.4 自定义栈

栈也是一种运算受限的线性表,其特点在于仅允许在一端进行元素的插入和删除操作,最后入栈的元素最先出栈,而最先入栈的元素最后出栈,即"先入后出(FILO)"或"后入先出(LIFO)"。

使用 Python 列表对象提供的 append()、pop()方法也可以模拟栈结构及其基本运算,但是无法限制栈的大小,并且在栈为空时尝试获取其元素时会引发异常,例如:

```
>>>s=[]
>>>s.append(3)                              #在尾部追加元素,模拟入栈操作
>>>s.append(5)
>>>s.append(7)
>>>s
[3, 5, 7]
>>>s.pop()                                  #在尾部弹出元素,模拟出栈操作
7
>>>s
[3, 5]
>>>s.pop()
5
>>>s
[3]
>>>s.pop()
3
>>>s.pop()
Traceback(most recent call last):
  File "<pyshell#74>", line 1, in <module>
    s.pop()
IndexError: pop from empty list
```

如同封装 Python 列表实现自定义队列类一样,也可以对 Python 列表进行封装来模拟栈结构。

小提示:使用 Python 列表对象的 insert(0,x)和 pop(0)也可以模拟栈的基本操作,实现在列表头部的元素插入和删除操作,但这两个方法会引发大量的元素移动操作,效率非常低,不建议使用。

例 4-5 设计自定义栈类,模拟入栈、出栈、判断栈是否为空、是否已满以及改变栈大小等操作。

```
class Stack:
    def __init__(self, size=10):
        self._content=[]                    #使用列表存放栈的元素
        self._size=size                     #初始栈大小
        self._current=0                     #栈中元素个数初始化为 0

    #析构函数
    def __del__(self):
        del self._content

    def empty(self):
        self._content=[]
        self._current=0
```

```python
    def isEmpty(self):
        return not self._content

    def setSize(self, size):
        #如果缩小栈空间,则删除指定大小之后的已有元素
        if size<self._current:
            for i in range(size, self._current)[::-1]:
                del self._content[i]
            self._current=size
        self._size=size

    def isFull(self):
        return self._current ==self._size

    def push(self, v):
        if self._current<self._size:
            self._content.append(v)
            self._current=self._current+1        #栈中元素个数加 1
        else:
            print('Stack Full!')

    def pop(self):
        if self._content:
            self._current=self._current-1       #栈中元素个数减 1
            return self._content.pop()
        else:
            print('Stack is empty!')

    def show(self):
        print(self._content)

    def showRemainderSpace(self):
        print('Stack can still PUSH ', self._size-self._current, ' elements.')

if __name__ =='__main__':
    print('Please use me as a module.')
```

将代码保存为 Stack.py 文件,下面的代码演示了自定义栈结构的用法。

```
>>>import Stack                              #导入模块
>>>s=Stack.Stack()                           #实例化对象
>>>s.isEmpty()                               #测试栈是否为空
True
>>>s.isFull()                                #测试栈是否已满
```

```
False
>>>s.push(5)                              #元素入栈
>>>s.push(8)
>>>s.push('a')
>>>s.pop()                                #元素出栈
'a'
>>>s.push('b')
>>>s.push('c')
>>>s.show()                               #查看栈的内容
[5, 8, 'b', 'c']
>>>s.showRemainderSpace()                 #查看栈的剩余大小
Stack can still PUSH  6  elements.
>>>s.setSize(3)                           #修改栈的大小
>>>s.isFull()
True
>>>s.show()
[5, 8, 'b']
>>>s.setSize(5)
>>>s.push('d')
>>>s.push('dddd')
>>>s.push(3)
Stack Full!
>>>s.show()
[5, 8, 'b', 'd', 'dddd']
```

4.2.5 自定义二叉树

二叉树是每个节点最多有两个子树(分别称为左子树和右子树)的树结构,二叉树的第 i 层最多有 $2^{(i-1)}$ 个节点,常用于排序或查找。

例 4-6 设计二叉树类,模拟二叉树创建、插入子节点以及前序遍历、中序遍历和后序遍历等遍历方式,同时还支持二叉树中任意子树的节点遍历。

```
class BinaryTree:
    def __init__(self, value):
        self.__left=None
        self.__right= None
        self.__data=value

    def __del__(self):                    #析构函数
        del self.__data

    def insertLeftChild(self, value):     #创建左子树
        if self.__left:
            print('Left child tree already exists.')
```

```
            else:
                self.__left=BinaryTree(value)
                return self.__left

        def insertRightChild(self, value):          #创建右子树
            if self.__right:
                print('Right child tree already exists.')
            else:
                self.__right=BinaryTree(value)
                return self.__right

        def show(self):
            print(self.__data)

        def preOrder(self):                         #前序遍历
            print(self.__data)                      #输出根节点的值
            if self.__left:
                self.__left.preOrder()              #遍历左子树
            if self.__right:
                self.__right.preOrder()             #遍历右子树

        def postOrder(self):                        #后序遍历
            if self.__left:
                self.__left.postOrder()
            if self.__right:
                self.__right.postOrder()
            print(self.__data)

        def inOrder(self):                          #中序遍历
            if self.__left:
                self.__left.inOrder()
            print(self.__data)
            if self.__right:
                self.__right.inOrder()

if __name__=='__main__':
    print('Please use me as a module.')
```

把上面的代码保存为 BinaryTree.py 文件,下面的代码创建了如图 4-3 所示的二叉树,并对该树进行遍历。

```
>>>import BinaryTree
>>>root=BinaryTree.BinaryTree('root')
>>>b=root.insertRightChild('B')
```

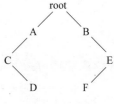

图 4-3 二叉树

```
>>>a=root.insertLeftChild('A')
>>>c=a.insertLeftChild('C')
>>>d=c.insertRightChild('D')
>>>e=b.insertRightChild('E')
>>>f=e.insertLeftChild('F')
>>>root.inOrder()
C D A root B F E
>>>root.postOrder()
D C A F E B root
>>>b.inOrder()
B F E
```

拓展知识：堆也是一种很重要的数据结构，在进行排序时使用较多，优先队列也是堆结构的一个重要应用。堆是一个二叉树，其中每个父节点的值都小于或等于其所有子节点的值。使用数组或列表来实现小根堆时，对于所有的 k(下标，从 0 开始)都满足 heap[k] <= heap[2*k+1]和 heap[k] <= heap[2*k+2]，并且整个堆中最小的元素总是位于二叉树的根节点，大根堆与小根堆正好相反。Python 在 heapq 模块中提供了对堆的支持，例如：

```
>>>import heapq                              #导入 heapq 模块
>>>import random
>>>data=random.sample(range(1000), 10)       #生成随机测试数据
>>>data
[638, 659, 212, 84, 737, 677, 553, 340, 526, 747]
>>>heapq.heapify(data)                       #堆化随机测试数据
>>>data
[84, 340, 212, 526, 737, 677, 553, 659, 638, 747]
>>>heapq.heappush(data, 30)                  #新元素入堆,自动调整堆结构
>>>data
[30, 84, 212, 526, 340, 677, 553, 659, 638, 747, 737]
>>>heapq.heappush(data, 5)
>>>data
[5, 84, 30, 526, 340, 212, 553, 659, 638, 747, 737, 677]
>>>heapq.heappop(data)                       #返回并删除最小元素,自动调整堆
5
>>>heapq.heappop(data)
30
>>>heapq.heappop(data)
84
>>>data
[212, 340, 553, 526, 737, 677, 747, 659, 638]
>>>heapq.heappushpop(data, 1000)             #弹出最小元素,同时新元素入堆
212
>>>data
```

```
[340, 526, 553, 638, 737, 677, 747, 659, 1000]
>>>heapq.heapreplace(data, 500)          #弹出最小元素,同时新元素入堆
340
>>>data
[500, 526, 553, 638, 737, 677, 747, 659, 1000]
>>>heapq.heapreplace(data, 700)
500
>>>data
[526, 638, 553, 659, 737, 677, 747, 700, 1000]
>>>heapq.nlargest(3, data)               #返回最大的前 3 个元素
[1000, 747, 737]
>>>heapq.nsmallest(2, data, key=str)     #返回指定排序规则下最小的 3 个元素
[1000, 526]
```

4.2.6　自定义有向图

有向图由若干节点和边组成,其中每条边都是有明确方向的,即从一个节点指向另一个节点。若有向图中两个节点之间存在若干条有向边则表示从起点可以到达终点,认为存在一条路径。

例 4-7　设计有向图类,模拟有向图的创建和路径搜索功能。

```
class DirectedGraph(object):
    def __init__(self, d):
        if isinstance(d, dict):
            self.__graph=d
        else:
            self.__graph=dict()
            print('Sth error')

    def __generatePath(self, graph, path, end, results):
        current=path[-1]
        if current ==end:
            results.append(path)
        else:
            for n in graph[current]:
                if n not in path:
                    self.__generatePath(graph, path+[n], end, results)

    def searchPath(self, start, end):
        self.__results=[]
        self.__generatePath(self.__graph, [start], end, self.__results)
        self.__results.sort(key=lambda x:len(x))      #按所有路径的长度进行排序
        print('The path from ',self.__results[0][0], ' to ', self.__results[0]
            [-1], ' is:')
```

```
            for path in self.__results:
                print(path)

d={'A':['B','C','D'],
   'B':['E'],
   'C':['D','F'],
   'D':['B','E','G'],
   'E':['D'],
   'F':['D','G'],
   'G':['E']}
g=DirectedGraph(d)
g.searchPath('A','D')
g.searchPath('A','E')
```

程序运行结果为

```
The path from  A  to  D  is:
['A', 'D']
['A', 'C', 'D']
['A', 'B', 'E', 'D']
['A', 'C', 'F', 'D']
['A', 'C', 'F', 'G', 'E', 'D']
The path from  A  to  E  is:
['A', 'B', 'E']
['A', 'D', 'E']
['A', 'C', 'D', 'E']
['A', 'D', 'B', 'E']
['A', 'D', 'G', 'E']
['A', 'C', 'D', 'B', 'E']
['A', 'C', 'D', 'G', 'E']
['A', 'C', 'F', 'D', 'E']
['A', 'C', 'F', 'G', 'E']
['A', 'C', 'F', 'D', 'B', 'E']
['A', 'C', 'F', 'D', 'G', 'E']
```

4.2.7 自定义集合

Python 内置了集合类型，支持并集、差集、交集等运算，并且不允许元素重复。本节案例通过封装 Python 列表模拟了集合类，并提供了有关的运算。

例 4-8 自定义集合类。

```
#自定义集合类
class Set(object):
    def __init__(self, data=None):
        if data==None:
```

```python
            self.__data=[]
    else:
        if not hasattr(data, '__iter__'):
            #提供的数据不可迭代,实例化失败
            raise Exception('必须提供可迭代的数据类型')
        temp=[]
        for item in data:
            #集合中的元素必须可哈希
            #这里的可哈希与第14章介绍的安全哈希算法并不一样
            hash(item)
            if not item in temp:
                temp.append(item)
        self.__data=temp

#析构函数
def __del__(self):
    del self.__data

#添加元素,要求元素必须可哈希
def add(self, value):
    hash(value)
    if value not in self.__data:
        self.__data.append(value)
    else:
        print('元素已存在,操作被忽略')

#删除元素
def remove(self, value):
    if value in self.__data:
        self.__data.remove(value)
        print('删除成功')
    else:
        print('元素不存在,删除操作被忽略')

#随机弹出并返回一个元素
def pop(self):
    if not self.__data:
        print('集合已空,弹出操作被忽略')
        return
    import random
    item=random.choice(self.__data)
    print(item)
    self.__data.remove(item)
```

```python
    #运算符重载,集合差集运算
    def __sub__(self, anotherSet):
        if not isinstance(anotherSet, Set):
            raise Exception('类型错误')
        #空集合
        result=Set()
        #如果一个元素属于当前集合而不属于另一个集合,添加
        for item in self.__data:
            if item not in anotherSet.__data:
                result.__data.append(item)
        return result

    #提供方法,集合差集运算
    def difference(self, anotherSet):
        if not isinstance(anotherSet, Set):
            raise Exception('类型错误')
        return self-anotherSet


    #|运算符重载,集合并集运算
    def __or__(self, anotherSet):
        if not isinstance(anotherSet, Set):
            raise Exception('类型错误')
        result=Set(self.__data)
        for item in anotherSet.__data:
            if item not in result.__data:
                result.__data.append(item)
        return result

    #提供方法,集合并集运算
    def union(self, anotherSet):
        if not isinstance(anotherSet, Set):
            raise Exception('类型错误')
        return self | anotherSet

    #& 运算符重载,集合交集运算
    def __and__(self, anotherSet):
        if not isinstance(anotherSet, Set):
            raise Exception('类型错误')
        result=Set()
        for item in self.__data:
            if item in anotherSet.__data:
                result.__data.append(item)
        return result
```

```python
#^运算符重载,集合对称差集
def __xor__(self, anotherSet):
    return(self-anotherSet)|(anotherSet-self)

#提供方法,集合对称差集运算
def symetric_difference(self, anotherSet):
    if not isinstance(anotherSet, Set):
        raise Exception('类型错误')
    return self ^ anotherSet

#==运算符重载,判断两个集合是否相等
def __eq__(self, anotherSet):
    if not isinstance(anotherSet, Set):
        raise Exception('类型错误')
    if sorted(self.__data)==sorted(anotherSet.__data):
        return True
    else:
        return False

#>运算符重载,集合包含关系
def __gt__(self, anotherSet):
    if not isinstance(anotherSet, Set):
        raise Exception('类型错误')
    if self !=anotherSet:
        flag1=True
        for item in self.__data:
            if item not in anotherSet.__data:
                #当前集合中有的元素不属于另一个集合
                flag1=False
                break
        flag2=True
        for item in anotherSet.__data:
            if item not in self.__data:
                #另一个集合中有的元素不属于当前集合
                flag2=False
                break
        if  not flag1 and flag2:
            return True
    return False

#>=运算符重载,集合包含关系
def __ge__(self, anotherSet):
    if not isinstance(anotherSet, Set):
```

```
            raise Exception('类型错误')
        return self==anotherSet or self>anotherSet

    #提供方法,判断当前集合是否为另一个集合的真子集
    def issubset(self, anotherSet):
        if not isinstance(anotherSet, Set):
            raise Exception('类型错误')
        if self<anotherSet:
            return True
        else:
            return False

    #提供方法,判断当前集合是否为另一个集合的超集
    def issuperset(self, anotherSet):
        if not isinstance(anotherSet, Set):
            raise Exception('类型错误')
        if self>anotherSet:
            return True
        else:
            return False


    #提供方法,清空集合所有元素
    def clear(self):
        while self.__data:
            del self.__data[-1]
        print('集合已清空')

    #运算符重载,使得集合可迭代
    def __iter__(self):
        return iter(self.__data)

    #运算符重载,支持 in 运算符
    def __contains__(self, value):
        if value in self.__data:
            return True
        else:
            return False

    #支持内置函数 len()
    def __len__(self):
        return len(self.__data)

    #直接查看该类对象时调用该函数
```

```
        def __repr__(self):
            return '{'+str(self.__data)[1:-1]+'}'

        #使用print()函数输出该类对象时调用该函数
        def __str__(self):
            return '{'+str(self.__data)[1:-1]+'}'
```

把这个文件保存成mySet.py,然后就可以像下面的代码这样使用自定义集合类了。

```
>>>from mySet import Set                    #导入自定义集合类
>>>x=Set(range(10))                         #创建集合对象
>>>y=Set(range(8, 15))
>>>z=Set([1, 2, 3, 4, 5])
>>>x
{0, 1, 2, 3, 4, 5, 6, 7, 8, 9}
>>>y
{8, 9, 10, 11, 12, 13, 14}
>>>z
{1, 2, 3, 4, 5}
>>>z.add(6)                                 #增加元素
>>>z
{1, 2, 3, 4, 5, 6}
>>>z.remove(3)                              #删除指定元素
删除成功
>>>z
{1, 2, 4, 5, 6}
>>>y.pop()                                  #随机删除一个元素
11
>>>x-y                                      #差集
{0, 1, 2, 3, 4, 5, 6, 7}
>>>x-z
{0, 3, 7, 8, 9}
>>>x.difference(y)
{0, 1, 2, 3, 4, 5, 6, 7}
>>>x | y                                    #并集
{0, 1, 2, 3, 4, 5, 6, 7, 8, 9, 10, 12, 13, 14}
>>>x.union(y)
{0, 1, 2, 3, 4, 5, 6, 7, 8, 9, 10, 12, 13, 14}
>>>x & z                                    #交集
{1, 2, 4, 5, 6}
>>>x ^ z                                    #对称差集
{0, 3, 7, 8, 9}
>>>x.symetric_difference(y)
{0, 1, 2, 3, 4, 5, 6, 7, 10, 12, 13, 14}
>>>(x-y)|(y-x)
```

```
{0, 1, 2, 3, 4, 5, 6, 7, 10, 12, 13, 14}
>>>x ==y                          #测试两个集合是否相等
False
>>>x> y                           #测试集合包含关系
False
>>>y> x
False
>>>x> z
True
>>>x >= z
True
>>>z.issubset(x)                  #测试 z 是否为 x 的子集
True
>>>x.issuperset(z)                #测试 x 是否为 z 的超集
True
>>>3 in x                         #测试集合中是否存在某个元素
True
>>>33 in x
False
>>>len(y)                         #计算集合中元素个数
6
>>>y.clear()
集合已清空
>>>y.pop()
集合已空,弹出操作被忽略
```

第 5 章 字符串与正则表达式

5.1 字　符　串

最早的字符串编码是美国标准信息交换码 ASCII,仅对 10 个数字、26 个大写英文字母、26 个小写英文字母及一些其他符号进行了编码。ASCII 采用 1 个字节来对字符进行编码,最多只能表示 256 个符号。

随着信息技术的发展和信息交换的需要,各国的文字都需要进行编码,不同的应用领域和场合对字符串编码的要求也略有不同,于是分别设计了不同的编码格式,常见的主要有 UTF-8、UTF-16、UTF-32、GB2312、GBK、CP936、base64、CP437 等。UTF-8 编码是国际通用的编码,以 1 个字节表示英语字符(兼容 ASCII),以 3 个字节表示中文,还有些语言的符号使用 2 个字节(如俄语和希腊语符号)或 4 个字节,UTF-8 对全世界所有国家需要用到的字符进行了编码。GB2312 是我国制定的中文编码,使用 1 个字节表示英语,2 个字节表示中文;GBK 是 GB2312 的扩充,而 CP936 是微软公司在 GBK 基础上开发的编码方式。GB2312、GBK 和 CP936 都是使用 2 个字节表示中文,UTF-8 使用 3 个字节表示中文。不同编码格式之间相差很大,采用不同的编码格式意味着不同的表示和存储形式,把同一字符存入文件时,写入的内容可能会不同,在理解其内容时必须了解编码规则并进行正确的解码。如果解码方法不正确就无法还原信息,从这个角度来讲,字符串编码也具有加密的效果。

Python 3.x 完全支持中文,使用 Unicode 编码格式,无论是一个数字、英文字母,还是一个汉字,都按一个字符对待和处理。例如,在 Python 3.5.1 中执行下面的代码,从代码中可以看到,在 Python 3.x 中甚至可以使用中文作为变量名。

```
>>>s='中国山东烟台'
>>>len(s)                    #字符串长度,或者包含的字符个数
6
>>>s='SDIBT'
>>>len(s)
5
>>>s='中国山东烟台 SDIBT'      #中文与英文字符同样对待,都算一个字符
>>>len(s)
11
>>>姓名='张三'                #使用中文作为变量名
```

```
>>>年龄=40
>>>print(姓名)                    #输出变量的值
张三
>>>print(年龄)
40
```

小提示：在 Windows 平台上使用 Python 2.x 时，input()函数从键盘输入的字符串默认为 GBK 编码，而 Python 程序中的字符串编码则使用#coding 显式地指定，常用的方式有：

```
#coding=utf-8
#coding:utf-8
#-*-coding:utf-8-*-
```

在 Python 中，字符串属于不可变序列类型，使用单引号（这是最常用的）、双引号、三单引号或三双引号作为界定符，并且不同的界定符之间可以互相嵌套。下面几种都是合法的 Python 字符串：

'abc'、'123'、'中国'、"Python"、'''Tom said,"Let's go"'''

除了支持序列通用方法（包括双向索引、比较大小、计算长度、元素访问、切片等操作）以外，字符串类型还支持一些特有的操作方法，如格式化、字符串查找、字符串替换（注意，不是原地替换）、排版等。但由于字符串属于不可变序列，不能直接对字符串对象进行元素增加、修改与删除等操作。另外，字符串对象提供的 replace()和 translate()方法也不是对原字符串直接进行修改替换，而是返回一个修改替换后的新字符串作为结果。

Python 支持短字符串驻留机制，对于短字符串，将其赋值给多个不同的对象时，内存中只有一个副本，多个对象共享该副本，与其他类型数具有相同的特点。然而，这一点并不适用于长字符串，长字符串不遵守驻留机制，下面的代码演示了短字符串和长字符串在这方面的区别。

```
>>>a='1234'
>>>b='1234'
>>>id(a)==id(b)                   #短字符串支持内存驻留机制
True
>>>a='1234'*50
>>>b='1234'*50
>>>id(a)==id(b)                   #长字符串不支持内存驻留机制
False
```

如果需要判断一个变量是否为字符串，可以使用内置方法 isinstance()或 type()。

```
>>>type('中国')
<class 'str'>
>>>type('中国'.encode('gbk'))      #编码成字节串，采用 GBK 编码格式
<class 'bytes'>
```

```
>>>bytes                                  #bytes 是 Python 的内置类
<class 'bytes'>
>>>isinstance('中国', str)
True
>>>type('中国')==str
True
>>>type('中国'.encode())==bytes
True
>>>type('中国')==bytes
False
```

拓展知识：转义字符。如果大家学习过其他语言，应该了解转义字符的概念，可以跳过这部分内容。转义字符是指，在字符串中某些特定的符号前加一个斜线之后该字符将被解释为另外一种含义，不再表示本来的字符。常见的转义字符如表 5-1 所示。

表 5-1 常见的转义字符

转义字符	含 义
\b	退格，把光标移动到前一列位置
\f	换页符
\n	换行符
\r	回车
\t	水平制表符
\v	垂直制表符
\\	一个\
\'	单引号
\"	双引号
\ooo	3 位八进制数对应的字符
\xhh	2 位十六进制数对应的字符
\uhhhh	4 位十六进制数表示的 Unicode 字符

下面的代码演示了转义字符的用法：

```
>>>print('Hello\nWorld')                  #包含转义字符的字符串
Hello
World
>>>oct(65)
'0o101'
>>>print('\101')                          #3 位八进制数对应的字符
A
>>>hex(65)
```

```
'0x41'
>>>print('\x41')                    #2 位十六进制数对应的字符
A
>>>ord('董')
33891
>>>hex(_)
'0x8463'
>>>print('\u8463')                  #4 位十六进制数表示的 Unicode 字符
董
```

5.1.1 字符串格式化的两种形式

如果需要将其他类型的数据转换为字符串,或者嵌入其他字符串或模板中再进行输出,就需要用到字符串格式化。Python 中字符串格式化的格式如图 5-1 所示,格式运算符%之前的部分为格式字符串,之后的部分为需要进行格式化的内容。

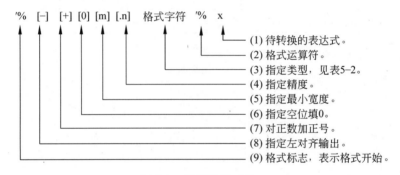

图 5-1 字符串格式化

Python 支持大量的格式字符,表 5-2 列出了比较常用的一部分。

表 5-2 格式字符

格式字符	说　明
%s	字符串(采用 str()的显示)
%r	字符串(采用 repr()的显示)
%c	单个字符
%b	二进制整数
%d	十进制整数
%i	十进制整数
%o	八进制整数
%x	十六进制整数
%e	指数(基底写为 e)

续表

格式字符	说　　明
%E	指数(基底写为E)
%f、%F	浮点数
%g	指数(e)或浮点数(根据显示长度)
%G	指数(E)或浮点数(根据显示长度)
%%	字符"%"

下面的代码演示了字符串格式化的用法：

```
>>>x=1235
>>>so="%o" %x
>>>so
'2323'
>>>sh="%x" %x
>>>sh
'4d3'
>>>se="%e" %x
>>>se
'1.235000e+03'
>>>"%s"%65                          #等价于 str()
'65'
>>>"%s"%65333
'65333'
>>>'%d,%c'%(65, 65)                 #使用元组对字符串进行格式化,按位置进行对应
'65,A'
>>>"%d"%"555"                       #试图将字符串转换为整数进行输出,抛出异常
Traceback(most recent call last):
  File "<pyshell#19>", line 1, in <module>
    "%d"%"555"
TypeError: %d format: a number is required, not str
>>>int('555')                       #可以使用int()函数将合法的数字字符串转换为整数
555
>>>'%s'%[1, 2, 3]
'[1, 2, 3]'
>>>str((1, 2, 3))                   #可以使用str()函数将任意类型数据转换为字符串
'(1, 2, 3)'
>>>str([1, 2, 3])
'[1, 2, 3]'
```

除了上面介绍的字符串格式化方法，目前 Python 社区更推荐使用 format()方法进行格式化，该方法更加灵活，不仅可以使用位置进行格式化，还支持使用与位置无关的参数名字来进行格式化，并且支持序列解包格式化字符串，为程序员提供了非常大的方便。

例如：

```
>>>print('{0:.3f}'.format(1/3))      #保留3位小数
0.333
>>>1/3
0.3333333333333333
>>>print("The number {0:,} in hex is: {0:#x}, in oct is {0:#o}".format(55))
The number 55 in hex is: 0x37, in oct is 0o67
>>>print("The number {0:,} in hex is: {0:x}, the number {1} in oct is {1:o}".
format(5555, 55))
The number 5,555 in hex is: 15b3, the number 55 in oct is 67
>>>print("The number {1} in hex is: {1:#x}, the number {0} in oct is {0:#o}".
format(5555, 55))
The number 55 in hex is: 0x37, the number 5555 in oct is 0o12663
>>>print("my name is {name}, my age is {age}, and my QQ is {qq}".format(name="
Dong", qq="306467355", age=38))
my name is Dong, my age is 38, and my QQ is 306467355
>>>position = (5, 8, 13)
>>>print("X:{0[0]};Y:{0[1]};Z:{0[2]}".format(position))
                      #使用元组同时格式化多个值
X:5;Y:8;Z:13
>>>weather=[("Monday", "rain"),("Tuesday", "sunny"),("Wednesday", "sunny"),
("Thursday", "rain"),("Friday", "Cloudy")]
>>>formatter="Weather of '{0[0]}' is '{0[1]}'".format
>>>for item in map(formatter, weather):
    print(item)
```

上面最后一段代码也可以改为下面的写法：

```
>>>for item in weather:
    print(formatter(item))
```

运行结果为

```
Weather of 'Monday' is 'rain'
Weather of 'Tuesday' is 'sunny'
Weather of 'Wednesday' is 'sunny'
Weather of 'Thursday' is 'rain'
Weather of 'Friday' is 'Cloudy'
```

拓展知识：在字符串格式化方法format()中常用的格式字符。在字符串格式化方法format()中可以使用的格式主要有b(二进制格式)、c(把整数转换成Unicode字符)、d(十进制格式)、o(八进制格式)、x(小写十六进制格式)、X(大写十六进制格式)、e/E(科学计数法格式)、f/F(固定长度的浮点数格式)、%(使用固定长度浮点数显示百分数)。

拓展知识：Python标准库string还提供了用于字符串格式化的模板类

Template。例如：

```
>>>from string import Template
>>>t=Template('My name is ${name}, and is ${age} years old.')
                                       #创建模板
>>>d={'name':'Dong', 'age':39}
>>>t.substitute(d)                     #替换
'My name is Dong, and is 39 years old.'
>>>tt=Template('My name is $name, and is $age years old.')
>>>tt.substitute(d)
'My name is Dong, and is 39 years old.'
```

5.1.2 字符串常用方法

1. find()、rfind()、index()、rindex()、count()

find()和rfind()方法分别用来查找一个字符串在另一个字符串指定范围（默认是整个字符串）中首次和最后一次出现的位置，如果不存在则返回-1；index()和rindex()方法用来返回一个字符串在另一个字符串指定范围中首次和最后一次出现的位置，如果不存在则抛出异常；count()方法用来返回一个字符串在另一个字符串中出现的次数，如果不存在则返回0。

```
>>>s="apple,peach,banana,peach,pear"
>>>s.find("peach")            #返回第一次出现的位置
6
>>>s.find("peach", 7)         #从指定位置开始查找
19
>>>s.find("peach", 7, 20)     #在指定范围中进行查找
-1
>>>s.rfind('p')               #从字符串尾部向前查找
25
>>>s.index('p')               #返回首次出现的位置
1
>>>s.index('pe')
6
>>>s.index('pear')
25
>>>s.index('ppp')             #指定子字符串不存在时抛出异常
Traceback(most recent call last):
  File "<pyshell#11>", line 1, in <module>
    s.index('ppp')
ValueError: substring not found
>>>s.count('p')               #统计子字符串出现的次数
5
```

```
>>>s.count('pp')
1
>>>s.count('ppp')                              #不存在时返回 0
0
```

 拓展知识：实际开发时应优先考虑使用 Python 内置函数和内置对象的方法，运行速度快，并且运行稳定。例如，下面的代码用来检查长字符串中哪些位置上的字母是 a，通过运行结果可以发现，使用字符串方法 find() 的速度明显要比逐个字符比较快很多。

```
from string import ascii_letters
from random import choice
from time import time

letters=''.join([choice(ascii_letters) for i in range(999999)])
def positions_of_character(sentence, ch):      #使用字符串对象的 find()方法
    result=[]
    index=0
    index=sentence.find(ch, index+1)
    while index!=-1:
        result.append(index)
        index=sentence.find(ch, index+1)
    return result

def demo(s, c):                                #普通方法,逐个字符比较
    result=[]
    for i,ch in enumerate(s):
        if ch==c:
            result.append(i)
    return result

start=time()
positions=positions_of_character(letters, 'a')
print(time()-start)

start=time()
p=demo(letters, 'a')
print(time()-start)
```

运行结果如下：

```
0.009000539779663086
0.08400487899780273
```

速度居然相差 10 倍左右，看来内置对象提供的方法还真是不错，简直是人见人爱，花见花开。但是不要高兴太早，一切都是相对的，这世间没有绝对得好，也没有绝对得坏，内

置对象的某些方法也不是在任何场合都能保证最优。例如把上面代码中的

letters=''.join([choice(ascii_letters)for i in range(999999)])

改为

letters=''.join([choice('ab')for i in range(999999)])

然后再次运行,会发现结果与上面的代码恰好相反,逐个比较的方法又比使用find()方法快了很多。稍加分析可以发现,上面两段代码是完全一样的,只是所查找数据的密度不一样,处理速度却有着翻天覆地的变化。所以说,首先要分析待处理的数据有什么样的特点(包括组成元素、分布情况等),然后才能设计最优的算法并采用最高效的方法。但一般情况下,Python 内置函数、内置对象的方法和标准库对象的效率要高于自己编写的代码。

2. split()、rsplit()、partition()、rpartition()

split()和 rsplit()方法分别用来以指定字符为分隔符,从字符串左端和右端开始将其分隔成多个字符串,并返回包含分隔结果的列表;partition()和 rpartition()用来以指定字符串为分隔符将原字符串分隔为3部分,即分隔符之前的字符串、分隔符字符串和分隔符之后的字符串,如果指定的分隔符不在原字符串中,则返回原字符串和两个空字符串。

```
>>>s="apple,peach,banana,pear"
>>>li=s.split(",")                    #使用逗号进行分隔
>>>li
["apple", "peach", "banana", "pear"]
>>>s.partition(',')                   #从左侧使用逗号进行切分
('apple', ',', 'peach,banana,pear')
>>>s.rpartition(',')                  #从右侧使用逗号进行切分
('apple,peach,banana', ',', 'pear')
>>>s.rpartition('banana')             #使用字符串作为分隔符
('apple,peach,', 'banana', ',pear')
>>>s="2014-10-31"
>>>t=s.split("-")                     #使用指定字符作为分隔符
>>>t
['2014', '10', '31']
>>>list(map(int, t))                  #将分隔结果转换为整数
[2014, 10, 31]
```

对于 split()和 rsplit()方法,如果不指定分隔符,则字符串中的任何空白符号(包括空格、换行符、制表符等)的连续出现都将被认为是分隔符,返回包含最终分隔结果的列表。

```
>>>s='hello world \n\n My name is Dong    '
>>>s.split()
['hello', 'world', 'My', 'name', 'is', 'Dong']
>>>s='\n\nhello world \n\n\n My name is Dong    '
```

```
>>>s.split()
['hello', 'world', 'My', 'name', 'is', 'Dong']
>>>s='\n\nhello\t\t world \n\n\n My name\t is Dong   '
>>>s.split()
['hello', 'world', 'My', 'name', 'is', 'Dong']
```

另外,split()和 rsplit()方法还允许指定最大分隔次数(注意,不是必须分隔这么多次),例如:

```
>>>s='\n\nhello\t\t world \n\n\n My name is Dong   '
>>>s.split(maxsplit=1)                   #分隔1次
['hello', 'world \n\n\n My name is Dong   ']
>>>s.rsplit(maxsplit=1)
['\n\nhello\t\t world \n\n\n My name is', 'Dong']
>>>s.split(maxsplit=2)
['hello', 'world', 'My name is Dong   ']
>>>s.rsplit(maxsplit=2)
['\n\nhello\t\t world \n\n\n My name', 'is', 'Dong']
>>>s.split(maxsplit=5)
['hello', 'world', 'My', 'name', 'is', 'Dong   ']
>>>s.split(maxsplit=6)
['hello', 'world', 'My', 'name', 'is', 'Dong']
>>>s.split(maxsplit=10)                  #最大分隔次数大于实际可分隔次数时,自动忽略
['hello', 'world', 'My', 'name', 'is', 'Dong']
```

小提示:调用 split()方法并且不传递任何参数时,将使用任何空白字符作为分隔符,如果字符串存在连续的空白字符,split()方法将自动忽略;明确传递参数指定 split()使用的分隔符时,情况略有不同。

```
>>>'a,,,bb,,ccc'.split(',')              #每个逗号都被作为独立的分隔符
['a', '', '', 'bb', '', 'ccc']
>>>'a\t\t\tbb\t\tccc'.split('\t')        #每个制表符都被作为独立的分隔符
['a', '', '', 'bb', '', 'ccc']
>>>'a\t\t\tbb\t\tccc'.split()            #连续多个制表符被作为一个分隔符
['a', 'bb', 'ccc']
```

3. join()

与 split()相反,join()方法用来将列表中多个字符串进行连接,并在相邻两个字符串之间插入指定字符。

```
>>>li=["apple", "peach", "banana", "pear"]
>>>sep=","
>>>s=sep.join(li)                        #使用逗号作为连接符
>>>s
```

```
"apple,peach,banana,pear"
>>>':'.join(li)                                    #使用冒号作为连接符
'apple:peach:banana:pear'
>>>''.join(li)                                     #使用空字符作为连接符
'applepeachbananapear'
```

小技巧：使用split()和join()方法可以删除字符串中多余的空白字符,如果有连续多个空白字符,只保留一个,例如:

```
>>>x='aaa     bb    c d e   fff    '
>>>' '.join(x.split())
'aaa bb c d e fff'
>>>def equavilent(s1, s2):                         #判断两个字符串在Python意义上是否等价
    if s1==s2:
        return True
    elif ' '.join(s1.split())==' '.join(s2.split()):
        return True
    elif ''.join(s1.split())==''.join(s2.split()):
        return True
    else:
        return False
>>>equavilent('pip list', 'pip    list')
True
>>>equavilent('[1, 2, 3]', '[1,2,3]')              #判断两个列表写法是否等价
True
>>>equavilent('[1, 2, 3]', '[1,2  ,3]')
True
>>>equavilent('[1, 2, 3]', '[1, 2   ,3 ]')
True
>>>equavilent('[1, 2, 3]', '[1, 2   ,3 ,4]')
False
```

注意：使用运算符"+"也可以连接字符串,但该运算符涉及大量数据的复制,效率非常低,不适合大量长字符串的连接。下面的代码演示了运算符"+"和字符串对象join()方法之间的速度差异。

```
import timeit

#使用列表推导式生成10000个字符串
strlist=['This is a long string that will not keep in memory.' for n in range(10000)]

#使用字符串对象的join()方法连接多个字符串
def use_join():
    return ''.join(strlist)
```

```python
#使用运算符"+"连接多个字符串
def use_plus():
    result=''
    for strtemp in strlist:
        result=result+strtemp
    return result

if __name__=='__main__':
    #重复运行次数
    times=1000
    jointimer=timeit.Timer('use_join()', 'from __main__ import use_join')
    print('time for join:', jointimer.timeit(number=times))
    plustimer=timeit.Timer('use_plus()', 'from __main__ import use_plus')
    print('time for plus:', plustimer.timeit(number=times))
```

该代码分别使用 join() 函数和"＋"对 10000 个字符串进行连接，并重复运行 1000 次，然后输出每种方法所使用的时间，运行结果为

```
time for join: 0.11133914429192587
time for plus: 1.6754796186748913
```

拓展知识：timeit 模块还支持下面代码演示的用法，从运行结果可以看出，当需要对大量数据进行类型转换时，内置函数 map() 可以提供非常高的效率。

```
>>>import timeit
>>>timeit.timeit('"-".join(str(n)for n in range(100))', number=10000)
                                        #重复运行 10000 次
0.3063435900577929
>>>timeit.timeit('"-".join([str(n) for n in range(100)])', number=10000)
0.27191914957273866
>>>timeit.timeit('"-".join(map(str, range(100)))', number=10000)
0.21119518171659024
```

4. lower()、upper()、capitalize()、title()、swapcase()

这几个方法分别用来将字符串转换为小写、大写字符串、将字符串首字母变为大写、将每个单词的首字母变为大写以及大小写互换，这几个方法都是生成新字符串，并不对原字符串做任何修改。

```
>>>s="What is Your Name?"
>>>s2=s.lower()                         #返回小写字符串
>>>s2
'what is your name?'
>>>s.upper()                            #返回大写字符串
```

```
'WHAT IS YOUR NAME?'
>>>s.capitalize()                       #字符串首字符大写
'What is your name?'
>>>s.title()                            #每个单词的首字母大写
'What Is Your Name?'
>>>s.swapcase()                         #大小写互换
'wHAT IS yOUR nAME?'
```

5. replace()

该方法用来替换字符串中指定字符或子字符串的所有重复出现,每次只能替换一个字符或一个字符串,类似于 Word、WPS、记事本等文本编辑器的查找与替换功能。该方法并不修改原字符串,而是返回一个新字符串。

```
>>>s="中国,中国"
>>>s
'中国,中国'
>>>print(s)
中国,中国
>>>print(s.replace("中国", "中华人民共和国"))
中华人民共和国,中华人民共和国
>>>print('abcdabc'.replace('abc', 'ABC'))
ABCdABC
```

6. maketrans()、translate()

maketrans()方法用来生成字符映射表,而 translate()方法则按映射表中定义的对应关系转换字符串并替换其中的字符,使用这两个方法的组合可以同时处理多个不同的字符,replace()方法则无法满足这一要求。下面的代码演示了这两个方法的用法,当然也可以定义自己的字符映射表,然后用来对字符串进行加密。

```
#创建映射表,将字符"abcdef123"一一对应地转换为"uvwxyz@#$"
>>>table=''.maketrans('abcdef123', 'uvwxyz@#$')
>>>s="Python is a greate programming language. I like it!"
>>>s.translate(table)                   #按映射表进行替换
'Python is u gryuty progrumming lunguugy. I liky it!'
#下面的代码模拟了恺撒加密算法
#每个英文字母替换为字母表中后面第 3 个字母,当然,3 也可以是其他数字
>>>import string
>>>lowerLetters=string.ascii_lowercase
>>>upperLetters=string.ascii_uppercase
>>>before=lowerLetters+upperLetters
>>>before
'abcdefghijklmnopqrstuvwxyzABCDEFGHIJKLMNOPQRSTUVWXYZ'
```

```
#循环移位
>>>after=lowerLetters[3:]+lowerLetters[:3]+upperLetters[3:]+upperLetters[:3]
>>>after
'defghijklmnopqrstuvwxyzabcDEFGHIJKLMNOPQRSTUVWXYZABC'
#创建字符转换表,将英文字母替换为该字母后面第3个字母
>>>table=''.maketrans(before, after)
>>>example='If the implementation is easy to explain, it may be a good idea.'
>>>example.translate(table)
'Li wkh lpsohphqwdwlrq lv hdvb wr hasodlq, lw pdb eh d jrrg lghd.'
```

拓展知识：Python 标准库中的 string 提供了英文字母大小写、数字字符、标点符号等常量,可以直接使用,下面的代码实现了随机密码生成功能。

```
>>>import string
>>>x=string.digits+string.ascii_letters+string.punctuation
                                                          #可能的字符集
>>>x
'0123456789abcdefghijklmnopqrstuvwxyzABCDEFGHIJKLMNOPQRSTUVWXYZ!"#$%&\'()*+,-./:;<=>?@[\\]^_`{|}~'
>>>import random
>>>''.join([random.choice(x)for i in range(8)])    #随机选择8个字符
'H\\{.#=)g'
>>>''.join([random.choice(x)for i in range(8)])
'(CrZ[44M'
>>>''.join([random.choice(x)for i in range(8)])
'o_?[M>iF'
>>>''.join([random.choice(x)for i in range(8)])
'n<[I)5V@'
```

拓展知识：在 Python 中,字符串属于不可变对象,不支持原地修改,如果需要修改其中的值,只能重新创建一个新的字符串对象。然而,如果确实需要一个支持原地修改的 unicode 数据对象,可以使用 io.StringIO 对象或 array 模块。

```
>>>from io import StringIO
>>>s="Hello world"
>>>sio=StringIO(s)                      #创建可变字符串对象
>>>sio
<_io.StringIO object at 0x0000000003096EE8>
>>>sio.tell()                           #返回当前位置
0
>>>sio.read()                           #从当前位置开始读取字符串
'Hello world'
>>>sio.getvalue()                       #返回可变字符串的全部内容
'Hello world'
>>>sio.tell()
```

```
11
>>>sio.seek(6)                              #重新定位当前位置
6
>>>sio.write('SDIBT')                       #从当前位置开始写入字符串
5
>>>sio.read()                               #从当前位置开始读取字符串
''
>>>sio.getvalue()
'Hello SDIBT'
>>>sio.tell()
11
>>>s="Hello world"
>>>from array import array
>>>sa=array('u', s)                         #创建可变字符串对象
>>>print(sa)
array('u', 'Hello world')
>>>print(sa.tostring())                     #查看可变字符串对象的内容
b'H\x00e\x00l\x00l\x00o\x00 \x00w\x00o\x00r\x00l\x00d\x00'
>>>print(sa.tounicode())                    #查看可变字符串对象的内容
Hello world
>>>sa[0]='F'                                #修改指定位置上的字符
>>>print(sa)
array('u', 'Fello world')
>>>sa.insert(5,'w')                         #在指定位置插入字符
>>>print(sa)
array('u', 'Fellow world')
>>>sa.remove('l')                           #删除指定字符的首次出现
>>>print(sa)
array('u', 'Felow world')
>>>sa.remove('w')
>>>print(sa)
array('u', 'Felo world')
```

7. strip()、rstrip()、lstrip()

这几个方法分别用来删除两端、右端或左端连续的空白字符或指定字符。

```
>>>s=" abc   "
>>>s2=s.strip()                             #删除空白字符
>>>s2
"abc"
>>>'\n\nhello world   \n\n'.strip()         #删除空白字符
'hello world'
>>>"aaaassddf".strip("a")                   #删除指定字符
"ssddf"
```

```
>>>"aaaassddf".strip("af")
"ssdd"
>>>"aaaassddfaaa".rstrip("a")          #删除字符串右端指定字符
'aaaassddf'
>>>"aaaassddfaaa".lstrip("a")          #删除字符串左端指定字符
'ssddfaaa'
```

> **注意**：这3个函数的参数指定的字符串并不作为一个整体对待,而是在原字符串的两侧、右侧、左侧删除参数字符串中包含的所有字符,例如：

```
>>>'aabbccddeeeffg'.strip('af')        #字母f不在字符串两侧,所以不删除
'bbccddeeeffg'
>>>'aabbccddeeeffg'.strip('gaf')
'bbccddeee'
>>>'aabbccddeeeffg'.strip('gaef')
'bbccdd'
>>>'aabbccddeeeffg'.strip('gbaef')
'ccdd'
>>>'aabbccddeeeffg'.strip('gbaefcd')
''
```

8. eval( )

内置函数eval()用来把任意字符串转化为Python表达式并进行求值。

```
>>>eval("3+4")                         #计算表达式的值
7
>>>a=3
>>>b=5
>>>eval('a+b')                         #这时候要求变量a和b已存在
8
>>>import math
>>>eval('help(math.sqrt)')
Help on built-in function sqrt in module math:
sqrt(...)
    sqrt(x)
    Return the square root of x.
>>>eval('math.sqrt(3)')
1.7320508075688772
>>>eval('aa')                          #当前作用域中不存在aa,抛出异常
Traceback (most recent call last):
  File "<pyshell#3>", line 1, in <module>
    eval('aa')
  File "<string>", line 1, in <module>
NameError: name 'aa' is not defined
```

在 Python 3.x 中，input()将用户的输入一律按字符串对待，如果需要将其还原为本来的类型，可以使用内置函数 eval()，有时候可能需要配合异常处理结构。

```
>>>x=input()
357
>>>x
'357'
>>>eval(x)
357
>>>x=input()
[3, 5, 7]
>>>x
'[3, 5, 7]'
>>>eval(x)                               #注意，这里不能使用 list(x)进行转换
[3, 5, 7]
>>>x=input()
abc
>>>x
'abc'
>>>try:                                  #当前作用域中不存在变量 abc
    print(eval(x))
except:
    print('wrong input')

wrong input
```

注意：虽然整体来说 Python 是一种非常安全的编程语言，不像 C 语言的 strcpy()、strcat()等库函数那样存在大量安全威胁，但这并不代表我们可以随意编写 Python 程序而不需要考虑任何安全性，就像安全系数再高的汽车在驾驶时也要遵守必要的交通规则一样。虽说智者千虑必有一失，但是如果我们不"虑"就会写出千疮百孔、漏洞百出的程序，那时候再想亡羊补牢就太难了。Python 的内置函数 eval()可以计算任意合法表达式的值，如果用户巧妙地构造输入的字符串，可以执行任意外部程序，例如，下面的代码运行后可以启动记事本程序：

```
>>>a=input('Please input a value:')
Please input a value:__import__('os').startfile(r'C:\Windows\\notepad.exe')
>>>eval(a)
```

是不是非常危险啊？如果觉得这没什么，再执行下面的代码试试：

```
>>>eval("__import__('os').system('md testtest')")
```

发生了什么？什么也没有，只是屏幕闪了一下，不知道各位读者会怎么想，我每次看到这样的情况都会心慌，因为我们不知道到底发生了什么，多年前有很多恶意软件都是屏幕一闪就悄悄地对我们的计算机做了很多事情，例如安装木马、窃取、删除或者破坏数据。

回到上面的代码,让我们看看当前工作目录中多了什么(太可怕了,居然多了个名字是testtest 的文件夹),当然可以调用 rd 命令来删除这个文件夹或其他文件,或者精心构造其他字符串来达到特殊目的。

因此,如果我们的程序中有使用内置函数 eval() 对用户输入的字符串求值的代码,一定要检查用户输入的字符串中是否有危险的字符串,例如"__import__('os')。",否则就很容易引发"血案"。不知道怎么检查?请继续往下看关键字 in 的介绍。

9. 关键字 in

与列表、元组、字典、集合一样,也可以使用关键字 in 和 not in 来判断一个字符串是否出现在另一个字符串中,返回 True 或 False。

```
>>>"a" in "abcde"                    #测试一个字符串是否存在于另一个字符串中
True
>>>'ab' in 'abcde'
True
>>>'ac' in 'abcde'                   #关键字 in 左边的字符串作为一个整体对待
False
>>>"j" in "abcde"
False
```

几乎所有论坛或社区都会对用户提交的输入进行检查,并过滤一些非法的敏感词,这极大地促进了网络文明和净化。这样的功能实际上就可以使用关键字 in 来实现,例如,下面的代码用来检测用户输入中是否有不允许的敏感字词,如果有就提示非法,否则提示正常。

```
>>>words = ('测试','非法','暴力')
>>>text=input('请输入:')
请输入:这句话里含有非法内容
>>>for word in words:
    if word in text:
        print('非法')
        break
else:
    print('正常')
```

下面的代码则可以用来测试用户输入中是否有敏感词,如果有就把敏感词替换为 3 个星号***。

```
>>>words = ('测试','非法','暴力','话')
>>>text='这句话里含有非法内容'
>>>for word in words:
    if word in text:
        text=text.replace(word, '***')
>>>text
```

'这句***里含有***内容'

10. startswith()、endswith()

这两个方法用来判断字符串是否以指定字符串开始或结束，可以接收两个整数参数来限定字符串的检测范围，例如：

```
>>>s='Beautiful is better than ugly.'
>>>s.startswith('Be')                    #检测整个字符串
True
>>>s.startswith('Be', 5)                 #指定检测范围的起始位置
False
>>>s.startswith('Be', 0, 5)              #指定检测范围的起始和结束位置
True
```

另外，这两个方法还可以接收一个字符串元组作为参数来表示前缀或后缀，例如，下面的代码可以列出指定文件夹下所有扩展名为 bmp、jpg 或 gif 的图片。

```
>>>import os
>>>[filename for filename in os.listdir(r'D:\\') if filename.endswith(('.bmp', '.jpg', '.gif'))]
```

11. isalnum()、isalpha()、isdigit()、isdecimal()、isnumeric()、isspace()、isupper()、islower()

用来测试字符串是否为数字或字母、是否为字母、是否为数字字符、是否为空白字符、是否为大写字母以及是否为小写字母。

```
>>>'1234abcd'.isalnum()
True
>>>'1234abcd'.isalpha()                  #全部为英文字母时返回 True
False
>>>'1234abcd'.isdigit()                  #全部为数字时返回 True
False
>>>'abcd'.isalpha()
True
>>>'1234.0'.isdigit()
False
>>>'1234'.isdigit()
True
>>>'九'.isnumeric()                      #isnumeric()方法支持汉字数字
True
>>>'九'.isdigit()
False
>>>'九'.isdecimal()
```

```
False
>>>'ⅣⅢⅩ'.isdecimal()
False
>>>'ⅣⅢⅩ'.isdigit()
False
>>>'ⅣⅢⅩ'.isnumeric()                    #支持罗马数字
True
```

拓展知识：Python 标准库 unicodedata 提供了不同形式数字字符到十进制数字的转换方法。

```
>>>import unicodedata
>>>unicodedata.numeric('2')
2.0
>>>unicodedata.numeric('九')             #汉字数字
9.0
>>>unicodedata.numeric('Ⅹ')             #罗马数字
10.0
```

12. center()、ljust()、rjust()、zfill()

center()、ljust()、rjust()返回指定宽度的新字符串,原字符串居中、左对齐或右对齐出现在新字符串中,如果指定的宽度大于字符串长度,则使用指定的字符进行填充,默认以空格进行填充。zfill()返回指定宽度的字符串,在左侧以字符 0 进行填充。

```
>>>'Hello world!'.center(20)              #居中对齐,以空格进行填充
'    Hello world!    '
>>>'Hello world!'.center(20, '=')         #居中对齐,以字符"="进行填充
'====Hello world!===='
>>>'Hello world!'.ljust(20, '=')          #左对齐
'Hello world!========'
>>>'Hello world!'.rjust(20, '=')          #右对齐
'========Hello world!'
>>>'abc'.zfill(5)                         #在左侧填充数字字符 0
'00abc'
>>>'abc'.zfill(2)                         #指定宽度小于字符串长度时,返回字符串本身
'abc'
>>>'uio'.zfill(20)
'00000000000000000uio'
```

拓展知识：Python 标准库 textwrap 提供了更加友好的排版函数。例如：

```
>>>import textwrap
>>>doc='''Beautiful is better than ugly.
Explicit is better than implicit.
```

```
Simple is better than complex.
Complex is better than complicated.
Flat is better than nested.
Sparse is better than dense.
Readability counts.
Special cases aren't special enough to break the rules.
Although practicality beats purity.'''
>>>print(textwrap.fill(doc, width=20))        #按指定宽度进行排版
'Beautiful is better
than ugly. Explicit
is better than
implicit. Simple is
better than complex.
Complex is better
than complicated.
Flat is better than
nested. Sparse is
better than dense.
Readability counts.
Special cases aren't
special enough to
break the rules.
Although
practicality beats
purity.
>>>print(textwrap.fill(doc, width=80))        #按指定宽度进行排版
'Beautiful is better than ugly. Explicit is better than implicit. Simple is
better than complex. Complex is better than complicated. Flat is better than
nested. Sparse is better than dense. Readability counts. Special cases aren't
special enough to break the rules. Although practicality beats purity.
>>>import pprint
>>>pprint.pprint(textwrap.wrap(doc))          #默认长度最大为 70
["'Beautiful is better than ugly. Explicit is better than implicit.",
 'Simple is better than complex. Complex is better than complicated.',
 'Flat is better than nested. Sparse is better than dense. Readability',
 "counts. Special cases aren't special enough to break the rules.",
 'Although practicality beats purity.']
```

13. 内置函数、切片

除了字符串对象提供的方法以外,很多 Python 内置函数也可以对字符串进行操作,例如:

```
>>>x='Hello world.'
```

```
>>>len(x)                                #字符串长度
12
>>>max(x)                                #最大字符
'w'
>>>min(x)
' '
>>>list(zip(x,x))                        #zip()也可以作用于字符串
[('H', 'H'),('e', 'e'),('l', 'l'),('l', 'l'),('o', 'o'),(' ', ' '),('w', 'w'),('o',
'o'),('r', 'r'),('l', 'l'),('d', 'd'),('.', '.')]
```

切片也适用于字符串,但仅限于读取其中的元素,不支持字符串修改。

```
>>>'Explicit is better than implicit.'[:8]
'Explicit'
>>>'Explicit is better than implicit.'[9:23]
'is better than'
```

另外,运算符"+"和"*"也支持字符串之间或者字符串与整数的运算,第1章已有介绍,不再赘述。

5.1.3 案例精选

例 5-1 编写函数实现字符串加密和解密,循环使用指定密钥,采用简单的异或算法。

```
def crypt(source, key):
    from itertools import cycle
    result=''
    temp=cycle(key)
    for ch in source:
        result=result+chr(ord(ch)^ord(next(temp)))
    return result

source='Shandong Institute of Business and Technology'
key='Dong Fuguo'

print('Before Encrypted:'+source)
encrypted=crypt(source, key)
print('After Encrypted:'+encrypted)
decrypted=crypt(encrypted, key)
print('After Decrypted:'+decrypted)
```

输出结果如图 5-2 所示。

```
Before Encrypted:Shandong Institute of Business and Technology
After Encrypted:╞•☼    D)← U&*┐╜T3 ┐U "0,IS/┐─d╜ └ ]┤ +└    Y
After Decrypted:Shandong Institute of Business and Technology
```

图 5-2 字符串加密与解密结果

例 5-2 编写程序，生成大量随机信息。

本例代码演示了如何使用 Python 标准库 random 来生成随机数据，这在需要获取大量数据来测试或演示软件功能的时候非常有用，不仅能真实展示软件功能或算法，还可以避免泄露真实数据或者引起不必要的争议。

```python
import random
import string
import codecs

#常用汉字 Unicode 编码表(部分)，完整列表详见配套源代码
StringBase='\u7684\u4e00\u4e86\u662f\u6211\u4e0d\u5728\u4eba'
#转换为汉字
StringBase=''.join(StringBase.split('\\u'))

def getEmail():
    #常见域名后缀，可以随意扩展该列表
    suffix=['.com', '.org', '.net', '.cn']
    characters=string.ascii_letters+string.digits+'_'
    username=''.join((random.choice(characters) for i in range(random.randint
                (6,12))))
    domain=''.join((random.choice(characters) for i in range(random.randint(3,
                6))))
    return username+'@'+domain+random.choice(suffix)

def getTelNo():
    return ''.join((str(random.randint(0,9)) for i in range(11)))

def getNameOrAddress(flag):
    '''flag=1 表示返回随机姓名,flag=0 表示返回随机地址'''
    result=''
    if flag==1:
        #大部分中国人姓名为 2~4 个汉字
        rangestart, rangeend=2, 5
    elif flag==0:
        #假设地址在 10~30 个汉字之间
        rangestart, rangeend=10, 31
    else:
        print('flag must be 1 or 0')
        return ''
    for i in range(rangestart, rangeend):
        result+=random.choice(StringBase)
    return result

def getSex():
```

```python
        return random.choice(('男','女'))

def getAge():
    return str(random.randint(18,100))

def main(filename):
    with codecs.open(filename, 'w', 'utf-8') as fp:
        fp.write('Name,Sex,Age,TelNO,Address,Email\n')
        #随机生成200个人的信息
        for i in range(200):
            name=getNameOrAddress(1)
            sex=getSex()
            age=getAge()
            tel=getTelNo()
            address=getNameOrAddress(0)
            email=getEmail()
            line=','.join([name, sex, age, tel, address, email])+'\n'
            fp.write(line)

def output(filename):
    with codecs.open(filename, 'r', 'utf-8') as fp:
        while True:
            line=fp.readline()
            if not line:
                return
            line=line.split(',')
            for i in line:
                print(i, end=',')
            print()

if __name__=='__main__':
    filename='information.txt'
    main(filename)
    output(filename)
```

拓展知识：Python 扩展库 jieba 和 snownlp 很好地支持了中文分词，可以使用 pip 命令进行安装。在自然语言处理领域经常需要对文字进行分词，分词的准确度直接影响了后续文本处理和挖掘算法的最终效果。

```
>>>import jieba                            #导入jieba模块
>>>x='分词的准确度直接影响了后续文本处理和挖掘算法的最终效果。'
>>>jieba.cut(x)                            #使用默认词库进行分词
<generator object Tokenizer.cut at 0x000000000342C990>
>>>list(_)
```

```
['分词', '的', '准确度', '直接', '影响', '了', '后续', '文本处理', '和', '挖掘',
'算法', '的', '最终', '效果', '。']
>>>list(jieba.cut('纸杯'))
['纸杯']
>>>list(jieba.cut('花纸杯'))
['花', '纸杯']
>>>jieba.add_word('花纸杯')                    #增加词条
>>>list(jieba.cut('花纸杯'))                   #使用新题库进行分词
['花纸杯']
>>>import snownlp                              导入snownlp模块
>>>snownlp.SnowNLP('学而时习之,不亦说乎').words
['学而', '时习', '之', ',', '不亦', '说乎']
>>>snownlp.SnowNLP(x).words
['分词', '的', '准确度', '直接', '影响', '了', '后续', '文本', '处理', '和', '挖掘',
'算法', '的', '最终', '效果', '。']
```

拓展知识：Python 扩展库 pypinyin 支持汉字到拼音的转换,并且可以和分词扩展库配合使用。

```
>>>from pypinyin import lazy_pinyin, pinyin
>>>lazy_pinyin('董付国')                       #返回拼音
['dong', 'fu', 'guo']
>>>lazy_pinyin('董付国', 1)                    #带声调的拼音
['dǒng', 'fù', 'guó']
>>>lazy_pinyin('董付国', 2)                    #另一种拼音风格
['do3ng', 'fu4', 'guo2']
>>>lazy_pinyin('董付国', 3)                    #只返回拼音首字母
['d', 'f', 'g']
>>>lazy_pinyin('重要', 1)                      #能够根据词组智能识别多音字
['zhòng', 'yào']
>>>lazy_pinyin('重阳', 1)
['chóng', 'yáng']
>>>pinyin('重阳')                              #返回拼音
[['chóng'], ['yáng']]
>>>pinyin('重阳节', heteronym=True)            #返回多音字的所有读音
[['zhòng', 'chóng', 'tóng'], ['yáng'], ['jié', 'jiē']]
>>>import jieba                                #其实不需要导入jieba,这里只是说明已安装
>>>x='中英文混合test123'
>>>lazy_pinyin(x)                              #自动调用已安装的jieba扩展库分词功能
['zhong', 'ying', 'wen', 'hun', 'he', 'test123']
>>>lazy_pinyin(jieba.cut(x))
['zhong', 'ying', 'wen', 'hun', 'he', 'test123']
>>>x='山东烟台的大樱桃真好吃啊'
>>>sorted(x, key=lambda ch: lazy_pinyin(ch))   #按拼音对汉字进行排序
```

['啊', '吃', '大', '的', '东', '好', '山', '台', '桃', '烟', '樱', '真']

5.2 正则表达式

正则表达式是字符串处理的有力工具和技术,正则表达式使用预定义的特定模式去匹配一类具有共同特征的字符串,主要用于字符串处理,可以快速、准确地完成复杂的查找、替换等处理要求。

5.2.1 正则表达式语法与子模式扩展语法

正则表达式由元字符及其不同组合来构成,通过巧妙地构造正则表达式可以匹配任意字符串,并完成复杂的字符串处理任务。常用的正则表达式元字符如表 5-3 所示。

表 5-3 常用的正则表达式元字符

元字符	功能说明
.	匹配除换行符以外的任意单个字符
*	匹配位于"*"之前的字符或子模式的 0 次或多次出现
+	匹配位于"+"之前的字符或子模式的 1 次或多次出现
-	用在[]之内用来表示范围
|	匹配位于"|"之前或之后的字符
^	匹配行首,匹配以^后面的字符开头的字符串
$	匹配行尾,匹配以$之前的字符结束的字符串
?	匹配位于"?"之前的 0 个或 1 个字符。当此字符紧随任何其他限定符(＊、＋、?、{n}、{n,}、{n,m})之后时,匹配模式是"非贪心的"。"非贪心的"模式匹配搜索到的、尽可能短的字符串,而默认的"贪心的"模式匹配搜索到的、尽可能长的字符串。例如,在字符串"oooo"中,"o＋?"只匹配单个 o,而"o＋"匹配所有 o
\	表示位于\之后的为转义字符
\num	此处的 num 是一个正整数。例如,"(.)\1"匹配两个连续的相同字符
\f	换页符匹配
\n	换行符匹配
\r	匹配一个回车符
\b	匹配单词头或单词尾
\B	与\b 含义相反
\d	匹配任何数字,相当于[0-9]
\D	与\d 含义相反,等效于[^0-9]
\s	匹配任何空白字符,包括空格、制表符、换页符,与 [\f\n\r\t\v] 等效

续表

元字符	功 能 说 明
\S	与\s含义相反
\w	匹配任何字母、数字以及下画线,相当于[a-zA-Z0-9_]
\W	与\w含义相反,与[^A-Za-z0-9_]等效
()	将位于()内的内容作为一个整体来对待
{}	按{}中的次数进行匹配
[]	匹配位于[]中的任意一个字符
[^xyz]	^放在[]内表示反向字符集,匹配除x、y、z之外的任何字符
[a-z]	字符范围,匹配指定范围内的任何字符
[^a-z]	反向范围字符,匹配除小写英文字母之外的任何字符

如果以"\"开头的元字符与转义字符相同,则需要使用"\\",或者使用原始字符串。在字符串前加上字符 r 或 R 之后表示原始字符串,字符串中任意字符都不再进行转义。原始字符串可以减少用户的输入,主要用于正则表达式和文件路径字符串的情况,但如果字符串以一个斜线"\"结束,则需要多写一个斜线,即以"\\"结束。

具体应用时,可以单独使用某种类型的元字符,但处理复杂字符串时,经常需要将多个正则表达式元字符进行组合,下面给出几个简单的示例。

(1) 最简单的正则表达式是普通字符串,只能匹配自身。

(2) '[pjc]ython'可以匹配'python'、'jython'、'cython'。

(3) '[a-zA-Z0-9]'可以匹配一个任意大小写字母或数字。

(4) '[^abc]'可以一个匹配任意除'a'、'b'、'c'之外的字符。

(5) 'python|perl'或'p(ython|erl)'都可以匹配'python'或'perl'。

(6) 子模式后面加上问号表示可选。r'(http://)?(www\.)? python\. org'只能匹配'http://www.python.org'、'http://python.org'、'www.python.org'和'python.org'。

(7) '^http'只能匹配所有以'http'开头的字符串。

(8) (pattern)*:允许模式重复0次或多次。

(9) (pattern)+:允许模式重复1次或多次。

(10) (pattern){m,n}:允许模式重复m~n次。

(11) '(a|b)*c':匹配多个(包含0个)a或b,后面紧跟一个字母c。

(12) 'ab{1,}':等价于'ab+',匹配以字母a开头后面带1个或多个字母b的字符串。

(13) '^[a-zA-Z]{1}([a-zA-Z0-9._]){4,19}$':匹配长度为5~20的字符串,必须以字母开头、可带数字、"_"、"."的字符串。

(14) '^(\w){6,20}$':匹配长度为6~20的字符串,可以包含字母、数字、下画线。

(15) '^\d{1,3}\.\d{1,3}\.\d{1,3}\.\d{1,3}$':检查给定字符串是否为合法IP地址。

(16) '^(13[4-9]\d{8})|(15[01289]\d{8})$':检查给定字符串是否为移动手机

号码。

(17) `'^[a-zA-Z]+$'`：检查给定字符串是否只包含英文字母大小写。

(18) `'^\w+@(\w+\.)+\w+$'`：检查给定字符串是否为合法电子邮件地址。

(19) `'^(\-)?\d+(\.\d{1,2})?$'`：检查给定字符串是否为最多带有 2 位小数的正数或负数。

(20) `'[\u4e00-\u9fa5]'`：匹配给定字符串中的所有汉字。

(21) `'^\d{18}|\d{15}$'`：检查给定字符串是否为合法身份证格式。

(22) `'\d{4}-\d{1,2}-\d{1,2}'`：匹配指定格式的日期，例如 2016-1-31。

(23) `'^(?=.*[a-z])(?=.*[A-Z])(?=.*\d)(?=.*[,._]).{8,}$'`：检查给定字符串是否为强密码，必须同时包含英文大写字母、英文小写字母、数字或特殊符号（如英文逗号、英文句号、下画线），并且长度必须至少 8 位。

(24) `"(?!.*[\"\/;=%?]).+"`：如果给定字符串中包含 \"、/、;、=、%、? 则匹配失败，关于子模式语法请参考表 5-4。

(25) `'(.)\\1+'`：匹配任意字符的一次或多次重复出现。

(26) `'((?P<f>\b\w+\b)\s+(?P=f))'`：匹配连续出现两次的单词。

注意：正则表达式只是进行形式上的检查，并不保证内容一定正确。例如，上面的例子中，正则表达式`'^\d{1,3}\.\d{1,3}\.\d{1,3}\.\d{1,3}$'`可以检查字符串是否为 IP 地址，字符串'888.888.888.888'这样的也能通过检查，但实际上并不是合法的 IP 地址。

正则表达式使用圆括号"()"表示一个子模式，圆括号内的内容作为一个整体出现，例如，"(red)+"可以匹配 redred、redredred 等多个重复 red 的情况。使用子模式扩展语法可以实现更加复杂的字符串处理，常用的扩展语法如表 5-4 所示。

表 5-4　常用子模式扩展语法

语　　法	功 能 说 明
(?P<groupname>)	为子模式命名
(?iLmsux)	设置匹配标志，可以是几个字母的组合，每个字母含义与编译标志相同
(?:…)	匹配但不捕获该匹配的子表达式
(?P=groupname)	表示在此之前的命名为 groupname 的子模式
(?#…)	表示注释
(?=…)	用于正则表达式之后，表示如果"="后的内容在字符串中出现则匹配，但不返回"="之后的内容
(?!…)	用于正则表达式之后，表示如果"!"后的内容在字符串中不出现则匹配，但不返回"!"之后的内容
(?<=…)	用于正则表达式之前，与(?=…)含义相同
(?<!…)	用于正则表达式之前，与(?!…)含义相同

小提示：正则表达式语法实在是太多了，很难全部记住（有读者说：Python 语法

也很多的,也很难记啊)。我个人的建议是,学会学习和学会思考比学会知识更加重要,大家要做个"知道分子",遇到问题之后要知道解决问题的大致思路和方向,具体的细节(比如函数用法和参数含义)可以通过搜索或查看帮助文档来解决,不需要(也不可能)把一切都记到脑子里。善于搜索能让我们更快地解决问题,站在巨人的肩膀上才能事半功倍。当然,必要的基础知识还是要掌握和熟悉的,不然很难判断搜索到的内容是否正确,毕竟网上的资料实在是太杂乱了。

5.2.2 re 模块方法与正则表达式对象

Python 标准库 re 提供了正则表达式操作所需要的功能,既可以直接使用 re 模块中的方法(见表 5-5)来实现字符串处理,也可以把模式编译成正则表达式对象再使用。

表 5-5 re 模块常用方法

方　　法	功 能 说 明
compile(pattern[, flags])	创建模式对象
search(pattern, string[, flags])	在整个字符串中寻找模式,返回 match 对象或 None
match(pattern, string[, flags])	从字符串的开始处匹配模式,返回 match 对象或 None
findall(pattern, string[, flags])	列出字符串中模式的所有匹配项
split(pattern, string[, maxsplit=0])	根据模式匹配项分割字符串
sub(pat, repl, string[, count=0])	将字符串中所有 pat 的匹配项用 repl 替换
escape(string)	将字符串中所有特殊正则表达式字符转义

其中,函数参数 flags 的值可以是 re.I(注意是大写字母 I,不是数字 1,表示忽略大小写)、re.L(支持本地字符集的字符)、re.M(多行匹配模式)、re.S(使元字符"."匹配任意字符,包括换行符)、re.U(匹配 Unicode 字符)、re.X(忽略模式中的空格,并可以使用#注释)的不同组合(使用"|"进行组合)。

1. 直接使用 re 模块中的方法

```
>>>import re                                      #导入 re 模块
>>>text='alpha. beta ...gamma delta'              #测试用的字符串
>>>re.split('[\. ]+', text)                       #使用指定字符作为分隔符进行分隔
['alpha', 'beta', 'gamma', 'delta']
>>>re.split('[\. ]+', text, maxsplit=2)           #最多分隔 2 次
['alpha', 'beta', 'gamma delta']
>>>re.split('[\. ]+', text, maxsplit=1)           #最多分隔 1 次
['alpha', 'beta ...gamma delta']
>>>pat='[a-zA-Z]+'
>>>re.findall(pat, text)                          #查找所有单词
['alpha', 'beta', 'gamma', 'delta']
>>>pat='{name}'
```

```
>>>text='Dear {name}…'
>>>re.sub(pat, 'Mr.Dong', text)          #字符串替换
'Dear Mr.Dong…'
>>>s='a s d'
>>>re.sub('a|s|d', 'good', s)            #字符串替换
'good good good'
>>>re.escape('http://www.python.org')    #字符串转义
'http\\:\\/\\/www\\.python\\.org'
>>>print(re.match('done|quit', 'done'))  #匹配成功,返回match对象
<_sre.SRE_Match object at 0x00B121A8>
>>>print(re.match('done|quit', 'done!')) #匹配成功
<_sre.SRE_Match object at 0x00B121A8>
>>>print(re.match('done|quit', 'doe!'))  #匹配不成功,返回空值None
None
>>>print(re.match('done|quit', 'd!one!')) #匹配不成功
None
>>>print(re.match('done|quit', 'd!one!done')) #匹配不成功
None
>>>print(re.search('done|quit', 'd!one!done')) #匹配成功
<_sre.SRE_Match object at 0x0000000002D03D98>
```

下面的代码使用不同的方法删除字符串中多余的空格,如果遇到连续多个空格则只保留一个,同时删除字符串两侧的所有空白字符。

```
>>>import re
>>>s='aaa     bb    c d e   fff   '
>>>' '.join(s.split())                   #不使用正则表达式,直接使用字符串对象的方法
'aaa bb c d e fff'
>>>re.split('[\s]+', s)
['aaa', 'bb', 'c', 'd', 'e', 'fff', '']
>>>re.split('[\s]+', s.strip())          #同时使用re模块中的方法和字符串对象的方法
['aaa', 'bb', 'c', 'd', 'e', 'fff']
>>>' '.join(re.split('[\s]+', s.strip()))
'aaa bb c d e fff'
>>>' '.join(re.split('\s+', s.strip()))
'aaa bb c d e fff'
>>>re.sub('\s+', ' ', s.strip())         #直接使用re模块的字符串替换方法
'aaa bb c d e fff'
```

下面的代码使用以"\"开头的元字符来实现字符串的特定搜索。

```
>>>import re
>>>example='ShanDong Institute of Business and Technology is a very beautiful school.'
>>>re.findall('\\ba.+?\\b', example)    #以字母a开头的完整单词,"?"表示非贪心模式
['and', 'a ']
```

```
>>>re.findall('\\ba.+\\b', example)        #贪心模式的匹配结果
['and Technology is a very beautiful school']
>>>re.findall('\\ba\w*\\b', example)
['and', 'a']
>>>re.findall('\\Bo.+?\\b', example)       #不以o开头且含有o字母的单词剩余部分
['ong', 'ology', 'ool']
>>>re.findall('\\b\w.+?\\b', example)  #所有单词
['ShanDong', 'Institute', 'of', 'Business', 'and', 'Technology', 'is', 'a ', 'very', 'beautiful', 'school']
>>>re.findall('\w+', example)              #所有单词
['ShanDong', 'Institute', 'of', 'Business', 'and', 'Technology', 'is', 'a', 'very', 'beautiful', 'school']
>>>re.findall(r'\b\w.+?\b', example)   #使用原始字符串
['ShanDong', 'Institute', 'of', 'Business', 'and', 'Technology', 'is', 'a ', 'very', 'beautiful', 'school']
>>>re.split('\s', example)                 #使用任何空白字符分隔字符串
['ShanDong', 'Institute', 'of', 'Business', 'and', 'Technology', 'is', 'a', 'very', 'beautiful', 'school.']
>>>re.findall('\d+\.\d+\.\d+', 'Python 2.7.11')
                                           #查找并返回x.x.x形式的数字
['2.7.11']
>>>re.findall('\d+\.\d+\.\d+', 'Python 2.7.11,Python 3.5.1')
['2.7.11', '3.5.1']
```

2. 使用正则表达式对象

首先使用re模块的compile()方法将正则表达式编译生成正则表达式对象,然后再使用正则表达式对象提供的方法进行字符串处理。使用编译后的正则表达式对象不仅可以提高字符串处理速度,还提供了更加强大的字符串处理功能。

正则表达式对象的match(string[, pos[, endpos]])方法用于在字符串开头或指定位置进行搜索,模式必须出现在字符串开头或指定位置;search(string[, pos[, endpos]])方法用于在整个字符串或指定范围中进行搜索;findall(string[, pos[, endpos]])方法用于在字符串中查找所有符合正则表达式的字符串并以列表形式返回。

```
>>>import re
>>>example='ShanDong Institute of Business and Technology'
>>>pattern=re.compile(r'\bB\w+\b')         #编译正则表达式对象,查找以B开头的单词
>>>pattern.findall(example)                #使用正则表达式对象的findall()方法
['Business']
>>>pattern=re.compile(r'\w+g\b')           #查找以字母g结尾的单词
>>>pattern.findall(example)
['ShanDong']
```

```
>>>pattern=re.compile(r'\b[a-zA-Z]{3}\b')    #查找 3 个字母长的单词
>>>pattern.findall(example)
['and']
>>>pattern.match(example)                     #从字符串开头开始匹配,失败返回空值
>>>pattern.search(example)                    #在整个字符串中搜索,成功
<_sre.SRE_Match object at 0x01228EC8>
>>>pattern=re.compile(r'\b\w*a\w*\b')         #查找所有含有字母 a 的单词
>>>pattern.findall(example)
['ShanDong', 'and']
>>>text="He was carefully disguised but captured quickly by police."
>>>re.findall(r"\w+ly", text)                 #查找所有以字母组合 ly 结尾的单词
['carefully', 'quickly']
```

正则表达式对象的 sub(repl，string[，count＝0])和 subn(repl，string[，count＝0])方法用来实现字符串替换功能。

```
>>>example='''Beautiful is better than ugly.
Explicit is better than implicit.
Simple is better than complex.
Complex is better than complicated.
Flat is better than nested.
Sparse is better than dense.
Readability counts.'''
>>>pattern=re.compile(r'\bb\w*\b', re.I)      #正则表达式对象,匹配以 b 或 B 开头的单词
>>>pattern.sub('*', example)                  #将符合条件的单词替换为 *
* is * than ugly.
Explicit is * than implicit.
Simple is * than complex.
Complex is * than complicated.
Flat is * than nested.
Sparse is * than dense.
Readability counts.
>>>pattern.sub('*', example, 1)               #只替换 1 次
* is better than ugly.
Explicit is better than implicit.
Simple is better than complex.
Complex is better than complicated.
Flat is better than nested.
Sparse is better than dense.
Readability counts.
>>>pattern=re.compile(r'\bb\w*\b')            #匹配以字母 b 开头的单词
>>>pattern.sub('*', example, 1)               #将符合条件的单词替换为 *,只替换 1 次
Beautiful is * than ugly.
Explicit is better than implicit.
Simple is better than complex.
```

```
Complex is better than complicated.
Flat is better than nested.
Sparse is better than dense.
Readability counts.
```

正则表达式对象的 split(string[，maxsplit＝0])方法用来实现字符串分隔。

```
>>>example=r'one,two,three.four/five\six?seven[eight]nine|ten'
>>>pattern=re.compile(r'[,./\\?[\]\|]')      #指定多个可能的分隔符
>>>pattern.split(example)
['one', 'two', 'three', 'four', 'five', 'six', 'seven', 'eight', 'nine', 'ten']
>>>example=r'one1two2three3four4five5six6seven7eight8nine9ten'
>>>pattern=re.compile(r'\d+')                #使用数字作为分隔符
>>>pattern.split(example)
['one', 'two', 'three', 'four', 'five', 'six', 'seven', 'eight', 'nine', 'ten']
>>>example=r'one two    three  four,five.six.seven,eight,nine9ten'
>>>pattern=re.compile(r'[\s,.\d]+')          #允许分隔符重复
>>>pattern.split(example)
['one', 'two', 'three', 'four', 'five', 'six', 'seven', 'eight', 'nine', 'ten']
```

3. match 对象

正则表达式模块或正则表达式对象的 match()方法和 search()方法匹配成功后都会返回 match 对象。match 对象的主要方法有 group()（返回匹配的一个或多个子模式内容）、groups()（返回一个包含匹配的所有子模式内容的元组）、groupdict()（返回包含匹配的所有命名子模式内容的字典）、start()（返回指定子模式内容的起始位置）、end()（返回指定子模式内容的结束位置的前一个位置）、span()（返回一个包含指定子模式内容起始位置和结束位置前一个位置的元组）等。下面的代码使用几种不同的方法来删除字符串中指定的内容：

```
>>>email="tony@tiremove_thisger.net"
>>>m=re.search("remove_this", email)         #使用 search()方法返回的 match 对象
>>>email[:m.start()]+email[m.end():]         #字符串切片
'tony@tiger.net'
>>>re.sub('remove_this', '', email)          #直接使用 re 模块的 sub()方法
'tony@tiger.net'
>>>email.replace('remove_this', '')          #也可以直接使用字符串替换方法
'tony@tiger.net'
```

下面的代码演示了 match 对象的 group()、groups()与 groupdict()以及其他方法的用法：

```
>>>m=re.match(r"(\w+) (\w+)", "Isaac Newton, physicist")
>>>m.group(0)                                #返回整个模式内容
'Isaac Newton'
```

```
>>>m.group(1)                                    #返回第1个子模式内容
'Isaac'
>>>m.group(2)                                    #返回第2个子模式内容.
'Newton'
>>>m.group(1, 2)                                 #返回指定的多个子模式内容
('Isaac', 'Newton')
```

下面的代码演示了子模式扩展语法的用法：

```
>>>m=re.match(r"(?P<first_name>\w+) (?P<last_name>\w+)", "Malcolm Reynolds")
>>>m.group('first_name')                         #使用命名的子模式
'Malcolm'
>>>m.group('last_name')
'Reynolds'
>>>m=re.match(r"(\d+)\.(\d+)", "24.1632")
>>>m.groups()                                    #返回所有匹配的子模式(不包括第0个)
('24', '1632')
>>>m=re.match(r"(?P<first_name>\w+) (?P<last_name>\w+)", "Malcolm Reynolds")
>>>m.groupdict()                                 #以字典形式返回匹配的结果
{'first_name': 'Malcolm', 'last_name': 'Reynolds'}
>>>exampleString='''There should be one—and preferably only one—obvious way
to do it.
Although that way may not be obvious at first unless you're Dutch.
Now is better than never.
Although never is often better than right now.'''
>>>pattern=re.compile(r'(?<=\w\s)never(?=\s\w)')
                                                 #查找不在句子开头和结尾的never
>>>matchResult=pattern.search(exampleString)
>>>matchResult.span()
(172, 177)
>>>pattern=re.compile(r'(?<=\w\s)never')         #查找位于句子末尾的单词
>>>matchResult=pattern.search(exampleString)
>>>matchResult.span()
(156, 161)
>>>pattern=re.compile(r'(?:is\s)better(\sthan)')
                                                 #查找前面是is的better than组合
>>>matchResult=pattern.search(exampleString)
>>>matchResult.span()
(141, 155)
>>>matchResult.group(0)                          #组0表示整个模式
'is better than'
>>>matchResult.group(1)
' than'
>>>pattern=re.compile(r'\b(?i)n\w+\b')           #查找以n或N字母开头的所有单词
>>>index=0
```

```
>>>while True:
    matchResult=pattern.search(exampleString, index)
    if not matchResult:
        break
    print(matchResult.group(0), ':', matchResult.span(0))
    index=matchResult.end(0)
not :(92, 95)
Now :(137, 140)
never :(156, 161)
never :(172, 177)
now :(205, 208)
>>>pattern=re.compile(r'(?<!not\s)be\b')     #查找前面没有单词 not 的单词 be
>>>index=0
>>>while True:
    matchResult=pattern.search(exampleString, index)
    if not matchResult:
        break
    print(matchResult.group(0), ':', matchResult.span(0))
    index=matchResult.end(0)
be :(13, 15)
>>>exampleString[13:20]                      #验证一下结果是否正确
'be one-'
>>>pattern=re.compile(r'(\b\w*(?P<f>\w+)(?P=f)\w*\b)')
                                             #匹配有连续相同字母的单词
>>>index=0
>>>while True:
    matchResult=pattern.search(exampleString, index)
    if not matchResult:
        break
    print(matchResult.group(0), ':', matchResult.group(2))
    index=matchResult.end(0)+1
unless : s
better : t
better : t
>>>s
'aabc abcd abbcd abccd abcdd'
>>>p=re.compile(r'(\b\w*(?P<f>\w+)(?P=f)\w*\b)')
>>>p.findall(s)
[('aabc', 'a'),('abbcd', 'b'),('abccd', 'c'),('abcdd', 'd')]
```

5.2.3 案例精选

例 5-3 使用正则表达式提取字符串中的电话号码。

```
import re

telNumber='''Suppose my Phone No. is 0535-1234567, yours is 010-12345678, his is
025-87654321.'''
pattern=re.compile(r'(\d{3,4})-(\d{7,8})')
index=0
while True:
    matchResult=pattern.search(telNumber, index)    #从指定位置开始匹配
    if not matchResult:
        break
    print('-'*30)
    print('Success:')
    for i in range(3):
        print('Searched content:', matchResult.group(i),\
        ' Start from:', matchResult.start(i), 'End at:', matchResult.end(i),\
        ' Its span is:', matchResult.span(i))
    index=matchResult.end(2)                        #指定下次匹配的开始位置
```

上面程序的运行结果如下：

```
------------------------------
Success:
Searched content: 0535-1234567   Start from: 24 End at: 36   Its span is:(24, 36)
Searched content: 0535   Start from: 24 End at: 28   Its span is:(24, 28)
Searched content: 1234567   Start from: 29 End at: 36   Its span is:(29, 36)
------------------------------
Success:
Searched content: 010-12345678   Start from: 47 End at: 59   Its span is:(47, 59)
Searched content: 010   Start from: 47 End at: 50   Its span is:(47, 50)
Searched content: 12345678   Start from: 51 End at: 59   Its span is:(51, 59)
------------------------------
Success:
Searched content: 025-87654321   Start from: 68 End at: 80   Its span is:(68, 80)
Searched content: 025   Start from: 68 End at: 71   Its span is:(68, 71)
Searched content: 87654321   Start from: 72 End at: 80   Its span is:(72, 80)
```

例 5-4　使用正则表达式提取 Python 程序中的类名、函数名以及变量名等标识符。

将下面的代码保存为 FindIdentifiersFromPyFile.py，在命令提示符环境中使用命令"Python FindIdentifiersFromPyFile.py 目标文件名"查找并输出目标文件中的标识符。

```
import re
import os
import sys

classes={}
```

```python
functions=[]
variables={'normal':{}, 'parameter':{}, 'infor':{}}

'''This is a test string:
atest, btest=3, 5
to verify that variables in comments will be ignored by this algorithm
'''

def _identifyClassNames(index, line):
    '''parameter index is the line number of line,
     parameter line is a line of code of the file to check'''
    pattern=re.compile(r'(?<=class\s)\w+(?=.*?:)')
    matchResult=pattern.search(line)
    if not matchResult:
        return
    className=matchResult.group(0)
    classes[className]=classes.get(className, [])
    classes[className].append(index)

def _identifyFunctionNames(index, line):
    pattern=re.compile(r'(?<=def\s)(\w+)\((.*?)\)(?=:)')
    matchResult=pattern.search(line)
    if not matchResult:
        return
    functionName=matchResult.group(1)
    functions.append((functionName, index))
    parameters=matchResult.group(2).split(r', ')
    if parameters[0]=='':
        return
    for v in parameters:
        variables['parameter'][v]=variables['parameter'].get(v, [])
        variables['parameter'][v].append(index)

def _identifyVariableNames(index, line):
    #find normal variables, including the case: a, b=3, 5
    pattern=re.compile(r'\b(.*?)(?=\s=)')
    matchResult=pattern.search(line)
    if matchResult:
        vs=matchResult.group(1).split(r', ')
        for v in vs:
            #consider the case 'if variable ==value'
            if 'if ' in v:
                v=v.split()[1]
            #consider the case: 'a[3]=3'
```

```python
            if '[' in v:
                v=v[0:v.index('[')]
            variables['normal'][v]=variables['normal'].get(v, [])
            variables['normal'][v].append(index)
    #find the variables in for statements
    pattern=re.compile(r'(?<=for\s)(.*?)(?=\sin)')
    matchResult=pattern.search(line)
    if matchResult:
        vs=matchResult.group(1).split(r', ')
        for v in vs:
            variables['infor'][v]=variables['infor'].get(v, [])
            variables['infor'][v].append(index)

def output():
    print('='*30)
    print('The class names and their line numbers are:')
    for key, value in classes.items():
        print(key, ':', value)
    print('='*30)
    print('The function names and their line numbers are:')
    for i in functions:
        print(i[0], ':', i[1])
    print('='*30)
    print('The normal variable names and their line numbers are:')
    for key, value in variables['normal'].items():
        print(key, ':', value)
    print('-'*20)
    print('The parameter names and their line numbers in functions are:')
    for key, value in variables['parameter'].items():
        print(key, ':', value)
    print('-'*20)
    print('The variable names and their line numbers in for statements are:')
    for key, value in variables['infor'].items():
        print(key, ':', value)

#suppose the lines of comments less than 50
def comments(index):
    for i in range(50):
        line=allLines[index+i].strip()
        if line.endswith('"""')or line.endswith("'''"):
            return i+1


if __name__=='__main__':
    fileName=sys.argv[1]                                    #命令行参数
```

```python
    if not os.path.isfile(fileName):
        print('Your input is not a file.')
        sys.exit(0)                                    #退出当前程序
    if not fileName.endswith('.py'):
        print('Sorry. I can only check Python source file.')
        sys.exit(0)
    allLines=[]
    with open(fileName, 'r')as fp:
        allLines=fp.readlines()
    index=0
    totalLen=len(allLines)
    while index<totalLen:
        line=allLines[index]
        #strip the blank characters at both end of line
        line=line.strip()
        #ignore the comments starting with '#'
        if line.startswith('#'):
            index +=1
            continue
        #ignore the comments between ''' or """
        if line.startswith('"""')or line.startswith("'''"):
            index +=comments(index)
            continue
        #identify identifiers
        _identifyClassNames(index+1, line)
        _identifyFunctionNames(index+1, line)
        _identifyVariableNames(index+1, line)
        index +=1
    output()
```

😊**温馨提示**：例 5-4 和例 5-5 的程序都需要在命令提示符环境中运行，并提供另一个文件名作为命令行参数。

例 5-5 使用正则表达式检查 Python 程序的代码风格是否符合规范。

本例代码主要检查 Python 程序的一些基本规范，例如，运算符两侧是否有空格，是否每次只导入一个模块，在不同的功能模块之间是否有空行，注释是否足够多，等等。

```python
import sys
import re

def checkFormats(lines, desFileName):
    fp=open(desFileName, 'w')
    for i, line in enumerate(lines):
        print('='* 30)
        print('Line:', i+1)
```

```python
            if line.strip().startswith('#'):
                print(' '*10+'Comments.Pass.')
                fp.write(line)
                continue
            flag=True
            #check operator symbols
            symbols=[',', '+', '-', '*', '/', '//', '**', '>>', '<<', '+=', '-=',
                    '*=', '/=']
            temp_line=line
            for symbol in symbols:
                pattern=re.compile(r'\s*'+re.escape(symbol)+r'\s*')
                temp_line=pattern.split(temp_line)
                sep=' '+symbol+' '
                temp_line=sep.join(temp_line)
            if line!=temp_line:
                flag=False
                print(' '*10+'You may miss some blank spaces in this line.')
            #check import statement
            if line.strip().startswith('import'):
                if ',' in line:
                    flag=False
                    print(' '*10+"You'd better import one module at a time.")
                    temp_line=line.strip()
                    modules=temp_line[temp_line.index(' ')+1:]
                    modules=modules.strip()
                    pattern=re.compile(r'\s*,\s*')
                    modules=pattern.split(modules)
                    temp_line=''
                    for module in modules:
                        temp_line +=line[:line.index('import')]+'import '+module+'\n'
                    line=temp_line
                pri_line=lines[i-1].strip()
                if pri_line and(not pri_line.startswith('import'))and \
                  (not pri_line.startswith('#')):
                    flag=False
                    print(' '*10+'You should add a blank line before this line.')
                    line='\n'+line
                after_line=lines[i+1].strip()
                if after_line and(not after_line.startswith('import')):
                    flag=False
                    print(' '*10+'You should add a blank line after this line.')
                    line =line+'\n'
            #check if there is a blank line before new funtional code block
            #including the class/function definition
```

```python
            if line.strip() and not line.startswith(' ') and i>0:
                pri_line=lines[i-1]
                if pri_line.strip() and pri_line.startswith(' '):
                    flag=False
                    print(' '*10+"You'd better add a blank line before this line.")
                    line='\n'+line
            if flag:
                print(' '*10+'Pass.')
            fp.write(line)
    fp.close()

if __name__=='__main__':
    fileName=sys.argv[1]                           #命令行参数
    fileLines=[]
    with open(fileName, 'r') as fp:
        fileLines=fp.readlines()
    desFileName=fileName[:-3]+'_new.py'
    checkFormats(fileLines, desFileName)
    #check the ratio of comment lines to all lines
    comments=[line for line in fileLines if line.strip().startswith('#')]
    ratio=len(comments)/len(fileLines)
    if ratio<=0.3:
        print('='*30)
        print('Comments in the file is less than 30%.')
        print('Perhaps you should add some comments at appropriate position.')
```

第 6 章 文件与文件夹操作

文件是长久保存信息并允许重复使用和反复修改的重要方式,同时也是信息交换的重要途径。数据库文件、图像文件、音频和视频文件、可执行文件、Office 文档、动态链接库文件等,都以文件的形式存储在不同形式的存储设备(如磁盘、U 盘、光盘、云盘等)上。按文件中数据的组织形式可以把文件分为文本文件和二进制文件两大类。

1. 文本文件

文本文件存储的是常规字符串,由若干文本行组成,通常每行以换行符'\n'结尾。常规字符串是指记事本之类的文本编辑器能正常显示、编辑并且人类能够直接阅读和理解的字符串,如英文字母、汉字、数字字符串。在 Windows 平台中,扩展名为 txt、log、ini 的文件都属于文本文件,可以使用字处理软件(如 gedit、记事本)进行编辑。

2. 二进制文件

常见的如图形图像文件、音频和视频文件、可执行文件、资源文件、各种数据库文件、各类 Office 文档等都属于二进制文件。二进制文件把信息以字节串(bytes)进行存储,无法用记事本或其他普通字处理软件直接进行编辑,通常也无法被人类直接阅读和理解,需要使用对应的软件进行解码后读取、显示、修改或执行。例如,图 6-1 中使用 Windows 记

图 6-1 二进制文件无法使用文本编辑器直接查看

事本打开 Python 主程序文件 pythonw.exe，由于这个文件是二进制可执行文件，无法使用记事本查看，所以显示乱码。当然，也可以使用 hexeditor、010Editor 之类的十六进制编辑器打开二进制文件进行查看和修改，但是这需要对不同类型的二进制文件结构有非常深入的理解才行，如图 6-2 所示。

图 6-2　使用 Winhex 十六进制编辑器打开可执行文件

6.1　文件对象常用方法与属性

无论是文本文件还是二进制文件，其操作流程基本都是一致的，即首先打开文件并创建文件对象，然后通过该文件对象对文件内容进行读取、写入、删除、修改等操作；最后关闭并保存文件内容。Python 内置了文件对象，通过 open() 函数即可以指定模式打开指定文件并创建文件对象，该函数用法为

```
open(file, mode='r', buffering=-1, encoding=None, errors=None, newline=None, closefd=True, opener=None)
```

该函数的主要参数含义如下。

（1）参数 file 指定要打开或创建的文件名称，如果该文件不在当前目录中，则需要指定完整路径，为了减少完整路径中"\"符号的输入，可以使用原始字符串。

（2）参数 mode（取值范围见表 6-1）指定打开文件后的处理方式，如"只读"、"只写"、"读写"、"追加"、"二进制只读"、"二进制读写"等，默认为"文本只读模式"。以不同方式打开文件时，文件指针的初始位置略有不同，例如，以"只读"和"只写"模式打开文件时文件指针的初始位置是文件头，而以"追加"模式打开文件时则文件指针的初始位置为文件尾。

（3）参数 buffering 指定读写文件的缓存模式，数值 0（只在二进制模式中可以用）表

示不缓存,数值 1(只在文本模式中可以用)表示使用行缓存模式,大于 1 的数字则表示缓冲区的大小,默认值是−1。当使用默认值−1 时,二进制文件和非交互式文本文件以固定大小的块为缓存单位,等价于 io.DEFAULT_BUFFER_SIZE,交互式文本文件(isatty()方法返回 True)采用行缓存模式。

(4) 参数 encoding 指定对文本进行编码和解码的方式,只适用于文本模式,可以使用 Python 支持的任何格式,详见标准库 codecs。

(5) 参数 newline 只适用于文本模式,取值可以是 None、''、'\n'、'\r'、'\r\n'中的任何一个,表示文件中新行的形式。

如果执行正常,open()函数返回一个可迭代的文件对象,通过该文件对象可以对文件进行读写操作,如果指定文件不存在、访问权限不够、磁盘空间不够或其他原因导致创建文件对象失败则抛出异常。下面的代码分别以读、写方式打开了两个文件并创建了与之对应的文件对象。

```
f1=open('file1.txt', 'r')
f2=open('file2.txt', 'w')
```

当对文件内容操作完以后,一定要关闭文件对象,这样才能保证所做的任何修改都确实被保存到文件中。

```
f1.close()
```

小提示:缓存机制使得修改文件时不需要频繁地进行磁盘文件的读写操作,而是等缓存满了以后再写入文件,或者调用 flush()方法强行将缓存中的内容写入磁盘文件,缓冲机制大幅度提高了文件操作速度,也延长了磁盘使用寿命。

注意:即使我们写了关闭文件的代码,也无法保证文件一定能够正常关闭,例如,在打开文件之后和关闭文件之前发生了错误导致程序崩溃。如果忘记关闭文件或者关闭文件的代码没有得到执行会怎么样呢?如果是以读模式打开的文件那一般没什么,但是如果是以写模式或追加模式打开的文件对象,有可能会导致数据并没有真正写入磁盘文件。如果被水平高超的黑客盯上,还有可能会造成内存数据的泄露。

表 6-1 文件打开模式

模 式	说 明
r	读模式(默认模式,可省略),如果文件不存在则抛出异常
w	写模式,如果文件已存在,先清空原有内容
x	写模式,创建新文件,如果文件已存在则抛出异常
a	追加模式,不覆盖文件中原有内容
b	二进制模式(可与其他模式组合使用)
t	文本模式(默认模式,可省略)
+	读、写模式(可与其他模式组合使用)

文件对象的常用属性如表 6-2 所示。

表 6-2 文件对象的常用属性

属性	说明
closed	判断文件是否关闭，若文件已关闭则返回 True
mode	返回文件的打开模式
name	返回文件的名称

文件对象的常用方法如表 6-3 所示。特别说明的是，文件读写操作相关的函数都会自动改变文件指针的位置。例如，以读模式打开一个文本文件，读取 10 个字符，会自动把文件指针移到第 11 个字符，再次读取字符的时候总是从文件指针的当前位置开始读取。写入文件的操作函数也具有相同的特点。

表 6-3 文件对象的常用方法

方法	功能说明
flush()	把缓冲区的内容写入文件，但不关闭文件
close()	把缓冲区的内容写入文件，同时关闭文件，并释放文件对象
read([size])	从文件中读取 size 个字节(Python 2.x)或字符(Python 3.x)的内容作为结果返回，如果省略 size 则表示读取所有内容
readline()	从文本文件中读取一行内容作为结果返回
readlines()	把文本文件中的每行文本作为一个字符串存入列表中，返回该列表
seek(offset[, whence])	把文件指针移到新的位置，offset 表示相对于 whence 的位置。whence 为 0 表示从文件头开始计算，1 表示从当前位置开始计算，2 表示从文件尾开始计算，默认为 0
tell()	返回文件指针的当前位置
truncate([size])	删除从当前指针位置到文件末尾的内容。如果指定了 size，则不论指针在什么位置都只留下前 size 个字节，其余的删除
write(s)	把字符串 s 的内容写入文件
writelines(s)	把字符串列表写入文本文件，不添加换行符
writable()	测试当前文件是否可写
readable()	测试当前文件是否可读

6.2 文本文件操作案例精选

例 6-1 向文本文件中写入内容。

```
s='Hello world\n 文本文件的读取方法 \n 文本文件的写入方法 \n'
f=open('sample.txt', 'a+')                    #打开文件
f.write(s)                                    #写入文件内容
```

```
f.close()                                                    #关闭文件
```

拓展知识:文件操作一般都要遵循"打开文件→读写文件→关闭文件"的标准套路,但是如果文件读写操作代码引发了异常,很难保证文件能够被正确关闭,使用上下文管理关键字 with 可以避免这个问题。关键字 with 可以自动管理资源,不论因为什么原因(哪怕是代码引发了异常)跳出 with 块,总能保证文件被正确关闭,并且可以在代码块执行完毕后自动还原进入该代码块时的现场,常用于文件操作、数据库连接、网络通信连接等场合。有了 with,再也不用担心文件没有关闭了。上面的代码改写如下:

```
s='Hello world\n文本文件的读取方法\n文本文件的写入方法\n'
with open('sample.txt', 'a+') as f:
    f.write(s)
```

另外,上下文管理语句 with 还支持下面的用法:

```
with open('test.txt', 'r') as src, open('test_new.txt', 'w') as dst:
    dst.write(src.read())
```

注意:下面的代码执行结束后不会自动关闭文件对象。

```
>>>for line in open('test.txt'):
    print(line)
```

拓展知识:在交互模式下使用文件对象的 write() 方法写入文件时,会显示成功写入的字符数量。如果想不显示这个数字,可以先导入 sys 模块,然后执行语句 sys.stdout=open('null', 'w'),这样再写入文件时就不会显示写入的字符数量了。

例 6-2 读取文本文件内容。

```
>>>fp=open('sample.txt')
>>>print(fp.read(4))                                        #从当前位置读取前 4 个字符
Hell
>>>print(fp.read(18))                                       #英文字符和汉字一样对待
o world
文本文件的读取方法
>>>print(fp.read())                                         #从当前位置读取后面的所有内容
文本文件的写入方法
>>>fp.close()                                               #关闭文件对象
```

小提示:Python 2.x 对中文支持不很好,文件对象的 read() 方法是读取文件中指定数量的字节。

拓展知识:JSON(JavaScript Object Notation)是一种轻量级的数据交换格式,易于阅读和编写,同时也易于机器解析和生成(一般用于提升网络传输速率),是一种比较理想的编码与解码格式。Python 标准库 json 提供对 JSON 的支持,例如:

```
>>>import json
```

```
>>>x=[1, 2, 3]
>>>json.dumps(x)                            #对列表进行编码
'[1, 2, 3]'
>>>json.loads(_)                            #解码
[1, 2, 3]
>>>type(_)
<class 'list'>
>>>x={'a':1, 'b':2, 'c':3}                  #对字典进行编码
>>>y=json.dumps(x)
>>>type(y)
<class 'str'>
>>>json.loads(y)
{'a': 1, 'b': 2, 'c': 3}
>>>type(_)
<class 'dict'>
>>>fp=open('test.txt', 'w')
>>>json.dump({'a':1, 'b':2, 'c':3}, fp)     #对字典进行编码并写入文件
>>>fp.close()
```

例 6-3 读取并显示文本文件的所有行。

```
with open('sample.txt')as fp:
    while True:
        line=fp.readline()
        if not line:
            break
        print(line)
```

聪明的读者是否想到了文件对象是可以迭代的呢？如果想到了，也就不难理解下面的代码了，是的，Python 就是可以这么简洁。

```
with open('sample.txt')as fp:
    for line in fp:                         #文件对象是可以迭代的
        print(line)
```

或者，也可以直接使用文件对象的 readlines() 方法来实现，但是操作大文件时不建议这样做，因为这会消耗大量的内存资源。

```
with open('sample.txt')as fp:
    lines=fp.readlines()                    #操作大文件时不建议这样使用
    print(''.join(lines))
```

例 6-4 移动文件指针。假设文件 sample.txt 中的内容原为"Hello world\n 文本文件的读取方法\n 文本文件的写入方法"。

```
>>>fp=open('sample.txt', 'r+')
>>>fp.tell()                                #返回文件指针的当前位置
```

```
0
>>>fp.read(20)                                    #读取20个字符
'Hello world\n 文本文件的读取方'
>>>fp.seek(13)                                    #重新定位文件指针的位置
13
>>>fp.read(5)
'文本文件的'
>>>fp.seek(13)
13
>>>fp.write('测试')                               #从文件指针当前位置写入内容
2
>>>fp.flush()                                     #把缓冲区内容写入磁盘文件
>>>fp.seek(0)
0
>>>fp.read()
'Hello world\n 测试文件的读取方法 \n 文本文件的写入方法 \n'
>>>fp.close()                                     #关闭文件
```

例 6-5 假设文件 data.txt 中有若干整数,整数之间使用英文逗号分隔,编写程序读取所有整数,将其按升序排序后再写入文本文件 data_asc.txt 中。

```
with open('data.txt', 'r')as fp:
    data=fp.readlines()                           #读取所有行
data=[line.strip()for line in data]               #删除每行两侧的空白字符
data=','.join(data)                               #合并所有行
data=data.split(',')                              #分割得到所有数字
data=[int(item)for item in data]                  #转换为数字
data.sort()                                       #升序排序
data=','.join(map(str,data))                      #将结果转换为字符串
with open('data_asc.txt', 'w')as fp:              #将结果写入文件
    fp.write(data)
```

拓展知识:CSV(Comma Separated Values)格式的文件常用于电子表格和数据库中内容的导入和导出。Python 标准库 csv 提供的 reader、writer 对象和 DictReader 和 DictWriter 类很好地支持了 CSV 格式文件的读写操作。另外,csvkit 支持命令行方式来实现更多关于 CSV 文件的操作以及与其他文件格式的转换,感兴趣的朋友可以参考 https://source.opennews.org/en-US/articles/eleven-awesome-things-you-can-do-csvkit/。

```
>>>import csv
>>>with open('test.csv', 'w', newline='')as fp:
    test_writer=csv.writer(fp, delimiter=' ', quotechar='"')
                                                  #创建 writer 对象
    test_writer.writerow(['red', 'blue', 'green'])    #写入一行内容
    test_writer.writerow(['test_string'] * 5)
>>>import csv
```

```
>>>with open('test.csv', newline='')as fp:
    test_reader=csv.reader(fp, delimiter=' ', quotechar='"')
                                            #创建 reader 对象
    for row in test_reader:                 #遍历所有行
        print(row)                          #每行作为一个列表返回
['red', 'blue', 'green']
['test_string', 'test_string', 'test_string', 'test_string', 'test_string']
>>>with open('test.csv', newline='')as fp:
    test_reader=csv.reader(fp, delimiter=':', quotechar='"')
                                            #使用不同的分隔符
    for row in test_reader:
        print(row)                          #注意,与上面的输出不同
['red blue green']
['test_string test_string test_string test_string test_string']
>>>with open('test.csv', newline='')as fp:
    test_reader=csv.reader(fp, delimiter=' ', quotechar='"')
    for row in test_reader:
        print(','.join(row))                #重新组织数据形式
red,blue,green
test_string,test_string,test_string,test_string,test_string
>>>import csv
>>>with open('names.csv', 'w')as fp:
    headers=['姓氏', '名字']
    test_dictWriter=csv.DictWriter(fp, fieldnames=headers)
                                            #创建 DictWriter 对象
    test_dictWriter.writeheader()           #写入表头信息
    test_dictWriter.writerow({'姓氏':'张', '名字':'三'})
                                            #写入数据
    test_dictWriter.writerow({'姓氏':'李', '名字':'四'})
    test_dictWriter.writerow({'姓氏':'王', '名字':'五'})
>>>import csv
>>>with open('names.csv')as fp:
    test_dictReader=csv.DictReader(fp)      #创建 DictReader 对象
    print(','.join(test_dictReader.fieldnames)) #读取表头信息
    for row in test_dictReader:             #遍历文件所有行
        print(row['姓氏'],',',row['名字'])
姓氏,名字
张 , 三
李 , 四
王 , 五
```

例 6-6 编写程序,保存为 demo.py,运行后生成文件 demo_new.py,其中的内容与 demo.py 一致,但是在每行的行尾加上了行号。

```
filename='demo.py'
```

```
with open(filename, 'r') as fp:
    lines=fp.readlines()                              #读取所有行
maxLength=max(map(len,lines))                         #最长行的长度
for index, line in enumerate(lines):                  #遍历所有行
    newLine=line.rstrip()                             #删除每行右侧的空白字符
    newLine=newLine+' '*(maxLength+5-len(newLine))
                                                      #在每行固定位置添加行号
    newLine=newLine+'#'+str(index+1)+'\n'             #添加行号
    lines[index]=newLine
with open(filename[:-3]+'_new.py', 'w') as fp:        #将结果写入文件
    fp.writelines(lines)
```

例 6-7 计算文本文件中最长行的长度和该行的内容。

```
with open('sample.txt') as fp:
    result=[0, '']
    for line in fp:
        t=len(line)
        if t>result[0]:
            result=[t,line]

print(result)
```

例 6-8 Python 程序代码复用度检查。

```
from os.path import isfile as isfile
from time import time as time

Result={}
AllLines=[]
FileName=r'XueshengKaoQin.pyw'
#FileName=input('Please input the file to check, including full path:')

#Read the content of given file
#Remove all the whitespace string of every line,
#preserving only one space character between words or operators
#note:The last line does not contain the '\n' character
def PreOperate():
    global AllLines
    with open(FileName, 'r', encoding='utf-8') as fp:
        for line in fp:
            line=' '.join(line.split())
            AllLines.append(line)

#Check if the current position is still the duplicated one
def IfHasDuplicated(Index1):
```

```python
        for item in Result.values():
            for it in item:
                if Index1==it[0]:
                    return it[1]                    #return the span
    return False

#If the current line Index2 is in a span of duplicated lines, return True,
#else False
def IsInSpan(Index2):
    for item in Result.values():
        for i in item:
            if i[0]<=Index2<i[0]+i[1]:
                return True
    return False


def MainCheck():
    global Result
    TotalLen=len(AllLines)
    Index1=0
    while Index1<TotalLen-1:
    #speed up
        span=IfHasDuplicated(Index1)
        if span:
            Index1 +=span
            continue
        Index2=Index1+1
        while Index2<TotalLen:
            #speed up, skip the duplicated lines
            if IsInSpan(Index2):
                Index2 +=1
                continue
            src=''
            des=''
            for i in range(10):
                if Index2+i >=TotalLen:
                    break
                src +=AllLines[Index1+i]
                des +=AllLines[Index2+i]
                if src ==des:
                    t=Result.get(Index1, [])
                    for tt in t:
                        if tt[0] ==Index2:
                            tt[1]=i+1
                            break
```

```
                    else:
                        t.append([Index2, i+1])
                    Result[Index1]=t
            else:
                break
        t=Result.get(Index1, [])
        for tt in t:
            if tt[0]==Index2:
                Index2 +=tt[1]
                break
            else:
                Index2 +=1

        #Optimize the Result dictionary, remove the items with span<3
        Result[Index1]=Result.get(Index1, [])
        for n in Result[Index1][::-1]:      #Note: here must use the reverse slice,
            if n[1]<3:
                Result[Index1].remove(n)
        if not Result[Index1]:
            del Result[Index1]

        #Compute the min span of duplicated codes of line Index1,modify the step
        #Index1
        a=[ttt[1] for ttt in Result.get(Index1, [[Index1, 1]])]
        if a:
            Index1 +=max(a)
        else:
            Index1 +=1

#Output the result
def Output():
    print('-'*20)
    print('Result:')
    for key, value in Result.items():
        print('The original line is: \n {0}'.format(AllLines[key]))
        print('Its line number is {0}'.format(key+1))
        print('The duplicated line numbers are:')
        for i in value:
            print('    Start:', i[0], '    Span:', i[1])
        print('-'*20)
    print('-'*20)

if isfile(FileName):
    start=time()
```

```
PreOperate()
MainCheck()
Output()
print('Time used:', time()-start)
```

6.3 二进制文件操作案例精选

数据库文件、图像文件、可执行文件、动态链接库文件、音频文件、视频文件、Office 文档等均属于二进制文件。对于二进制文件，不能使用记事本或其他文本编辑软件直接进行正常读写，也不能通过 Python 的文件对象直接读取和理解二进制文件的内容。必须正确理解二进制文件的结构和序列化规则，然后设计正确的反序列化规则，才能准确地理解二进制文件内容。

所谓序列化，简单地说就是把内存中的数据在不丢失其类型信息的情况下转换成对象的二进制形式的过程，对象序列化后的数据经过正确的反序列化过程应该能够准确无误地恢复为原来的对象。Python 中常用的序列化模块有 struct、pickle、shelve、marshal 和 json，其中，json 常用于文本信息的序列化，在 6.2 节中已经介绍了。

6.3.1 使用 pickle 模块读写二进制文件

Python 标准库 pickle 提供的 dump()方法用于将数据进行序列化并写入文件（dump()方法的 protocol 参数为 True 时可以实现压缩的效果），而 load()用于读取二进制文件内容并进行反序列化，还原为原来的信息。

例 6-9 使用 pickle 模块写入二进制文件。

```
import pickle

n=7
i=13000000
a=99.056
s='中国人民 123abc'
lst=[[1, 2, 3], [4, 5, 6], [7, 8, 9]]
tu = (-5, 10, 8)
coll={4, 5, 6}
dic={'a':'apple', 'b':'banana', 'g':'grape', 'o':'orange'}
f=open('sample_pickle.dat', 'wb')          #以写模式打开二进制文件
try:
    pickle.dump(n, f)                       #对象个数
    pickle.dump(i, f)                       #写入整数
    pickle.dump(a, f)                       #写入实数
    pickle.dump(s, f)                       #写入字符串
    pickle.dump(lst, f)                     #写入列表
    pickle.dump(tu, f)                      #写入元组
```

```
            pickle.dump(coll, f)             #写入集合
            pickle.dump(dic, f)              #写入字典
    except:
        print('写文件异常')
    finally:
        f.close()
```

例 6-10 使用 pickle 模块读取例 6-9 中写入二进制文件的内容。

```
import pickle

f=open('sample_pickle.dat', 'rb')
n=pickle.load(f)                             #读出文件的数据个数
for i in range(n):
    x=pickle.load(f)
    print(x)
f.close()
```

💡**小提示**：pickle 模块还提供了一个 dumps() 方法，可以返回对象序列化之后的字节形式，例如：

```
>>>pickle.dumps([1,2,3])                    #序列化列表
b'\x80\x03]q\x00(K\x01K\x02K\x03e.'
>>>pickle.dumps([1,2,3,4])
b'\x80\x03]q\x00(K\x01K\x02K\x03K\x04e.'
>>>pickle.dumps({1,2,3,4})                  #序列化集合
b'\x80\x03cbuiltins\nset\nq\x00]q\x01(K\x01K\x02K\x03K\x04e\x85q\x02Rq\x03.'
>>>pickle.dumps({1,2,3})
b'\x80\x03cbuiltins\nset\nq\x00]q\x01(K\x01K\x02K\x03e\x85q\x02Rq\x03.'
>>>pickle.dumps((1,2,3))                    #序列化元组
b'\x80\x03K\x01K\x02K\x03\x87q\x00.'
>>>pickle.dumps(123)                        #序列化数字
b'\x80\x03K{.'
```

📁**拓展阅读**：下面的代码可以用来把文本文件转换为二进制文件，其中，test.txt 是包含若干文本信息的源文件，test_pickle.dat 是转换后的二进制文件，注意 with 语句的用法。

```
>>>import pickle
>>>with open('test.txt')as src, open('test_pickle.dat', 'wb')as dest:
    lines=src.readlines()
    pickle.dump(len(lines), dest)
    for line in lines:
        pickle.dump(line, dest)
>>>with open('test_pickle.dat', 'rb')as fp:
```

```
    n=pickle.load(fp)
    for i in range(n):
        print(pickle.load(fp))
```

6.3.2 使用 struct 模块读写二进制文件

使用 struct 模块需要使用 pack()方法把对象按指定个数进行序列化,然后使用文件对象的 write()方法将序列化的结果写入二进制文件;读取时需要使用文件对象的 read()方法读取二进制文件内容,然后再使用 struct 模块的 unpack()方法反序列化得到原来的信息。

例 6-11 使用 struct 模块写入二进制文件。

```
import struct

n=1300000000
x=96.45
b=True
s='a1@中国'
sn=struct.pack('if?', n, x, b)              #序列化,i 表示整数,f 表示实数,?表示逻辑值
f=open('sample_struct.dat', 'wb')
f.write(sn)
f.write(s.encode())                          #字符串需要编码为字节串再写入文件
f.close()
```

例 6-12 使用 struct 模块读取例 6-11 中二进制文件的内容。

```
import struct

f=open('sample_struct.dat', 'rb')
sn=f.read(9)
tu=struct.unpack('if?', sn)                  #使用指定格式反序列化
print(tu)
n, x, b1=tu
print('n=',n, 'x=',x, 'b1=',b1)
s=f.read(9)
s=s.decode()                                 #字符串解码
print('s=', s)
```

小提示:在上面的代码中,可能读者会疑惑如何确定要读取几个字节,为什么是 9 而不是其他数字呢?看完下面的代码应该能够明白了。

```
>>>import struct
>>>struct.pack('if?', 13000, 56.0, True)
b'\xc82\x00\x00\x00`B\x01'
>>>len(_)
```

```
9
>>>len(struct.pack('if?', 9999, 5336.0, False))
9
>>>x='a1@中国'
>>>len(x.encode())
9
```

6.3.3 使用 shelve 模块操作二进制文件

Python 标准库 shelve 也提供了二进制文件操作的功能,可以像字典赋值一样来写入二进制文件,也可以像字典一样读取二进制文件,有点类似于后面第 8 章介绍的 NoSQL 数据库 MongoDB。

```
>>>import shelve                                  #导入 shelve 模块
>>>fp=shelve.open('shelve_test.dat')              #创建或打开二进制文件
>>>zhangsan={'age':38, 'sex':'Male', 'address':'SDIBT'}
>>>fp['zhangsan']=zhangsan                        #写入文件内容
>>>lisi={'age':40, 'sex':'Male', 'qq':'1234567', 'tel':'7654321'}
>>>fp['lisi']=lisi                                #写入文件内容
>>>fp.close()                                     #关闭文件
>>>fp=shelve.open('shelve_test.dat')
>>>print(fp['zhangsan']['age'])                   #查看文件内容
38
>>>print(fp['lisi']['qq'])
1234567
>>>fp.close()
```

6.3.4 使用 marshal 模块操作二进制文件

Python 标准库 marshal 也可以进行对象的序列化和反序列化,下面的代码进行了简单演示。

```
>>>import marshal                                 #导入模块
>>>x1=30                                          #待序列化的对象
>>>x2=5.0
>>>x3=[1, 2, 3]
>>>x4 = (4, 5, 6)
>>>x5={'a':1, 'b':2, 'c':3}
>>>x6={7, 8, 9}
>>>x=[eval('x'+str(i))for i in range(1,7)]       #把需要序列化的对象放到一个列表中
>>>x
[30, 5.0, [1, 2, 3],(4, 5, 6), {'a': 1, 'b': 2, 'c': 3}, {8, 9, 7}]
>>>with open('test.dat', 'wb')as fp:             #创建二进制文件
    marshal.dump(len(x), fp)                     #先写入对象个数
```

```
    for item in x:
        marshal.dump(item,fp)              #把列表中的对象依次序列化并写入文件
>>>with open('test.dat','rb')as fp:        #打开二进制文件
    n=marshal.load(fp)                     #获取对象个数
    for i in range(n):
        print(marshal.load(fp))            #反序列化,输出结果
30
5.0
[1, 2, 3]
(4, 5, 6)
{'a': 1, 'b': 2, 'c': 3}
{8, 9, 7}
```

6.4 文件与文件夹操作

6.4.1 标准库 os、os.path 与 shutil 简介

os 模块除了提供使用操作系统功能和访问文件系统的简便方法之外,还提供了大量文件与文件夹操作的方法,如表 6-4 所示。os.path 模块提供了大量用于路径判断、切分、连接以及文件夹遍历的方法,如表 6-5 所示。shutil 模块也提供了大量的方法支持文件和文件夹操作,常用方法如表 6-6 所示。

表 6-4 os 模块常用成员

方　　法	功　能　说　明
access(path, mode)	按照 mode 指定的权限访问文件
chdir(path)	把 path 设为当前工作目录
chmod(path, mode, *, dir_fd=None, follow_symlinks=True)	改变文件的访问权限
extsep	当前操作系统所使用的文件扩展名分隔符
fstat(path)	返回打开的文件的所有属性
get_exec_path()	返回可执行文件的搜索路径
getcwd()	返回当前工作目录
listdir(path)	返回 path 目录下的文件和目录列表
mkdir(path[, mode=0777])	创建目录
makedirs(path1/path2…, mode=511)	创建多级目录
open(path, flags, mode = 0o777, *, dir_fd=None)	按照 mode 指定的权限打开文件,默认权限为可读、可写、可执行

续表

方法	功能说明
rmdir(path)	删除目录，目录中不能有文件或子文件夹
remove(path)	删除指定的文件
removedirs(path1/path2…)	删除多级目录，目录中不能有文件
rename(src, dst)	重命名文件或目录，可实现文件的移动
scandir(path='.')	返回包含指定文件夹中所有 DirEntry 对象的迭代对象
sep	当前操作系统所使用的路径分隔符
startfile(filepath [, operation])	使用关联的应用程序打开指定文件或启动指定应用程序
stat(path)	返回文件的所有属性
truncate(path, length)	将文件截断，只保留指定长度的内容
walk(top, topdown=True, onerror=None)	遍历目录树，该方法返回一个元组，包括 3 个元素：所有路径名、所有目录列表与文件列表
write(fd, data)	将 bytes 对象 data 写入文件 fd

表 6-5 os.path 模块常用成员

方法	功能说明
abspath(path)	返回给定路径的绝对路径
basename(path)	返回指定路径的最后一个组成部分
commonpath(paths)	返回给定的多个路径的最长公共路径
commonprefix(paths)	返回给定的多个路径的最长公共前缀
dirname(p)	返回给定路径的文件夹部分
exists(path)	判断文件是否存在
getatime(filename)	返回文件的最后访问时间
getctime(filename)	返回文件的创建时间
getmtime(filename)	返回文件的最后修改时间
getsize(filename)	返回文件的大小
isabs(path)	判断 path 是否为绝对路径
isdir(path)	判断 path 是否为文件夹
isfile(path)	判断 path 是否为文件
join(path, *paths)	连接两个或多个 path

续表

方 法	功 能 说 明
split(path)	对路径进行分隔，以列表形式返回
splitext(path)	从路径中分隔文件的扩展名
splitdrive(path)	从路径中分隔驱动器的名称

表6-6 shutil模块常用成员

方 法	功 能 说 明
copyfile(src, dst)	复制文件
copytree(src, dst)	递归复制文件夹
disk_usage(path)	查看磁盘使用情况
move(src, dst)	移动文件或递归移动文件夹
rmtree(path)	递归删除文件夹
make_archive(base_name, format, root_dir=None, base_dir=None)	创建tar或zip格式的压缩文件
unpack_archive(filename, extract_dir=None, format=None)	解压缩文件

下面通过几个示例来演示os、os.path以及shutil模块的基本用法。

```
>>>import os
>>>import os.path
>>>os.path.basename('C:\\windows\\notepad.exe')
'notepad.exe'
>>>os.path.basename('C:\\windows')              #获取路径的最后一个组成部分
'windows'
>>>os.path.exists('test1.txt')                  #测试文件是否存在
False
>>>os.rename('C:\\test1.txt', 'D:\\test2.txt')  #源文件不存在,重命名失败
  File "<pyshell#150>", line 1, in <module>
    os.rename('C:\\test1.txt', 'D:\\test2.txt')
FileNotFoundError: [WinError 2] 系统找不到指定的文件。: 'C:\\test1.txt' -> 'D:\\test2.txt'
>>>os.rename('C:\\dfg.txt', 'D:\\test2.txt')
                                                #os.rename()可以实现文件的改名和移动
>>>os.path.exists('C:\\dfg.txt')
False
>>>os.path.exists('D:\\test2.txt')
True
>>>path='D:\\mypython_exp\\new_test.txt'
>>>os.path.dirname(path)                        #返回路径的文件夹名
```

```
'D:\\mypython_exp'
>>>os.path.split(path)                    #切分文件路径和文件名
('D:\\mypython_exp', 'new_test.txt')
>>>os.path.splitdrive(path)
('D:', '\\mypython_exp\\new_test.txt')
>>>os.path.splitext(path)                 #切分文件扩展名
('D:\\mypython_exp\\new_test', '.txt')
>>>print([fname for fname in os.listdir(os.getcwd()) if os.path.isfile(fname) and fname.endswith('.pyc')])
['consts.pyc', 'database_demo.pyc', 'nqueens.pyc']
>>>os.getcwd()                            #返回当前工作目录
'C:\\Python35'
>>>os.mkdir(os.getcwd()+'\\temp')         #创建目录
>>>os.chdir(os.getcwd()+'\\temp')         #改变当前工作目录
>>>os.getcwd()
'C:\\Python35\\temp'
>>>os.mkdir(os.getcwd()+'\\test')
>>>os.listdir('.')
['test']
>>>os.rmdir('test')                       #删除目录
>>>os.listdir('.')
[]
>>>os.path.commonpath([r'C:\windows\notepad.exe', r'C:\windows\system'])
'C:\\windows'
>>>os.path.commonpath([r'a\b\c\d', r'a\b\c\e'])
'a\\b\\c'
>>>os.path.commonprefix([r'a\b\c\d', r'a\b\c\e'])
'a\\b\\c\\'
>>>import shutil                          #导入 shutil 模块
>>>shutil.copyfile('C:\\dir.txt', 'C:\\dir1.txt')   #复制文件
```

下面的代码将 C:\Python34\Dlls 文件夹以及该文件夹中所有文件压缩至 D:\a.zip 文件：

```
>>>shutil.make_archive('D:\\a', 'zip', 'C:\\Python34', 'Dlls')
'D:\\a.zip'
```

下面的代码则将刚压缩得到的文件 D:\a.zip 解压缩至 D:\a_unpack 文件夹：

```
>>>shutil.unpack_archive('D:\\a.zip', 'D:\\a_unpack')
```

下面的代码使用 shutil 模块的方法删除刚刚解压缩得到的文件夹：

```
>>>shutil.rmtree('D:\\a_unpack')
```

Python 标准库 shutil 的 rmtree()函数还支持更多的参数，例如，可以使用 onerror 参数指定回调函数来处理删除文件或文件夹失败的情况：

```
>>> import os
>>> import stat
>>> import shutil
>>> def remove_readonly(func, path, _):      #定义回调函数
        os.chmod(path, stat.S_IWRITE)        #删除文件的只读属性
        func(path)                            #再次执行删除操作
>>> shutil.rmtree('D:\\des_test')            #文件夹中有一个只读文件
Traceback (most recent call last):
  File "<pyshell#21>", line 1, in <module>
    shutil.rmtree('D:\\des_test')
  File "C:\Python35\lib\shutil.py", line 488, in rmtree
    return _rmtree_unsafe(path, onerror)
  File "C:\Python35\lib\shutil.py", line 383, in _rmtree_unsafe
    onerror(os.unlink, fullname, sys.exc_info())
  File "C:\Python35\lib\shutil.py", line 381, in _rmtree_unsafe
    os.unlink(fullname)
PermissionError: [WinError 5] 拒绝访问。: 'D:\\des_test\\test1.txt'
>>> shutil.rmtree('D:\\des_test', onerror=remove_readonly)
                                              #指定回调函数,删除成功
```

下面的代码用来递归复制文件夹,并忽略扩展名为 pyc 的文件和以"新"开头的文件和子文件夹:

```
>>> from shutil import copytree, ignore_patterns
>>> copytree('C:\\python35\\test', 'D:\\des_test', ignore=ignore_patterns('*.pyc', '新*'))
```

如果需要遍历指定目录下的所有子目录和文件,可以使用递归的方法,例如:

```
import os

def visitDir(path):
    if not os.path.isdir(path):
        print('Error:"', path, '" is not a directory or does not exist.')
        return
    for lists in os.listdir(path):
        sub_path=os.path.join(path, lists)
        print(sub_path)
        if os.path.isdir(sub_path):
            visitDir(sub_path)                #递归调用

visitDir('E:\\test')
```

下面的代码使用 os 模块的 walk() 方法进行指定目录的遍历。

```
import os
```

```
def visitDir2(path):
    if not os.path.isdir(path):
        print('Error:"', path, '" is not a directory or does not exist.')
        return
    list_dirs=os.walk(path)
    for root, dirs, files in list_dirs:        #遍历该元组的目录和文件信息
        for d in dirs:
            print(os.path.join(root, d))       #获取完整路径
        for f in files:
            print(os.path.join(root, f))       #获取文件的绝对路径

visitDir2('h:\\music')
```

拓展知识：除了用于文件操作和文件夹操作的方法之外，os 模块还提供了大量其他方法。例如，system()方法可以用来执行外部程序或系统内置命令，popen()和 startfile()也可以用来启动外部程序。另外，Python 标准库 subprocess 也提供了大量与进程创建与管理有关的对象，pywin32 工具包提供的 ShellExecute()和 CreateProcess()函数，这些对象和函数也可以实现启动外部程序的目的。如果对 Python 标准库 ctypes 和系统 API 函数熟悉，也可以直接调用底层 API 函数实现更加高级的功能。

```
>>>import subprocess
>>>h=subprocess.Popen('', executable='C:\\windows\\notepad.exe')
                                                #打开记事本程序
>>>h.terminate()                                #结束进程
>>>h=subprocess.Popen('', executable='C:\\windows\\notepad.exe')
                                                #打开记事本程序
>>>h.kill()                                     #结束进程
>>>os.popen('C:\\windows\\notepad.exe')         #打开记事本程序
>>>import os
>>>os.startfile(r'C:\windows\notepad.exe')      #打开记事本程序
>>>os.startfile(r'test.py')                     #执行 Python 程序
>>>import win32api
>>>win32api.ShellExecute(0, 'open', 'notepad.exe', '', '',0)
                                                #0 表示后台运行程序
42
>>>win32api.ShellExecute(0, 'open', 'notepad.exe', '', '',1)
                                                #1 表示前台运行程序
42
>>>win32api.ShellExecute(0, 'open', 'notepad.exe', 'C:\\dir.txt', '',1)
                                                #打开指定文件
42
>>>win32api.ShellExecute(0, 'open', 'www.python.org', '', '',1)
                                                #打开网址
42
```

```
>>>win32api.ShellExecute(0, 'open',r'C:\dir.txt', '', '',1)
                                                       #相当于双击文件
42
>>>import win32process
>>>handle=win32process.CreateProcess(r'C:\windows\notepad.exe', '', None,
None, 0, win32process.CREATE_NO_WINDOW, None, None, win32process.STARTUPINFO())
                                                       #打开记事本程序
>>>win32process.TerminateProcess(handle[0], 0)    #关闭程序
>>>handle=win32process.CreateProcess(r'C:\windows\notepad.exe', '', None,
None, 0, win32process.CREATE_NO_WINDOW, None, None, win32process.STARTUPINFO())
>>>import win32event
>>>win32event.WaitForSingleObject(handle[0],-1)
                                                       #需要手动关闭记事本
0
```

6.4.2 案例精选

例 6-13 将当前目录的所有扩展名为 html 的文件重命名为扩展名为 htm 的文件。

```
import os

file_list=os.listdir(".")
for filename in file_list:
    pos=filename.rindex(".")
    if filename[pos+1:]=="html":
        newname=filename[:pos+1]+"htm"
        os.rename(filename, newname)
        print(filename+"更名为:"+newname)
```

当然,也可以改写为下面的简洁而等价的代码:

```
import os

file_list=[filename for filename in os.listdir(".")if filename.endswith('.html')]
for filename in file_list:
    newname=filename[:-4]+'htm'
    os.rename(filename, newname)
    print(filename+"更名为:"+newname)
```

例 6-14 计算文件的 CRC32 值。

```
import sys
import zlib
import os.path

filename=sys.argv[1]
```

```
if os.path.isfile(filename):
    fp=open(filename, 'rb')
    contents=fp.read()
    fp.close()
    print(zlib.crc32(contents.encode()))
else:
    print('file not exists')
```

拓展知识：CRC 又称为循环冗余检验码，常用于数据存储和通信领域，具有极强的检错能力。CRC32 产生校验值时源数据块的每一个 bit（位）都参与了计算，所以数据块中即使只有一位发生了变化，也会得到不同的 CRC32 值，也可用于文件完整性保护。

例 6-15 判断一个文件是否为 GIF 图像文件。任何一种文件都具有专门的文件头结构，在文件头中存放了大量的信息，其中就包括该文件的类型。通过文件头信息来判断文件类型的方法可以得到更加准确的信息，而不依赖于文件扩展名。

```
>>>def is_gif(fname):
    f=open(fname, 'r')
    first4=tuple(f.read(4))
    f.close()
    return first4==('G', 'I', 'F', '8')
>>>is_gif('C:\\test.gif')
True
>>>is_gif('C:\\dir.txt')
False
```

例 6-16 使用 xlwt 模块写入 Excel 文件。

```
from xlwt import *

book=Workbook()                              #创建新的 Excel 文件
sheet1=book.add_sheet("First")               #添加新的 worksheet
al=Alignment()
al.horz=Alignment.HORZ_CENTER                #对齐方式
al.vert=Alignment.VERT_CENTER
borders=Borders()
borders.bottom=Borders.THICK                 #边框样式
style=XFStyle()
style.alignment=al
Style.borders=borders
row0=sheet1.row(0)                           #获取第 0 行
row0.write(0, 'test', style=style)           #写入单元格
book.save(r'D:\test.xls')                    #保存文件
```

小提示：xlwt 和下面一个案例用到的 xlrd 都是用来操作 Excel 2003 之前版本的 Python 扩展库，默认没有安装，可以使用 pip 进行安装。

例 6-17 使用 xlrd 模块读取 Excel 文件。

```
>>>import xlrd
>>>book=xlrd.open_workbook(r'D:\test.xls')   #打开 Excel 文件
>>>sheet1=book.sheet_by_name('First')        #打开 worksheet
>>>row0=sheet1.row(0)                        #获取第 0 行
>>>print(row0[0])                            #查看该行第 0 个单元格信息
text:u'test'
>>>print(row0[0].value)                      #查看单元格中的内容
test
```

例 6-18 使用 Pywin32 操作 Excel 文件。

```
xlApp=win32com.client.Dispatch('Excel.Application')
                                             #打开 Excel 程序
xlBook=xlApp.Workbooks.Open('D:\\1.xls')     #打开 Excel 文件
xlSht=xlBook.Worksheets('sheet1')            #打开 worksheet
aaa=xlSht.Cells(1, 2).Value                  #访问单元格的内容
xlSht.Cells(2, 3).Value=aaa
xlBook.Close(SaveChanges=1)
del xlApp
```

拓展知识：Pywin32 模块需要单独安装，这是一个功能非常强大的模块，提供了 Windows 底层 API 函数的封装，使得可以在 Python 中直接调用 Windows API 函数，支持大量的 Windows 底层操作。

拓展知识：Python 标准库 ctypes 提供了访问 DLL 动态链接库的功能，很好地支持了与 C/C++ 等语言混合编程的需求，也可以调用系统底层 API 函数。下面给出三段代码来演示如何通过 Python 标准库 ctypes 来调用 Windows API 实现特定功能，这部分内容涉及 Windows 系统底层的知识，感觉有难度的读者可以选择跳过。

1. 监视用户计算机桌面窗口焦点的变化情况

```
from ctypes import *
from time import sleep
from datetime import datetime

#方便调用 Windows 底层 API 函数
user32=windll.user32
kernel32=windll.kernel32
psapi=windll.psapi

#实时查看当前窗口
def getProcessInfo():
    global windows
```

```python
        #获取当前位于桌面最顶端的窗口句柄
        hwnd=user32.GetForegroundWindow()
        pid=c_ulong(0)
        #获取进程ID
        user32.GetWindowThreadProcessId(hwnd, byref(pid))
        processId=str(pid.value)
        #获取可执行文件名称
        executable=create_string_buffer(512)
        h_process=kernel32.OpenProcess(0x400|0x10, False, pid)
        psapi.GetModuleBaseNameA(h_process, None, byref(executable), 512)
        #获取窗口标题
        windowTitle=create_string_buffer(512)
        user32.GetWindowTextA(hwnd, byref(windowTitle), 512)
        #关闭句柄
        kernel32.CloseHandle(hwnd)
        kernel32.CloseHandle(h_process)
        #更新最近两个窗口列表
        windows.pop(0)
        windows.append([executable.value.decode('gbk'),windowTitle.value.decode
        ('gbk')])

def main():
    global windows
    windows=[None, None]
    while True:
        getProcessInfo()
        #如果用户切换窗口则进行提示
        if windows[0] !=windows[1]:
            print('='* 30)
            print(str(datetime.now())[:19],windows[0],'==>',windows[1])
        sleep(0.2)
if __name__ == '__main__':
    main()
```

2. 查杀Windows系统中指定的进程

```python
from ctypes.wintypes import *
from ctypes import *

kernel32=windll.kernel32

class tagPROCESSENTRY32(Structure):              #定义结构体
    _fields_=[('dwSize',              DWORD),
              ('cntUsage',            DWORD),
```

```
                ('th32ProcessID',      DWORD),
                ('th32DefaultHeapID',  POINTER(ULONG)),
                ('th32ModuleID',       DWORD),
                ('cntThreads',         DWORD),
                ('th32ParentProcessID',DWORD),
                ('pcPriClassBase',     LONG),
                ('dwFlags',            DWORD),
                ('szExeFile',          c_char * 260)]
def killProcess(processNames):
    #创建进程快照
    hSnapshot=kernel32.CreateToolhelp32Snapshot(15, 0)
    fProcessEntry32=tagPROCESSENTRY32()
    if hSnapshot:
        fProcessEntry32.dwSize=sizeof(fProcessEntry32)
        hasmore=kernel32.Process32First(hSnapshot, byref(fProcessEntry32))
        #枚举进程
        while hasmore:
            #可执行文件
            processName = (fProcessEntry32.szExeFile)
            #进程 ID
            processID=fProcessEntry32.th32ProcessID
            if processName.decode().lower()in processNames:
                #获取进程句柄
                hProcess=kernel32.OpenProcess(1, False, processID)
                #结束进程
                kernel32.TerminateProcess(hProcess,0)
            #获取下一个进程
            hasmore=kernel32.Process32Next(hSnapshot, byref(fProcessEntry32))

#待查杀的进程列表
processNames =('notepad.exe', 'mspaint.exe')
killProcess(processNames)
```

3. 实时显示鼠标所处窗口的文本

```
from ctypes import *
from ctypes import wintypes
from time import sleep

#调用 Windows 系统动态链接库 user32.dll
user32=windll.user32
p=wintypes.POINT()
buffer=create_string_buffer(255)
```

```
while True:
    sleep(0.5)
    #获取鼠标的位置
    user32.GetCursorPos(byref(p))
    #获取鼠标所处位置的窗口句柄
    HWnd=user32.WindowFromPoint(p)
    #注释掉的代码本来是可以用星号密码查看的,但在Windows 7以后的系统中失效了
    #dwStyle=user32.GetWindowLongA(HWnd, -16)          #-16是GWL_STYLE消息的值
    #user32.SetWindowWord(HWnd, -16, 0)
    sleep(0.2)
    #获取窗口文本
    user32.SendMessageA(HWnd, 13, 255, byref(buffer)) #13是WM_GETTEXT消息的值
    #user32.SetWindowLongA(HWnd, -16, dwStyle)
    print(buffer.value.decode('gbk'))
```

例 6-19 检查Word文档的连续重复字。在Word文档中,经常会由于键盘操作不小心而使得文档中出现连续的重复字,例如,"用户的的资料"或"需要需要用户输入"之类的情况。本例使用Pywin32模块中win32com对Word文档进行检查并提示类似的重复汉字。

```
import sys
from win32com import client

filename=sys.argv[1]
word=client.Dispatch('Word.Application')              #打开Word程序
doc=word.Documents.Open(filename)                      #打开Word文件
content=str(doc.Content)                               #获取文件内容
doc.Close()
word.Quit()

repeatedWords=[]

lens=len(content)
for i in range(lens-2):
    ch=content[i]
    ch1=content[i+1]
    ch2=content[i+2]
    #\u4e00至\u9fa5是汉字Unicode编码范围
    if(u'\u4e00'<=ch<=u'\u9fa5' or ch in(',','。','、')):
        if ch==ch1 and ch+ch1 not in repeatedWords:
            print(ch+ch1)
            repeatedWords.append(ch+ch1)
        elif ch==ch2 and ch+ch1+ch2 not in repeatedWords:
            print(ch+ch1+ch2)
```

```
            repeatedWords.append(ch+ch1+ch2)
```

例 6-20　编写程序,进行文件夹增量备份。

程序功能与用法:指定源文件夹与目标文件夹,自动检测自上次备份以来源文件夹中内容的改变,包括修改的文件、新建的文件、新建的文件夹等,自动复制新增或修改过的文件到目标文件夹中,自上次备份以来没有修改过的文件将被忽略而不复制,从而实现增量备份。本例属于系统运维的范畴。

```
import os
import filecmp
import shutil
import sys

def autoBackup(scrDir, dstDir):
    if((not os.path.isdir(scrDir))or(not os.path.isdir(dstDir))or
      (os.path.abspath(scrDir)!=scrDir)or(os.path.abspath(dstDir)!=dstDir)):
        usage()
    for item in os.listdir(scrDir):
        scrItem=os.path.join(scrDir, item)
        dstItem=scrItem.replace(scrDir,dstDir)
        if os.path.isdir(scrItem):
            #创建新增的文件夹,保证目标文件夹的结构与原始文件夹一致
            if not os.path.exists(dstItem):
                os.makedirs(dstItem)
                print('make directory'+dstItem)
            #递归调用自身函数
            autoBackup(scrItem, dstItem)
        elif os.path.isfile(scrItem):
            #只复制新增或修改过的文件
            if((not os.path.exists(dstItem))or
              (not filecmp.cmp(scrItem, dstItem, shallow=False))):
                shutil.copyfile(scrItem, dstItem)
                print('file:'+scrItem+'==>'+dstItem)

def usage():
    print('scrDir and dstDir must be existing absolute path of certain directory')
    print('For example:{0} c:\\olddir c:\\newdir'.format(sys.argv[0]))
    sys.exit(0)

if __name__=='__main__':
    if len(sys.argv)!=3:
        usage()
    scrDir, dstDir=sys.argv[1], sys.argv[2]
    autoBackup(scrDir, dstDir)
```

例 6-21　编写程序，统计指定文件夹大小以及文件和子文件夹数量。本例也属于系统运维范畴，可用于磁盘配额的计算，例如 E-mail、博客、FTP、快盘等系统中每个账号所占空间大小的统计。

```python
import os

totalSize=0
fileNum=0
dirNum=0

def visitDir(path):
    global totalSize
    global fileNum
    global dirNum
    for lists in os.listdir(path):
        sub_path=os.path.join(path, lists)
        if os.path.isfile(sub_path):
            fileNum=fileNum+1                                       #统计文件数量
            totalSize=totalSize+os.path.getsize(sub_path)           #统计文件总大小
        elif os.path.isdir(sub_path):
            dirNum=dirNum+1                                         #统计文件夹数量
            visitDir(sub_path)                                      #递归遍历子文件夹

def main(path):
    if not os.path.isdir(path):
        print('Error:"', path, '" is not a directory or does not exist.')
        return
    visitDir(path)

def sizeConvert(size):                                              #单位换算
    K, M, G=1024, 1024**2, 1024**3
    if size >=G:
        return str(size/G)+'G Bytes'
    elif size >=M:
        return str(size/M)+'M Bytes'
    elif size >=K:
        return str(size/K)+'K Bytes'
    else:
        return str(size)+'Bytes'

def output(path):
    print('The total size of '+path+' is:'+sizeConvert(totalSize)+' ('+str(totalSize)+' Bytes)')
    print('The total number of files in '+path+' is:',fileNum)
```

```
        print('The total number of directories in '+path+' is:',dirNum)

if __name__=='__main__':
    path=r'd:\idapro6.5plus'
    main(path)
    output(path)
```

例 6-22 编写程序,统计指定目录所有 C++ 源程序文件中不重复代码行数。

本例只考虑 C++ 源程序文件(扩展名为 cpp),并且只认为严格相等的两行为重复行。

```
from os.path import isdir, join
from os import listdir

NotRepeatedLines=[]                                    #保存非重复的代码行
file_num=0                                             #文件数量
code_num=0                                             #代码总行数

def LinesCount(directory):
    global NotRepeatedLines
    global file_num
    global code_num

    for filename in listdir(directory):
        temp=join(directory, filename)
        if isdir(temp):                                #递归遍历子文件夹
            LinesCount(temp)
        elif temp.endswith('.cpp'):                    #只考虑.cpp文件
            file_num +=1
            with open(temp, 'r') as fp:
                while True:
                    line=fp.readline()
                    if not line:
                        break
                    if line not in NotRepeatedLines:
                        NotRepeatedLines.append(line)  #记录非重复行
                    code_num +=1                       #记录所有代码行
path=r'C:\Users\Dong\Desktop\VC++6.0'
print('总行数:{0},非重复行数:{1}'.format(code_num, len(NotRepeatedLines)))
linescount(path)
print('文件数量:{0}'.format(file_num))
```

例 6-23 编写程序,递归删除指定文件夹中指定类型的文件。

本例代码也属于系统运维范畴,可用于清理系统中的临时垃圾文件或其他指定类型的文件,稍加扩展还可以删除大小为 0 字节的文件,大家可以自行补充和完成。

```python
from os.path import isdir, join, splitext
from os import remove, listdir
import sys

filetypes=['.tmp', '.log', '.obj', '.txt']        #指定要删除的文件类型

def delCertainFiles(directory):
    if not isdir(directory):
        return
    for filename in listdir(directory):
        temp=join(directory, filename)
        if isdir(temp):
            delCertainFiles(temp)                  #递归调用
        elif splitext(temp)[1] in filetypes:       #检查文件类型
            remove(temp)                           #删除文件
            print(temp, ' deleted…')

def main():
    directory=r'E:\new'
    #directory=sys.argv[1]
    delCertainFiles(directory)

main()
```

如果文件夹中有带特殊属性的文件或子文件夹,上面的代码可能会无法删除带特殊属性的文件,利用 Python 扩展库 pywin32 可以解决这一问题。

```python
import win32con
import win32api
import os
from win32con import FILE_ATTRIBUTE_NORMAL

def del_dir(path):
    for file in os.listdir(path):
        file_or_dir=os.path.join(path,file)
        if os.path.isdir(file_or_dir) and not os.path.islink(file_or_dir):
            del_dir(file_or_dir)                   #递归删除子文件夹及其文件
        else:
            try:
                os.remove(file_or_dir)             #尝试删除该文件
            except:                                #无法删除,很可能是文件拥有特殊属性
                win32api.SetFileAttributes(file_or_dir, FILE_ATTRIBUTE_NORMAL)
                os.remove(file_or_dir)             #修改文件属性,设置为普通文件,再次删除
    os.rmdir(path)                                 #删除文件夹
del_dir("E:\\old")
```

拓展知识：系统运维涵盖的内容非常多，还包括电力系统维护、数据库维护、磁盘配额、用户账号与权限、网络设备与带宽分配、病毒防护与入侵检测、系统资源分配等。前面几个案例都属于系统运维的范畴，另外，跨平台的 Python 扩展库 psutil 可以用来查询进程或 CPU、内存、硬盘以及网络等系统资源占用率等信息，常用于系统运行状态检测和维护。可以使用 pip 工具安装该库。

(1) 查看 CPU 信息。

```
>>>psutil.cpu_count()                    #查看 CPU 核数
>>>psutil.cpu_count(logical=False)       #查看物理 CPU 个数
>>>psutil.cpu_percent()                  #查看 CPU 使用率
>>>psutil.cpu_percent(percpu=True)       #查看每个 CPU 的使用率
>>>psutil.cpu_times()                    #查看 CPU 时间分配情况
```

(2) 查看开机时间。

```
>>>import datetime
>>>t=psutil.boot_time()
>>>datetime.datetime.fromtimestamp(t).strftime('%Y-%m-%d %H:%M:%S')
'2015-12-26 11:32:17'
```

(3) 查看内存信息。

```
>>>virtual_memory=psutil.virtual_memory()
>>>virtual_memory.total /1024/1024/1024    #内存总大小
>>>virtual_memory.used/1024/1024/1024      #已使用内存
>>>virtual_memory.free/1024/1024/1024      #空闲内存
>>>virtual_memory.percent                  #内存使用率
```

(4) 查看磁盘信息。

```
>>>psutil.disk_partitions()                       #查看所有分区信息
>>>psutil.disk_usage('C:\\')                      #查看指定分区的磁盘空间情况
>>>psutil.disk_io_counters(perdisk=True)          #查看硬盘读写操作情况
```

(5) 查看与网络连接有关的信息。

```
>>>psutil.net_connections()              #查看网络连接情况
>>>psutil.net_io_counters()              #查看收发包情况
>>>psutil.net_if_addrs()                 #查看网络地址
```

(6) 查看当前登录用户信息。

```
>>>psutil.users()
```

(7) 查看进程信息。

```
>>>psutil.pids()                         #查看当前所有进程 ID
>>>p=psutil.Process(4204)                #获取指定 ID 的进程
>>>p.name()                              #进程名
```

```
>>>p.username()                          #查看创建该进程的用户名
>>>p.cmdline()                           #查看该进程对应的exe文件
>>>p.cwd()                               #查看该进程的工作目录
>>>p.exe()                               #进程对应的可执行文件名
>>>p.cpu_affinity()                      #该进程CPU占用情况(运行在哪个CPU上)
>>>p.num_threads()                       #该进程包含的线程数量
>>>p.threads()                           #该进程所有线程对象
>>>p.status()                            #进程状态
>>>p.is_running()                        #进程是否正在运行
>>>p.suspend()                           #挂起进程
>>>p.resume()                            #恢复运行
>>>p.kill()                              #结束进程
```

(8) 检查记事本程序是否在运行,如果在运行则返回记事本程序对应的进程ID。

```
>>>for pid in psutil.pids():
    try:
        p=psutil.Process(pid)
        if os.path.basename(p.exe())=='notepad.exe':
            print(pid)
    except:
        pass
```

(9) 查看指定进程的信息,如线程数量、打开的文件、用户账户等。

```
>>>psutil._psutil_windows.proc_username(7660)
                                                 #7660是进程ID
'dfg-PC\\dfg'
>>>psutil._psutil_windows.proc_open_files(7660)
['\\Device\\HarddiskVolume1\\Windows\\System32', '\\Device\\KsecDD']
>>>psutil._psutil_windows.proc_threads(4636)
                                                 #查看ID为4636的进程中的线程信息
>>>psutil._psutil_windows.proc_exe(5424)         #查看ID为5424的进程可执行文件
'\\Device\\HarddiskVolume1\\Python35\\pythonw.exe'
```

(10) 查看指定进程的线程信息,包括线程数量和所用CPU时间。

```
import psutil
import os

#遍历系统当前运行的所有进程ID
for pid in psutil.pids():
    try:
        proc=psutil.Process(pid)
        #获取进程对应的可执行文件
        exeFile=os.path.basename(proc.exe())
        #获取该进程的所有线程信息
```

```
            threads=psutil._psutil_windows.proc_threads(pid)
            times=0
            #遍历该进程的所有线程
            for thread in threads:
                #下标为 0 的元素是线程 ID,后面是每个 CPU 上的运行时间
                for timeUsed in thread[1:]:
                    times+=timeUsed
            print('='*20)
            print('Exe file:', os.path.basename(proc.exe()))
            print('Number of threads:', len(threads))
            print('Time used:', times)
        except:
            pass
```

拓展知识：正如前面所说,系统运维涉及面非常广,也包括系统中进程的创建与结束、系统服务状态等。下面再通过 Python 扩展库 wmi(该模块提供了对 Windows Management Instrumentation 的访问)来演示一下这两个方面的内容。

(1) 监视 Windows 系统中进程创建情况。

```
import wmi

c=wmi.WMI()
process_watcher=c.Win32_Process.watch_for('creation')
while True:
    try:
        new_process=process_watcher()
        proc_owner='{0[0]}\\{0[1]}'.format(new_process.GetOwner())
        temp_creation_date=new_process.CreationDate
        creation_date=temp_creation_date[:4]
        for i in(4, 6):
            creation_date+='-'+temp_creation_date[i:i+2]
        creation_date+=' '
        for i in(8, 10, 12):
            creation_date+=temp_creation_date[i:i+2]+':'
        creation_date=creation_date[:-1]
        executable=new_process.ExecutablePath
        cmdline=new_process.CommandLine
        pid=new_process.ProcessId
        parent_pid=new_process.ParentProcessId
        print('='*30)
        print('Process owner:'.ljust(18), proc_owner)
        print('Creation Time:'.ljust(18), str(creation_date))
        print('Executable:'.ljust(18), executable)
        print('Cmdline:'.ljust(18), cmdline)
```

```python
            print('ProcessId:'.ljust(18), pid)
            print('Parent ProcessId:'.ljust(18), parent_pid)
    except:
        pass
```

（2）查看 Windows 系统中的服务状态。

```python
import itertools
import wmi

def group(service):
    if service.State=='Stopped':
        return 'Stopped'
    elif service.State=='Running':
        return 'Running'
    else:
        return 'Others'


result=dict()

c=wmi.WMI()
for service in c.Win32_Service():
    state=service.State
    caption=service.Caption
    t=result.get(state,[])
    t.append(caption)
    result[state]=t

for state, captions in result.items():
    print('=' * 30)
    print(state)
    print('\n'.join(sorted(captions)))
```

例 6-24 使用扩展库 openpyxl 读写 Excel 2007 及更高版本的 Excel 文件。

```python
import openpyxl
from openpyxl import Workbook
fn=r'f:\test.xlsx'                          #文件名
wb=Workbook()                               #创建工作簿
ws=wb.create_sheet(title='你好,世界')        #创建工作表
ws['A1']='这是第一个单元格'                   #单元格赋值
ws['B1']=3.1415926
wb.save(fn)                                 #保存 Excel 文件
wb=openpyxl.load_workbook(fn)               #打开已有的 Excel 文件
ws=wb.worksheets[1]                         #打开指定索引的工作表
print(ws['A1'].value)                       #读取并输出指定单元格的值
```

```python
ws.append([1,2,3,4,5])                                    #添加一行数据
ws.merge_cells('F2:F3')                                   #合并单元格
ws['F2']="=sum(A2:E2)"                                    #写入公式
for r in range(10,15):
    for c in range(3,8):
        _=ws.cell(row=r, column=c, value=r*c)             #写入单元格数据
wb.save(fn)
```

假设某学校所有课程每学期允许多次考试,学生可随时参加考试,系统自动将每次成绩添加到 Excel 文件(包含 3 列:姓名、课程、成绩)中,现期末要求统计所有学生每门课程的最高成绩。下面的代码首先模拟生成随机成绩数据,然后进行统计分析。

```python
import openpyxl
from openpyxl import Workbook
import random

#生成随机数据
def generateRandomInformation(filename):
    workbook=Workbook()
    worksheet=workbook.worksheets[0]
    worksheet.append(['姓名','课程','成绩'])
    #中文名字中的第一、第二、第三个字
    first=tuple('赵钱孙李')
    middle=tuple('伟昀琛东')
    last=tuple('坤艳志')
    #课程名称
    subjects = ('语文','数学','英语')
    #随机生成 200 个数据
    for i in range(200):
        line=[]
        r=random.randint(1,100)
        name=random.choice(first)
        #按一定概率生成只有两个字的中文名字
        if r>50:
            name=name+random.choice(middle)
        name=name+random.choice(last)
        #依次生成姓名、课程名称和成绩
        line.append(name)
        line.append(random.choice(subjects))
        line.append(random.randint(0,100))
        worksheet.append(line)
    #保存数据,生成 Excel 2007 格式的文件
    workbook.save(filename)

def getResult(oldfile, newfile):
```

```python
    #用于存放结果数据的字典
    result=dict()
    #打开原始数据
    workbook=openpyxl.load_workbook(oldfile)
    worksheet=workbook.worksheets[0]
    #遍历原始数据
    #跳过第 0 行的表头
    for row in worksheet.rows[1:]:
        #姓名、课程名称、本次成绩
        name, subject, grade=row[0].value, row[1].value, row[2].value
        #获取当前姓名对应的课程名称和成绩信息
        #如果 result 字典中不包含,则返回空字典
        t=result.get(name, {})
        #获取当前学生当前课程的成绩,若不存在,返回 0
        f=t.get(subject, 0)
        #只保留该学生该课程的最高成绩
        if grade>f:
            t[subject]=grade
            result[name]=t
    #创建 Excel 文件
    workbook1=Workbook()
    worksheet1=workbook1.worksheets[0]
    worksheet1.append(['姓名','课程','成绩'])
    #将 result 字典中的结果数据写入 Excel 文件
    for name, t in result.items():
        for subject, grade in t.items():
            worksheet1.append([name, subject, grade])
    workbook1.save(newfile)

if __name__ =='__main__':
    oldfile=r'D:\test.xlsx'
    newfile=r'D:\result.xlsx'
    generateRandomInformation(oldfile)
    getResult(oldfile, newfile)
```

例 6-25 编写代码,查看指定 ZIP 和 RAR 压缩文件中的文件列表。

Python 标准库 zipfile 提供了对 ZIP 和 APK 文件的访问。

```
>>>import zipfile
>>>fp=zipfile.ZipFile(r'D:\Jakstab-0.8.3.zip')
>>>for f in fp.namelist():
    print(f)                                    #如果中文显示乱码可以参考例 6-29 进行修改
>>>fp.close()
```

Python 扩展库 rarfile(可通过 pip 工具进行安装)提供了对 RAR 文件的访问。

```
>>>import rarfile
>>>r=rarfile.RarFile(r'D:\asp网站.rar')
>>>for f in r.namelist():
    print(f)
>>>r.close()
```

例 6-26 小学口算题库生成器。

```
import random
import os
import tkinter
import tkinter.ttk
from docx import Document

columnsNumber=4

def main(rowsNumber=20, grade=4):
    if grade<3:
        operators='＋－'
        biggest=20
    elif grade <=4:
        operators='＋－×÷'
        biggest=100
    elif grade ==5:
        operators='＋－×÷('
        biggest=100

    document=Document()
    #创建表格
    table=document.add_table(rows=rowsNumber, cols=columnsNumber)
    #遍历每个单元格
    for row in range(rowsNumber):
        for col in range(columnsNumber):
            first=random.randint(1, biggest)
            second=random.randint(1, biggest)
            operator=random.choice(operators)
            if operator !='(':
                if operator =='－':
                    #如果是减法口算题,确保结果为正数
                    if first<second:
                        first, second=second, first
                r=str(first).ljust(2, ' ')+' '+operator\
                    +str(second).ljust
                    (2, ' ')+'='
            else:
```

```python
            #生成带括号的口算题,需要3个数字和2个运算符
            third=random.randint(1, 100)
            while True:
                o1=random.choice(operators)
                o2=random.choice(operators)
                if o1 !='(' and o2 !='(':
                    break
            rr=random.randint(1, 100)
            if rr>50:
                if o2 =='-':
                    if second<third:
                        second, third=third, second
                r=str(first).ljust(2, ' ')+o1+'(' \
                    +str(second).ljust(2, ' ')+o2+str(third).ljust(2, ' ')+')='
            else:
                if o1 =='-':
                    if first<second:
                        first, second=second, first
                r='('+str(first).ljust(2, ' ')+o1 \
                    +str(second).ljust(2, ' ')+')' \
                    +o2+str(third).ljust(2, ' ')+'='
        #获取指定单元格并写入口算题
            cell=table.cell(row, col)
            cell.text=r
    document.save('kousuan.docx')
    os.startfile('kousuan.docx')

if __name__ =='__main__':
    #tkinter编程知识请参考第15章
    app=tkinter.Tk()
    app.title('KouSuan——by Dong Fuguo')
    app['width']=300
    app['height']=150
    labelNumber=tkinter.Label(app, text='Number:', justify=tkinter.RIGHT,
                              width=50)
    labelNumber.place(x=10, y=40, width=50,height=20)
    comboNumber=tkinter.ttk.Combobox(app, values=(100,200,300,400,500),
                                     width=50)
    comboNumber.place(x=70, y=40, width=50, height=20)

    labelGrade=tkinter.Label(app, text='Grade:', justify=tkinter.RIGHT,
                             width=50)
    labelGrade.place(x=130, y=40, width=50,height=20)
    comboGrade=tkinter.ttk.Combobox(app, values=(1,2,3,4,5), width=50)
```

```
            comboGrade.place(x=200, y=40, width=50, height=20)

            def generate():
                number=int(comboNumber.get())
                grade=int(comboGrade.get())
                main(number, grade)
            buttonGenerate=tkinter.Button(app, text='GO', width=40, command=generate)
            buttonGenerate.place(x=130, y=90, width=40, height=30)

            app.mainloop()
```

小提示：例 6-26 的代码需要先使用 pip 工具安装扩展库 python-docx。

拓展知识：Python 程序编译与打包。学到这里，相信很多读者已经能够编写出一些程序来实现自己需要的功能了，然后应该会有一个很大的疑问：难道 Python 程序只能以源代码的方式来运行吗？能不能通过某种方式来保护自己的源代码呢？答案是肯定的。下面我们就来补充一下这方面的技术。一种方法是把 Python 程序伪编译成扩展名为 pyc 的字节码文件，一种是通过 py2exe 或 pyinstaller 对 Python 程序进行打包。

(1) Python 程序伪编译。

可以使用 py_compile 模块的 compile()函数对 Python 程序进行编译得到扩展名为 pyc 的字节码以提高加载和运行速度，同时还可以隐藏源代码。另外，Python 还提供了 compileall 模块，其中包含 compile_dir()、compile_file()和 compile_path()等方法，用来支持批量 Python 源程序文件的编译。另外，也可以在命令提示符环境中使用"python -m py_compile file.py"来编译 Python 程序文件 file.py，或者"python -O -m py_compile file.py"或"python -OO -m py_compile file.py"进行优化编译。

(2) Python 程序打包。

把 Python 程序转换为 exe 版本可执行程序之后再发布，可以在没有安装 Python 环境和扩展库的 Windows 平台上运行，这个功能极大地方便了用户。为了将 Python 程序转换为 exe 可执行文件，需要用到 py2exe 和 distutils 模块。当然，首先应保证 Python 程序可以正常运行，并且本机已安装了所有需要的扩展模块和相关的动态链接库文件。

例如，假设有 Python 源程序文件 test.py，然后编写 setup.py 文件，内容为

```
import distutils
import py2exe
distutils.core.setup(console=['test.py'])
```

然后在命令提示符下执行下面的命令：

```
python setup.py py2exe
```

接下来就会看到控制台窗口中大量的提示内容飞快地闪过，这个过程会自动搜集 test.py 程序执行所需要的所有支持文件，如果创建成功则会在当前文件夹下生成一个 dist 子文件夹，其中包含了最终程序执行所需要的所有内容。等待编译完成以后，将 dist

文件中的文件打包发布即可。

py2exe 模块的详细用法可以查阅有关资料,但是对于一般应用而言,上面的代码已经足够了。唯一要注意的问题是,对于控制台应用程序,要想转换为 exe 可执行程序直接套用上面的代码框架即可,只需要把

```
distutils.core.setup(console=['test.py'])
```

这行代码中的文件名替换为自己的 Python 程序文件名即可。而对于 GUI 应用程序,还需要将上面代码中的关键字 console 修改为 windows。

另外一个比较好用的 Python 程序打包工具是 pyinstaller,可以通过 pip 工具进行安装。安装之后在命令提示符环境中使用命令"pyinstaller -F -w kousuan.pyw"即可将 Python 程序 kousuan.pyw 及其所有依赖包打包成为 kousuan.exe 可执行文件,从而脱离 Python 解释器环境而独立运行于 Windows 系统。

例 6-27 将 docx 文档中的题库导入 SQLite 数据库。

```python
#docx 文档题库包含很多段,每段一个题目,格式为:问题。(答案)
#数据库 datase.db 中 tiku 表包含 kechengmingcheng、zhangjie、timu、daan 四个字段
#数据库有关知识可以查看本书第 8 章
import sqlite3
from docx import Document

doc=Document('《Python 程序设计》题库.docx')

#连接数据库
conn=sqlite3.connect('database.db')
cur=conn.cursor()

#先清空原来的题,可选操作
cur.execute('delete from tiku')
conn.commit()

for p in doc.paragraphs:
    text=p.text
    if '(' in text and ')' in text:
        index=text.index('(')
        #分离问题和答案
        question=text[:index]
        if '___' in question:
            question='填空题:'+question
        else:
            question='判断题:'+question
        answer=text[index+1:-1]
        #将数据写入数据库
        sql='insert into tiku(kechengmingcheng,zhangjie,timu,daan)values
```

```
              ("Python程序设计","未分类","'+question+'","'+answer+'")'
        cur.execute(sql)
conn.commit()
#关闭数据库连接
conn.close()
```

例 6-28 提取 docx 文档中例题、插图和表格清单。

```
from docx import Document
import re

result={'li':[], 'fig':[], 'tab':[]}
doc=Document(r'C:\Python可以这样学.docx')

for p in doc.paragraphs:                            #遍历文档所有段落
    t=p.text                                        #获取每一段的文本
    if re.match('例\d+-\d+', t):                    #例题
        result['li'].append(t)
    elif re.match('图\d+-\d+', t):                  #插图
        result['fig'].append(t)
    elif re.match('表\d+-\d+', t):                  #表格
        result['tab'].append(t)

for key in result.keys():                           #输出结果
    print('=' * 30)
    for value in result[key]:
        print(value)
```

例 6-29 将指定文件夹中的文件压缩至已有压缩包。

可能会有朋友问,为什么不把这个例子和下面的例子放在例 6-25 一起介绍呢?其实我是故意这样组织的,同类知识集中讲解和分散介绍相结合,学起来就不会太累,更容易加深记忆。就像书店里的书籍和超市里商品的摆放位置集中和分散相结合一样,也类似于报纸内容的横排和竖排相结合,都是有讲究的,后面还有类似的情况。什么也别说了,我们一起看代码吧!

```
from zipfile import ZipFile
from os import listdir
from os.path import isfile, isdir, join

def addFileIntoZipfile(srcDir, fp):
    for subpath in listdir(srcDir):
        subpath=join(srcDir, subpath)
        if isfile(subpath):
            fp.write(subpath)                       #写入文件
        elif isdir(subpath):
```

```
            fp.write(subpath)                        #写入文件夹
            addFileIntoZipfile(subpath, fp)          #递归调用

def zipCompress(srcDir, desZipfile):
    fp=ZipFile(desZipfile, mode='a')                 #以追加模式打开或创建 zip 文件
    addFileIntoZipfile(srcDir, fp)
    fp.close()

paths=[r'C:\python35\Scripts', r'C:\python35\Dlls', r'D:\tc']
for path in paths:
    zipCompress(path, 'test.zip')
```

例 6-30　使用密码字典暴力破解 RAR 或 ZIP 文件密码。

众所周知,使用字典暴力破解的成功率取决于密码字典文件中密码是否足够多。如果想保证非常高的成功率,可以参考本书 14.2 节中介绍的 MD5 破解方法来生成所有可能的密码,然后再逐个尝试。另外,如果下面的代码不能运行,需要做以下几个操作:①到 http://www.rarlab.com/rar/UnRARDLL.exe 下载并安装 unrardll 库,然后根据需要把安装文件夹中的 UnRAR.dll 或 x64\UnRAR64.dll 文件复制到 unrar 安装文件夹(例如,C:\Python 3.5\Lib\site-packages\unrar)中;②打开 unrar 安装文件夹中的 unrarlib.py 文件,把 lib_path=lib_path or find_library("unrar.dll")直接改为 unrarlib=ctypes.WinDLL(r"C:\Python 3.5\Lib\site-packages\unrar\UnRAR64.dll"),并把接下来的两行代码删除或注释掉。另外,如果想提高破解速度,可以参考第 10 章的多线程编程知识稍微改写一下本例代码。

```
import os
import sys
import zipfile                                       #zipfile 是标准库
try:
    from unrar import rarfile                        #尝试导入扩展库,如果没有就临时安装
except:
    path='"'+os.path.dirname(sys.executable)\
        +'\\scripts\\pip" install --upgrade pip'
    os.system(path)
    path='"'+os.path.dirname(sys.executable)+'\\scripts\\pip" install unrar'
    os.system(path)
    from unrar import rarfile

def decryptRarZipFile(filename):
    if filename.endswith('.zip'):
        fp=zipfile.ZipFile(filename)
    elif filename.endswith('.rar'):
        fp=rarfile.RarFile(filename)
    desPath=filename[:-4]                            #解压缩的目标文件夹
```

```
            if not os.path.exists(desPath):
                os.mkdir(desPath)
            try:                                           #尝试不用密码解压缩
                fp.extractall(desPath)
                fp.close()
                print('No password')
                return
            except:                                        #使用密码字典进行暴力破解
                try:
                    fpPwd=open('pwddict.txt')
                except:
                    print('No dict file pwddict.txt in current directory.')
                    return
                for pwd in fpPwd:
                    pwd=pwd.rstrip()
                    try:
                        if filename.endswith('.zip'):
                            for file in fp.namelist():    #重新编码再解码,避免中文乱码
                                fp.extract(file, path=desPath, pwd=pwd.encode())
                                os.rename(desPath+'\\'+file,
                                    desPath+'\\'+file.encode('cp437').decode('gbk'))
                            print('Success! ====>'+pwd)
                            fp.close()
                            break
                        elif filename.endswith('.rar'):
                            fp.extractall(path=desPath, pwd=pwd)
                            print('Success! ====>'+pwd)
                            fp.close()
                            break
                    except:
                        pass
                fpPwd.close()

if __name__=='__main__':
    filename=sys.argv[1]
    if os.path.isfile(filename)and filename.endswith(('.zip', '.rar')):
        decryptRarZipFile(filename)
    else:
        print('Must be RAR or ZIP file')
```

例 6-31 把 Excel 2007$^+$ 文件中的多个同结构 worksheet 中的内容合并到新文件中的一个 worksheet 中。

> **注意**：Excel 2007$^+$ 表示 Excel 2007 或更高版本。

感谢中国石油大学(华东)计算机与通信工程学院李昕老师提供本例的问题和第一版

代码,并和我一起反复修改和完善得到了最终版本的代码,关于基于 tkinter 的 GUI 编程知识请参考本书第 15 章。

```python
import os
import sys
from tkinter import Tk, Button
from tkinter import filedialog
from tkinter import simpledialog

try:
    import openpyxl
except:
    #先把 pip 升级到最新版本
    path='"'+os.path.dirname(sys.executable)\
        +'\\scripts\\pip" install --upgrade pip'
    os.system(path)
    #安装 openpyxl 扩展库
    path='"'+os.path.dirname(sys.executable)\
        +'\\scripts\\pip" install openpyxl'
    os.system(path)
    import openpyxl

def merge(start):
    #显示打开文件对话框,打开要合并的 Excel 2007+文件
    opts={'filetypes':[('Excel 2007', '.xlsx')]}
    filename=filedialog.askopenfilename(**opts)
    #如果没有选择文件,不再执行后面的代码
    if not filename:
        return
    #分隔路径和文件名
    filepath, tempfilename=os.path.split(filename)
    shotname=os.path.splitext(tempfilename)[0]
    #生成的新文件名
    newFile=filepath+'\\'+shotname+'_merge.xlsx'
    #创建新的 Excel 2007+文件
    workbook=openpyxl.Workbook()
    #添加新的 worksheet
    worksheet=workbook.worksheets[0]
    data=openpyxl.load_workbook(filename)
    for sheetnum, sheet in enumerate(data.worksheets):
        #根据设定的表头行数,设置读取的起始行
        #第一个 sheet 读取表头,后面的 sheet 忽略表头
        if sheetnum==0:
            rowStart=0
```

```
        else:
            rowStart=start
    #遍历原sheet,根据情况忽略表头
    for row in sheet.rows[rowStart:]:
        line=[col.value for col in row]
        worksheet.append(line)
#保存新文件
workbook.save(newFile)
#打开刚刚创建的新文件
os.startfile(newFile)

#单击按钮后执行的函数,参数a表示Excel文件中每个worksheet预期表头行数
def callback():
    kw={'initialvalue':1, 'minvalue':0, 'maxvalue':10}
    headerNum=simpledialog.askinteger('表头行数', '请输入表头行数',**kw)
    if headerNum !=None:
        merge(headerNum)

root=Tk()
root.title("合并 sheet")
Button(root, text="合并 WorkSheets", bg='blue', bd=2, width=28, command=
callback).pack()

root.mainloop()
```

例 6-32 把记事本文件 test.txt 转换成 Excel 2007$^+$ 文件。假设 test.txt 文件中第一行为表头,从第二行开始是实际数据,并且表头和数据行中的不同字段信息都是用逗号分隔。

可能会有朋友疑惑这样的代码有什么意义,毕竟使用 Excel 打开 txt 文件时只需要简单单击几下鼠标就可以实现数据导入。是的,确实是这样的,如果只有几个文件是完全可以手动完成的,但是如果有几百、几千甚至更多的文件需要转换呢?想象一下,我们写一段代码然后一边喝咖啡一边看着程序批量处理大量的文件,一杯咖啡还没喝完任务就已经完成了,那该是多么的惬意啊!

```
from openpyxl import Workbook

def main(txtFileName):
    new_XlsxFileName=txtFileName[:-3]+'xlsx'
    wb=Workbook()
    ws=wb.worksheets[0]
    with open(txtFileName) as fp:
        for line in fp:
            line=line.strip().split(',')
            ws.append(line)
    wb.save(new_XlsxFileName)

main('test.txt')
```

第 7 章 异常处理结构、代码测试与调试

再牛的程序员也无法提前预见代码运行时可能会遇到的所有情况,几乎每个程序员(也包括我,虽然我不是职业程序员,但是也开发过不少系统)都被用户说过"你编的那个软件不好用啊",而程序员经过反复检查以后发现问题的原因是用户操作不规范或者输入了错误类型的数据,于是一边修改代码加强类型检查一边抱怨用户为什么不按套路出牌。其实呢,我个人认为这样的问题的根源还是在程序员而不在用户,程序员编写代码时有义务也有必要考虑这些特殊情况,因为大多时候恰恰是一些特殊情况影响了整个系统的美感和成就感。虽然软件在发布前一般都经过了充分的测试,然而再充分的测试也很难枚举所有可能出现的情况,这时候异常处理结构则是避免特殊情况下软件崩溃的利器。

每种高级编程语言都提供了不同形式的异常处理结构,大幅度提高了代码的健壮性。简单地说,异常是指程序运行时引发的错误,引发错误的原因有很多,例如除零、下标越界、文件不存在、网络异常等。如果这些错误得不到正确的处理将会导致程序崩溃并终止运行,合理地使用异常处理结构可以使得程序更加健壮,具有更高的容错性,不会因为用户不小心的错误输入而造成程序崩溃,也可以使用异常处理结构为用户提供更加友好的提示。有效的软件测试方法能够在软件发布之前发现尽可能多的 Bug,而软件发布之后再出现错误时是否能够调试程序并快速定位和解决存在的问题则是程序员综合水平和能力的重要体现。

7.1 异常处理结构

7.1.1 异常是什么

首先,"异常"这个高大上的词语所描述的现象实际上在前面章节的代码中已经多次出现了,只是没有详细解释这个概念,让我们一起回顾一下:

```
>>>2 / 0                                    #除 0 错误
Traceback(most recent call last):
  File "<pyshell#9>", line 1, in <module>
    2 / 0
ZeroDivisionError: division by zero
>>>'a'+2                                    #操作数类型不支持
Traceback(most recent call last):
```

```
    File "<pyshell#10>", line 1, in <module>
        'a'+2
TypeError: Can't convert 'int' object to str implicitly
>>>{3, 4, 5} * 3                          #操作数类型不支持
Traceback(most recent call last):
    File "<pyshell#11>", line 1, in <module>
        {3, 4, 5} * 3
TypeError: unsupported operand type(s)for *:'set' and 'int'
>>>print(testStr)                         #变量名不存在
Traceback(most recent call last):
    File "<pyshell#12>", line 1, in <module>
        print(testStr)
NameError: name 'testStr' is not defined
>>>fp=open(r'D:\test.data', 'rb')         #文件不存在
Traceback(most recent call last):
    File "<pyshell#13>", line 1, in <module>
        fp=open(r'D:\test.data', 'rb')
FileNotFoundError: [Errno 2] No such file or directory: 'D:\\test.data'
>>>len(3)                                 #参数类型不匹配
Traceback(most recent call last):
    File "<pyshell#16>", line 1, in <module>
        len(3)
TypeError: object of type 'int' has no len()
>>>list(3)                                #参数类型不匹配
Traceback(most recent call last):
    File "<pyshell#17>", line 1, in <module>
        list(3)
TypeError: 'int' object is not iterable
```

通过前面章节的学习，大家应该已经注意到，如果类似于上面这些的错误得不到正确的处理将会导致程序崩溃并终止运行。合理地使用异常处理结构可以使得程序更加健壮，具有更高的容错性，不会因为用户不小心的错误输入而造成程序终止，也可以使用异常处理结构为用户提供更加友好的提示。

异常处理是因为程序执行过程中出错而在正常控制流之外采取的行为。严格来说，语法错误和逻辑错误不属于异常，但有些语法错误往往会导致异常，例如，由于大小写拼写错误而试图访问不存在的对象，或者试图访问不存在的文件，等等。当 Python 检测到一个错误时，解释器就会指出当前程序流已无法继续执行下去，这时候就出现了异常。当程序执行过程中出现错误时 Python 会自动引发异常，程序员也可以通过 raise 语句显式地引发异常。

> **注意**：尽管异常处理机制非常重要也非常有效，但是不建议使用异常来代替常规的检查，例如必要的 if…else 判断等。在编程时应避免过多依赖于异常处理机制来提高程序的健壮性。

🔖 **小常识**：异常和错误这两个概念并不完全一样。异常一般是指运行时由于某些条件不符合而引发的错误，一旦引发异常并且没有得到有效的处理，一般是直接导致程序崩溃。错误一般又可以分为语法错误和逻辑错误两种。拼写错误、缩进不一致、引号或括号不闭合等都属于语法错误，一般来说存在语法错误的代码是无法运行的，这类错误很容易发现和解决；而存在逻辑错误的代码通常可以运行，只是非常可能会得到一个错误的结果，这类错误非常难发现。

🔖 **常见疑问解答**：为啥要使用异常处理结构呢？简单地说，可以理解为"请求原谅比请求允许要容易"。也就是说，有些代码执行可能会出现错误，也可能不会出现错误，这主要由运行时的各种客观因素决定，此时建议使用异常处理结构。如果使用大量的选择结构来提前判断，仅当满足相应条件时才执行该代码，这些条件判断可能会严重干扰正常的业务逻辑，也会严重降低代码的可读性。

7.1.2 Python内置异常类层次结构

下面全面展示了Python内建异常类的继承层次，其中BaseException是所有内置异常类的基类。

```
BaseException
 +--SystemExit
 +--KeyboardInterrupt
 +--GeneratorExit
 +--Exception
      +--StopIteration
      +--ArithmeticError
      |    +--FloatingPointError
      |    +--OverflowError
      |    +--ZeroDivisionError
      +--AssertionError
      +--AttributeError
      +--BufferError
      +--EOFError
      +--ImportError
      +--LookupError
      |    +--IndexError
      |    +--KeyError
      +--MemoryError
      +--NameError
      |    +--UnboundLocalError
      +--OSError
      |    +--BlockingIOError
      |    +--ChildProcessError
      |    +--ConnectionError
```

```
      |      |      +--BrokenPipeError
      |      |      +--ConnectionAbortedError
      |      |      +--ConnectionRefusedError
      |      |      +--ConnectionResetError
      |      +--FileExistsError
      |      +--FileNotFoundError
      |      +--InterruptedError
      |      +--IsADirectoryError
      |      +--NotADirectoryError
      |      +--PermissionError
      |      +--ProcessLookupError
      |      +--TimeoutError
      +--ReferenceError
      +--RuntimeError
      |      +--NotImplementedError
      +--SyntaxError
      |      +--IndentationError
      |             +--TabError
      +--SystemError
      +--TypeError
      +--ValueError
      |      +--UnicodeError
      |             +--UnicodeDecodeError
      |             +--UnicodeEncodeError
      |             +--UnicodeTranslateError
      +--Warning
             +--DeprecationWarning
             +--PendingDeprecationWarning
             +--RuntimeWarning
             +--SyntaxWarning
             +--UserWarning
             +--FutureWarning
             +--ImportWarning
             +--UnicodeWarning
             +--BytesWarning
             +--ResourceWarning
```

7.1.3 常见异常处理结构形式

Python 提供了多种不同形式的异常处理结构，基本思路都是一致的：先尝试运行代码，然后处理可能发生的错误。在实际使用时，可以根据需要来选择使用哪一种。

1．try…except…

Python 异常处理结构中最基本的结构是 try…except…结构。其中 try 子句中的代码块包含可能会引发异常的语句,而 except 子句用来捕捉相应的异常。如果 try 子句中的代码引发异常并被 except 子句捕捉,则执行 except 子句的代码块;如果 try 中的代码块没有出现异常则继续往下执行异常处理结构后面的代码;如果出现异常但没有被 except 捕获,则继续往外层抛出;如果所有层都没有捕获并处理该异常,则程序崩溃并将该异常呈现给最终用户,这是我们最不希望发生的事情。该结构语法如下:

```
try:
    #可能会引发异常的代码,先执行一下试试
except Exception[ as reason]:
    #如果 try 中的代码抛出异常并被 except 捕捉,就执行这里的代码
```

例如,下面的代码用来接收用户输入,并且要求用户必须输入整数,而不接收其他类型的输入。

```
>>>while True:
    x=input('Please input:')
    try:
        x=int(x)
        print('You have input {0}'.format(x))
        break
    except Exception as e:
        print('Error.')
Please input:a
Error.
Please input:b
Error.
Please input:234c
Error.
Please input:5
You have input 5
```

小提示:一般而言,应避免捕捉 Python 异常类的基类 BaseException,而应该明确指定捕捉哪种异常,然后有针对性地编写相应的处理代码。

拓展知识:回调函数原理。6.4.1 节介绍 shutil 模块时曾经提到过回调函数的概念,不知道大家有没有想过什么是回调函数,回调函数又是怎么实现的呢?回调函数的定义与普通函数并没有本质的区别,但一般不直接调用,而是作为参数传递给另一个函数,当另一个函数中触发了某个事件、满足了某个条件时就会自动调用回调函数。前面例 6-23 介绍了如何删除包含只读属性文件的文件夹,下面的代码实现同样的功能,主要用来演示回调函数的原理。

```
import os
import stat
```

```
def remove_readonly(func, path):            #定义回调函数
    os.chmod(path, stat.S_IWRITE)           #删除文件的只读属性
    func(path)                              #再次调用刚刚失败的函数

def del_dir(path, onerror=None):
    for file in os.listdir(path):
        file_or_dir=os.path.join(path,file)
        if os.path.isdir(file_or_dir) and not os.path.islink(file_or_dir):
            del_dir(file_or_dir)            #递归删除子文件夹及其文件
        else:
            try:
                os.remove(file_or_dir)      #尝试删除该文件
            except:                         #删除失败
                if onerror and callable(onerror):
                    onerror(os.remove, file_or_dir)
                                            #自动调用回调函数
                else:
                    print('You have an exception but did not capture it.')
    os.rmdir(path)                          #删除文件夹

del_dir("E:\\old", remove_readonly)         #调用函数,指定回调函数
```

2. try…except…else…

带有 else 子句的异常处理结构可以看作是一种特殊的选择结构,如果 try 中的代码抛出了异常并且被某个 except 语句捕捉则执行相应的异常处理代码,这种情况下就不会执行 else 中的代码;如果 try 中的代码没有抛出异常,则执行 else 块的代码。该结构的语法如下:

```
try:
    #可能会引发异常的代码
except Exception [ as reason]:
    #用来处理异常的代码
else:
    #如果 try 子句中的代码没有引发异常,就继续执行这里的代码
```

例如,前面要求用户必须输入整数的代码也可以这样写:

```
>>>while True:
    x=input('Please input:')
    try:
        x=int(x)
    except Exception as e:
        print('Error.')
    else:
```

```
            print('You have input {0}'.format(x))
            break
Please input:a
Error.
Please input:b
Error.
Please input:888c
Error.
Please input:888
You have input 888
```

3. try…except…finally…

在这种结构中,无论 try 中的代码是否发生异常,也不管抛出的异常有没有被 except 语句捕获,finally 子句中的代码总是会得到执行。因此,finally 中的代码常用来做一些清理工作以释放 try 子句中申请的资源。该结构语法为

```
try:
    #可能会引发异常的代码
except Exception [ as reason]:
    #处理异常的代码
finally:
    #无论 try 子句中的代码是否引发异常,都会执行这里的代码
```

例如下面的代码,不论是否发生异常,finally 子句中的代码总是被执行。

```
>>>def div(a, b):
    try:
        print(a/b)
    except ZeroDivisionError:
        print('The second parameter cannot be 0.')
    finally:
        print(-1)
>>>div(3, 5)
0.6
-1
>>>div(3, 0)
The second parameter cannot be 0.
-1
```

注意:如果 try 子句中的异常没有被 except 语句捕捉和处理,或者 except 子句或 else 子句中的代码抛出了异常,那么这些异常将会在 finally 子句执行完后再次抛出,例如:

```
>>>def div(a, b):
```

```
        try:
            print(a/b)
        except ZeroDivisionError:
            print('The second parameter cannot be 0.')
        finally:
            print(-1)
>>>div('3', 5)
-1
Traceback(most recent call last):
  File "<pyshell#60>", line 1, in <module>
    div('3', 5)
  File "<pyshell#59>", line 3, in div
    print(a/b)
TypeError: unsupported operand type(s)for /: 'str' and 'int'
```

注意：finally 子句中的代码也可能会引发异常。下面代码的本意是使用异常处理结构来避免文件对象没有关闭的情况发生，但是由于指定的文件不存在而导致打开失败，结果在 finally 子句中关闭文件时引发了异常。

```
>>>try:
    f1=open('test1.txt', 'r')           #文件不存在,抛出异常,不会创建文件对象 f1
    line=f1.readline()                  #后面的代码不会被执行
    print(line)
except SyntaxError:                     #这个 except 并不能捕捉上面的异常
    print('Sth wrong')
finally:
    f1.close()                          #f1 不存在,再次引发异常

Traceback(most recent call last):
  File "<pyshell#75>", line 2, in <module>
    f1=open('test1.txt', 'r')
FileNotFoundError: [Errno 2] No such file or directory: 'test1.txt'

During handling of the above exception, another exception occurred:
Traceback(most recent call last):
  File "<pyshell#75>", line 8, in <module>
    f1.close()
NameError: name 'f1' is not defined
```

注意：如果在函数中使用异常处理结构，尽量不要在 finally 子句中使用 return 语句，以免发生非常难以发现的逻辑错误。例如下面的代码，不管参数是否符合函数要求，调用函数时都得到了同样的错误信息。

```
>>>def div(a, b):
    try:
```

```
        return a/b
    except ZeroDivisionError:
        return 'The second parameter cannot be 0.'
    finally:
        return 'Error'
>>>div(3,5)
'Error'
>>>div('3', 5)
'Error'
>>>div(3, 0)
'Error'
```

4．可以捕捉多种异常的异常处理结构

在实际开发中，同一段代码可能会抛出多种异常，并且需要针对不同的异常类型进行相应的处理。为了支持多种异常的捕捉和处理，Python 提供了带有多个 except 的异常处理结构，一旦某个 except 捕捉到了异常，则其他的 except 子句将不会再尝试捕捉异常。该结构类似于多分支选择结构，语法格式为

```
try:
    #可能会引发异常的代码
except Exception1:
    #处理异常类型 1 的代码
except Exception2:
    #处理异常类型 2 的代码
except Exception3:
    #处理异常类型 3 的代码
 ⋮
```

下面的代码演示了这种异常处理结构的用法，连续运行 3 次并输入不同的数据，结果如下：

```
>>>try:
    x=float(input('请输入被除数：'))
    y=float(input('请输入除数：'))
    z=x / y
except ZeroDivisionError:
    print('除数不能为零')
except TypeError:
    print('被除数和除数应为数值类型')
except NameError:
    print('变量不存在')
else:
    print(x, '/', y, '=', z)
```

```
请输入被除数：30                    #第一次运行
请输入除数：5
30.0 / 5.0=6.0

请输入被除数：30                    #第二次运行，略去重复代码
请输入除数：abc
Traceback(most recent call last):
  File "<pyshell#95>", line 3, in <module>
    y=float(input('请输入除数：'))
ValueError: could not convert string to float: 'abc'

请输入被除数：30                    #第三次运行，略去重复代码
请输入除数：0
除数不能为零
```

在实际开发中，有时候可能会为几种不同的异常设计相同的异常处理代码（虽然这种情况很少）。为了减少代码量，Python 允许把多个异常类型放到一个元组中，然后使用一个 except 子句同时捕捉多种异常，并且共用同一段异常处理代码，例如：

```
>>>try:
    x=float(input('请输入被除数：'))
    y=float(input('请输入除数：'))
    z=float(x)/y
except(ZeroDivisionError, TypeError, NameError):
    print('捕捉到了异常')
else:
    print(x, '/', y, '=', z)

请输入被除数：30
请输入除数：0
捕捉到了异常
```

5. 同时包含 else 子句、finally 子句和多个 except 子句的异常处理结构

Python 异常处理结构中可以同时包含多个 except 子句、else 子句和 finally 子句，例如：

```
>>>def div(x, y):
    try:
        print(x / y)
    except ZeroDivisionError:
        print('ZeroDivisionError')
    except TypeError:
        print('TypeError')
    else:
```

```
        print('No Error')
    finally:
        print("executing finally clause")

>>>div(3,5)
0.6
No Error
executing finally clause
>>>div('3',5)
TypeError
executing finally clause
>>>div(3,0)
ZeroDivisionError
executing finally clause
```

拓展知识：断言语句 assert 也是一种比较常用的技术，常用来在程序的某个位置确认指定条件必须满足，常和异常处理结构一起使用。断言语句 assert 仅当脚本的__debug__属性值为 True 时有效，一般只在开发和测试阶段使用。当使用-O 选项把 Python 程序编译为字节码文件时，assert 语句将被删除。

```
>>>a=3
>>>b=5
>>>assert a==b, 'a must be equal to b'
Traceback(most recent call last):
  File "<pyshell#17>", line 1, in <module>
    assert a==b, 'a must be equal to b'
AssertionError: a must be equal to b
>>>try:
    assert a==b, 'a must be equal to b'
except AssertionError as reason:
    print('%s:%s'%(reason.__class__.__name__, reason))
AssertionError:a must be equal to b
```

7.2 代 码 测 试

代码测试不仅是测试团队的任务，更是开发人员自己的义务。尽管错误无法完全避免，一个优秀的程序员在编写代码时还是应该尽量减少代码中潜在的错误（优秀的程序员还会考虑代码的易测试性），测试人员一旦发现问题一定要经过反复确认之后再和开发人员沟通，并且双方要注意沟通的方式和语气。敢于质疑不代表可以粗鲁无礼，正如哈佛大学著名心理学家威廉·詹姆斯所说："思考可以随心所欲，表达想法则必须谨慎小心。"毕竟每个人都爱面子，开发人员不希望被人说自己写的代码不好，测试人员也不希望自己的努力被人说没意义。良言一句三冬暖，恶语伤人六月寒。必须要"以德服人"(每当看到这个词,我就想起来电影《方世玉》中的雷老虎),就事论事,千万不要说些伤感情的话。

建议:作为一个小建议,团队成员之间进行沟通时应该记住以下几点:①言之有物,切忌胡搅蛮缠、无理取闹,一定要针对核心问题进行有效沟通并争取尽快达成一致;②适当赞美、保持微笑,即使确实发现了对方代码或思路存在问题,也不可一副小人得志的样子,得理不饶人,要善于发现对方的优点和可取之处;③注意场合、话留三分,即使和对方私交非常好,也尽量不要在正式场合开没有分寸的玩笑,要考虑其他人的感受;④适当沉默,让对方把话说完,切不可随意打断别人的话,一定要让对方清晰、完整地表达完自己的想法之后再进一步沟通。

7.2.1 doctest

Python 标准库 doctest 可以搜索程序中类似于交互式 Python 代码的文本片段,并运行这些交互式代码来验证是否符合预期结果和功能,常用于 Python 程序的模块测试。

例 7-1 使用 doctest 模块测试 Python 代码。

下面的代码演示了 doctest 模块的用法,定义了一个函数,预期功能为可以对整数或实数相加,或连接 2 个字符串、列表、元素,或对两个集合求并集,并返回结果。

```
def add(value1, value2):
    #下面三个单引号之间是测试代码,doctest 会搜索这些代码并执行
    #并且根据执行结果与预期结果的匹配程度来测试代码是否正确
    '''return the addition of two numbers or the concatenation of two string/list/
    tuple
    >>>add(3, 5)
    8
    >>>add(3.0, 5.0)
    8.0
    >>>add([1,2], [3, 4])
    [1, 2, 3, 4]
    >>>add((1,),(2, 3, 4))
    (1, 2, 3, 4)
    >>>add(1, [3])
    Traceback(most recent call last):
        ...
    TypeError: value1 and value2 must be of the same type
    >>>add(1, '2')
    Traceback(most recent call last):
        ...
    TypeError: value1 and value2 must be of the same type
    >>>add([1],(2,))
    Traceback(most recent call last):
```

```
        ...
    TypeError: value1 and value2 must be of the same type
    >>>add('1234', [1,2,3,4])
    Traceback(most recent call last):
        ...
    TypeError: value1 and value2 must be of the same type
    >>>add({1,2,3}, {3,4,5})
    {1, 2, 3, 4, 5}
    >>>add({1:1}, {2:2})
    Traceback(most recent call last):
        ...
    TypeError: value1 and value2 must be the type of int,float,str,list,tuple
    or set
    '''
    #下面是正式的功能代码
    if type(value1)not in(int, float, str, list, tuple, set):
        raise TypeError('value1 and value2 must be the type of int,float,str,
        list,tuple or set')
    if type(value1)!=type(value2):
        raise TypeError('value1 and value2 must be of the same type')
    if type(value1)==set:
        return value1 | value2
    else:
        return value1+value2

if __name__=="__main__":
    import doctest
    doctest.testmod()
    print(add(3,5))
```

把上面的代码保存成 Python 程序文件 doctest_demo.py，在 IDLE 中直接运行，如果函数功能完全符合预期的功能要求就会输出正确的结果，如果有不符合预期结果的代码就会给出相应的提示。在命令提示符环境中使用带-v 参数的方式执行，可以看到详细的测试过程，如图 7-1 所示。

7.2.2 单元测试

软件测试对于保证软件质量非常重要，尤其是升级过程中对代码的改动不应该影响系统的原有功能，是未来重构代码的信心保证。一般来说稍微有些规模的软件公司都有专门的测试团队来保证软件质量，但作为程序员，首先应该保证自己编写的代码准确无误地实现了预定功能。

软件测试方法有很多，从软件工程角度来讲，可以分为白盒测试和黑盒测试两大类。其中，白盒测试主要通过阅读程序源代码来判断是否符合功能要求，对于复杂的业务逻辑

第 7 章 异常处理结构、代码测试与调试　265

```
C:\Python 3.5>python doctest_demo.py
8

C:\Python 3.5>python doctest_demo.py -v
Trying:
    add(3, 5)
Expecting:
    8
ok
Trying:
    add(3.0, 5.0)
Expecting:
    8.0
ok
Trying:
    add([1,2], [3, 4])
Expecting:
    [1, 2, 3, 4]
ok
Trying:
    add((1,), (2, 3, 4))
Expecting:
    (1, 2, 3, 4)
ok
Trying:
    add(1, [3])
Expecting:
    Traceback (most recent call last):
        ...
    TypeError: value1 and value2 must be of the same type
ok
Trying:
    add(1, '2')
Expecting:
```

图 7-1　doctest 测试过程示意图

白盒测试难度非常大,一般以黑盒测试为主,白盒测试为辅。黑盒测试不关心模块的内部实现方式,只关心其功能是否正确,通过精心设计一些测试用例来检验模块的输入和输出是否正确,最终判断其是否符合预定的功能要求。

单元测试是保证模块质量的重要手段之一,通过单元测试来管理设计好的测试用例,不仅可以避免测试过程中人工反复输入可能引入的错误,还可以重复利用设计好的测试用例,具有很好的可扩展性,大幅度缩短代码的测试时间。Python 标准库 unittest 提供了大量用于单元测试的类和方法,其中最常用的是 TestCase 类,其常用方法如表 7-1 所示。

表 7-1　TestCase 类的常用方法

方 法 名 称	功 能 说 明	方 法 名 称	功 能 说 明
assertEqual(a, b)	a == b	assertNotEqual(a, b)	a != b
assertTrue(x)	bool(x)is True	assertFalse(x)	bool(x)is False
assertIs(a, b)	a is b	assertIsNot(a, b)	a is not b
assertIsNone(x)	x is None	assertIsNotNone(x)	x is not None
assertIn(a, b)	a in b	assertNotIn(a, b)	a not in b
assertIsInstance(a, b)	isinstance(a, b)	assertNotIsInstance(a, b)	not isinstance(a, b)

续表

方 法 名 称	功 能 说 明	方 法 名 称	功 能 说 明
assertAlmostEqual(a, b)	round(a−b, 7)==0	assertNotAlmostEqual(a, b)	round(a−b, 7)!=0
assertGreater(a, b)	a > b	assertGreaterEqual(a, b)	a >= b
assertLess(a, b)	a < b	assertLessEqual(a, b)	a <= b
assertRegex(s, r)	r.search(s)	assertNotRegex(s, r)	not r.search(s)
setUp()	每项测试开始之前自动调用该函数	tearDown()	每项测试完成之后自动调用该函数

其中，setUp()和 tearDown()这两个方法比较特殊，分别在每个测试之前和之后自动调用，常用来执行数据库连接的创建与关闭、文件的打开与关闭等操作，避免编写过多的重复代码。

例 7-2 编写单元测试程序。

以第 4 章自定义栈的代码为例，演示如何利用 unittest 库对 Stack 类中入栈、出栈、改变大小以及满/空测试等方法进行测试，并将测试结果写入文件 test_Stack_result.txt。

```python
#要测试的模块,在本书第 4 章
import Stack
#Python 单元测试标准库
import unittest

class TestStack(unittest.TestCase):
    def setUp(self):
        #测试之前以追加模式打开指定文件
        self.fp=open('D:\\test_Stack_result.txt', 'a+')

    def tearDown(self):
        #测试结束后关闭文件
        self.fp.close()

    def test_isEmpty(self):
        try:
            s=Stack.Stack()
            #确保函数返回结果为 True
            self.assertTrue(s.isEmpty())
            self.fp.write('isEmpty passed\n')
        except Exception as e:
            self.fp.write('isEmpty failed\n')

    def test_empty(self):
        try:
```

```python
            s=Stack.Stack(5)
            for i in ['a', 'b', 'c']:
                s.push(i)
            #测试清空栈操作是否工作正常
            s.empty()
            self.assertTrue(s.isEmpty())
            self.fp.write('empty passed\n')
        except Exception as e:
            self.fp.write('empty failed\n')

    def test_isFull(self):
        try:
            s=Stack.Stack(3)
            s.push(1)
            s.push(2)
            s.push(3)
            self.assertTrue(s.isFull())
            self.fp.write('isFull passed\n')
        except Exception as e:
            self.fp.write('isFull failed\n')

    def test_pushpop(self):
        try:
            s=Stack.Stack()
            s.push(3)
            #确保入栈后立刻出栈得到原来的元素
            self.assertEqual(s.pop(), 3)
            s.push('a')
            self.assertEqual(s.pop(), 'a')
            self.fp.write('push and pop passed\n')
        except Exception as e:
            self.fp.write('push or pop failed\n')

    def test_setSize(self):
        try:
            s=Stack.Stack(8)
            for i in range(8):
                s.push(i)
            self.assertTrue(s.isFull())
            #测试扩大栈空间是否正常工作
            s.setSize(9)
            s.push(8)
            self.assertTrue(s.isFull())
            self.assertEqual(s.pop(), 8)
```

```
            #测试缩小栈空间是否正常工作
            s.setSize(4)
            self.assertTrue(s.isFull())
            self.assertEqual(s.pop(), 3)
            self.fp.write('setSize passed\n')
        except Exception as e:
            self.fp.write('setSize failed\n')

if __name__ == '__main__':
    unittest.main()
```

注意：①测试用例的设计应该是完备的，应保证覆盖尽可能多的情况，尤其是要覆盖边界条件，对目标模块的功能进行充分测试，避免漏测；②测试用例以及测试代码本身也可能会存在 Bug，通过测试并不代表目标代码没有错误，但是一般而言，不能通过测试的模块代码是存在问题的；③再好的测试方法和测试用例也无法保证能够发现所有错误，只能通过改进和综合多种测试方法并且精心设计测试用例来发现尽可能多的潜在问题；④除了功能测试，还应对程序进行性能测试与安全性测试，甚至还需要进行规范性测试以保证代码可读性和可维护性。

拓展知识：在工程界，不管是安全专家还是恶意攻击者，最常使用的漏洞发现和挖掘方法是 Fuzz，属于"灰"盒测试技术，也可以说是一种特殊的黑盒测试技术。Fuzz 的主要目的是 crash、break 和 destroy，Fuzz 的测试用例往往是带有攻击性的畸形数据，用来触发各种类型的潜在漏洞。

拓展知识：有时候可能需要把代码执行过程中的一些调试信息、出错信息或其他信息记录下来而不影响正常的输出，这时可以使用 Python 标准库 logging 提供的功能。该模块提供的几个输出方法默认把信息输出到标准控制台 sys.stderr，可以修改这个值使得信息能够输出到文件中。

```
import sys
import logging


old=sys.stderr                                      #记下原输出目的地
fp=open('log_test.txt', 'a')                        #创建日志文件
sys.stderr=fp                                       #把信息输出到指定文件

logging.debug('Debugging information')              #输出信息到文件
logging.info('Informational message')               #debug()和info()的信息一般会被忽略
logging.warning('Warning:config file %s not found', 'server.conf')
logging.error('Error occurred')
logging.critical('Critical error—shutting down')


sys.stderr=old                                      #恢复标准控制台
fp.close()                                          #关闭文件
```

拓展知识：软件性能测试。在本书 2.4.1 节、2.4.3 节、3.2.3 节、5.1.2 节等多个章节中都演示过使用 Python 标准库 time 提供的 time() 函数来测试代码运行时间，在 5.1.2 节还演示过 timeit 模块的有关用法。除此之外，还可以使用下面的方法来测试代码的运行时间：

```python
from time import time

class Timer(object):
    def __enter__(self):
        self.start=time()
        return self

    def __exit__(self, *args):
        self.end=time()
        self.seconds=self.end-self.start

def isPrime(n):
    if n ==2:
        return True
    for i in range(2, int(n**0.5)+2):
        if n%i ==0:
            return False
    return True

with Timer() as t:
    for i in range(1000):
        isPrime(999999999999999999999)
print(t.seconds)
```

运行上面的程序，会输出 isPrime() 函数运行 1000 次所需要的时间。另外，在很多时候，除了要测试代码运行所需要的时间，还需要检测代码运行过程中的内存占用情况，这时候需要使用 pip 安装 Python 扩展库 memory_profiler，然后编写下面的代码：

```python
from memory_profiler import profile

@profile                                      #修饰器
def isPrime(n):
    if n ==2:
        return True
    for i in range(2, int(n**0.5)+2):
        if n%i ==0:
            return False
    return True
```

```
isPrime(99999999999999999999)
```

运行上面的程序,会得到下面的输出,从中可以清楚地看到代码对内存的使用情况:

```
Line #    Mem usage    Increment   Line Contents
================================================
    3     33.9 MiB     0.0 MiB     @profile
    4                              def isPrime(n):
    5     33.9 MiB     0.0 MiB         if n==2:
    6                                      return True
    7     33.9 MiB     0.0 MiB         for i in range(2, int(n**0.5)+2):
    8     33.9 MiB     0.0 MiB             if n%i==0:
    9     33.9 MiB     0.0 MiB                 return False
   10                                  return True
```

7.3 代码调试

7.3.1 使用 IDLE 调试

当程序运行发生错误或者得到了非预期的结果时,是否能够熟练地对程序进行调试并快速定位和解决问题是体现程序员综合能力的重要标准之一。

几乎任何一种集成开发环境都提供了代码调试功能,Python 标准开发环境 IDLE 也不例外。使用 IDLE 的调试功能时,首先单击 IDLE 的菜单 Debug→Debugger 打开调试器窗口,然后打开并运行要调试的程序,最后切换到调试器窗口使用其中的控制按钮进行调试。图 7-2 为 IDLE 调试窗口及其功能简要介绍,可以使用调试按钮对程序进行单步

图 7-2 IDLE 调试器窗口

执行,实时查看变量的当前值并跟踪其变化过程,对于理解程序内部的工作原理和发现程序中存在的问题非常有帮助。

例 7-3 使用 IDLE 调试 Python 程序。

假设有 Python 程序 demo.py,其功能为生成 1000 个随机字符(英语字母大小写或数字),然后查看某个字符的出现次数,代码如下:

```
import string
from random import choice

characters=string.ascii_letters+string.digits
selected=[choice(characters) for i in range(1000)]
ch=choice(selected)
print(ch, ':', selected.count(ch))
```

然后使用 IDLE 对该程序进行单步调试(使用 Step 按钮),调试过程中的部分截图如图 7-3～图 7-6 所示。可以发现,在调试过程中执行了很多不属于 demo.py 程序的代码,这是正常的,因为调用标准库函数时会自动进入标准库并执行其中的代码。如果不想进入和执行标准库代码,可以使用 Over 按钮。另外,上面代码的调试过程显示,列表推导式在本质上还是循环,只是形式比较简单而已。

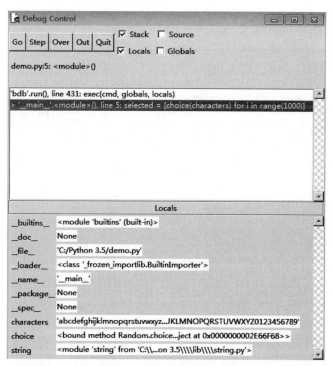

图 7-3 程序调试截图(一)

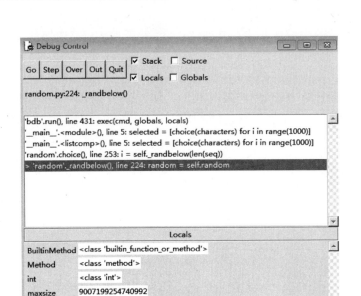

图 7-4　程序调试截图(二)

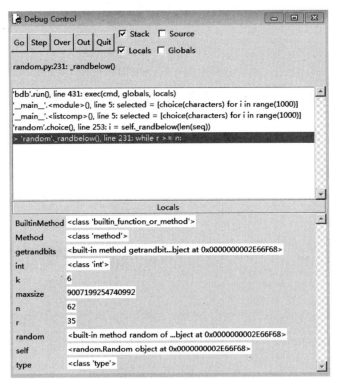

图 7-5　程序调试截图(三)

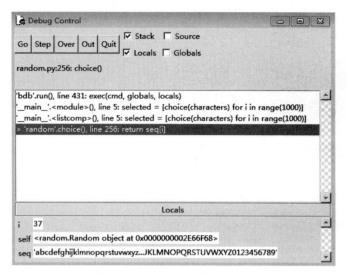

图 7-6　程序调试截图（四）

7.3.2　使用 pdb 调试

pdb 是 Python 自带的交互式源代码调试模块，源代码文件为 pdb.py，感兴趣的读者可以在 Python 安装目录下找到该文件进行阅读并理解其工作原理。pdb 模块提供了代码调试所需要的绝大部分功能，包括设置/清除(条件)断点、启用/禁用断点、单步执行、查看栈帧、查看变量值、查看当前执行位置、列出源代码、执行任意 Python 代码或表达式等。pdb 还支持事后调试，可在程序控制下被调用，并且可以通过 pdb 和 cmd 接口对该调试器进行扩展。pdb 模块常用调试命令如表 7-2 所示。

表 7-2　常用 pdb 调试命令

简写/完整命令	用法示例	解　释
a(rgs)		显示当前函数中的参数
b(reak) [[filename:]lineno | function [, condition]]	b 173	在 173 行设置断点
	b function	在 function 函数第一条可执行语句位置设置断点
	b	不带参数则列出所有断点，包括每个断点的触发次数、当前忽略计数以及与之关联的条件
	b 175，condition	设置条件断点，仅当 condition 的值为 True 时该断点有效
cl(ear) [filename:lineno | bpnumber [bpnumber …]]	cl	清除所有断点
	cl file:line	删除指定文件中指定行的所有断点
	cl 3 5 9	删除第 3、5、9 个断点

续表

简写/完整命令	用 法 示 例	解　　释
condition bpnumber [condition]	condition 3 a<b	仅当 a<b 时 3 号断点有效
	condition 3	将 3 号断点设置为无条件断点
continue		继续运行至下一个断点或脚本结束
disable [bpnumber [bpnumber …]]	disable 3 5	禁用第 3、5 个断点,禁用后断点仍存在,可以再次被启用
d(own)		在栈跟踪器中向下移动一个栈帧
enable [bpnumber [bpnumber …]]	enable n	启用第 n 个断点
h(elp)[command]		查看 pdb 帮助
ignore bpnumber [count]		为断点设置忽略计数,count 的默认值为 0。若某断点的忽略计数不为 0,则每次触发时自动减 1,当忽略计数为 0 时该断点处于活动状态
j(ump)	j 20	跳至第 20 行继续运行
l(ist)[first [, last]]	l	列出脚本清单,默认 11 行
	l m, n	列出从第 m 行到第 n 之间的脚本代码
	l m	列出从第 m 行开始的 11 行代码
n(ext)		执行下一条语句,遇到函数时不进入其内部
p(rint)	p i	打印变量 i 的值
q(uit)		退出 pdb 调试环境
r(eturn)		一直运行至当前函数返回
tbreak		设置临时断点,该类型断点只被中断一次,触发后该断点自动删除
step		执行下一条语句,遇到函数时进入其内部
u(p)		在栈跟踪器中向上移动一个栈帧
w(here)		查看当前栈帧
[!]statement		在 pdb 中执行语句,!与要执行的语句之间不需要空格,任何非 pdb 命令都被解释为 Python 语句并执行,甚至可以调用函数或修改当前上下文中变量的值
		直接回车则默认执行上一个命令

使用 pdb 模块调试 Python 代码的形式常见的有 3 种:在交互模式下调试特定的代码块,在程序中显式插入断点,把 pdb 作为模块来调试程序。

(1) 在交互模式下使用 pdb 模块提供的功能可以直接调试语句块、表达式、函数等多种脚本,常用的调试方法有 4 个。

① pdb.run(statement[, globals[, locals]])：调试指定语句，可选参数 globals 和 locals 用来指定代码执行的环境，默认是__main__模块的字典。

② pdb.runeval(expression[, globals[, locals]])：返回表达式的值，可选参数 globals 和 locals 的含义与上面的 run()函数一样。

③ pdb.runcall(function[, argument, …])：调试指定函数。

④ pdb.post_mortem([traceback])：进入指定 traceback 对象的事后调试模式，如果没有指定 traceback 对象，则使用当前正在处理的一个异常。

例如，下面的代码演示了如何调试一个函数，其中"(Pdb)"为提示符，在后面输入并执行前面表 7-2 中介绍的命令即可。

```
>>>import pdb
>>>def demo():
    from random import randint
    x=[randint(1,10)for i in range(20)]      #随机生成 20 个介于 1~10 的整数
    m=max(x)                                 #最大数
    r=[index for index, value in enumerate(x)if value ==m]
    print(r)                                 #输出最大数所在的下标

>>>pdb.runcall(demo)                         #调试函数
><pyshell#13>(2)demo()
(Pdb)n                                       #执行下一条语句
><pyshell#13>(3)demo()
(Pdb)n
><pyshell#13>(4)demo()
(Pdb)p x                                     #查看变量值
[9, 5, 8, 9, 7, 8, 10, 3, 8, 5, 2, 8, 8, 4, 9, 10, 8, 7, 5, 1]
(Pdb)p m
*** NameError: name 'm' is not defined
(Pdb)n
><pyshell#13>(5)demo()
(Pdb)p m
10
(Pdb)r                                       #运行函数直至结束
[6, 15]
--Return--
><pyshell#13>(6)demo()->None
(Pdb)l
[EOF]
(Pdb)p r
[6, 15]
(Pdb)p m
10
(Pdb)q                                       #退出调试模式
```

(2) 在程序中嵌入断点来实现调试功能。

在程序中首先导入 pdb 模块，然后使用 pdb.set_trace() 在需要的位置设置断点。如果程序中存在通过该方法调用显式插入的断点，那么在命令提示符环境下执行该程序或双击执行程序时将自动进行 pdb 调试模式，即使该程序当前不处于调试状态。例如，下面的程序 IsPrime.py：

```
import pdb

n=37
pdb.set_trace()
for i in range(2, n):
    if n%i==0:
        print('No')
        break
else:
    print('Yes')
```

由于使用 pdb 设置了断点，在 IDLE 中运行该程序时会自动打开调试模式，如图 7-7 所示。在命令提示符环境中运行该程序时也会自动进入 pdb 调试模式，如图 7-8 所示。

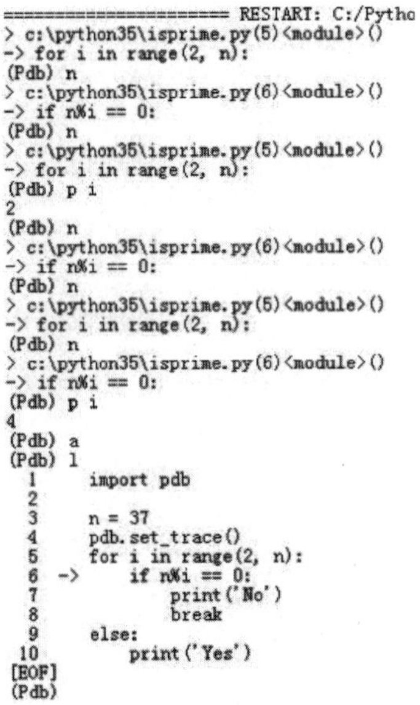

图 7-7 运行程序自动进行 pdb 调试模式　　图 7-8 在命令提示符环境运行程序

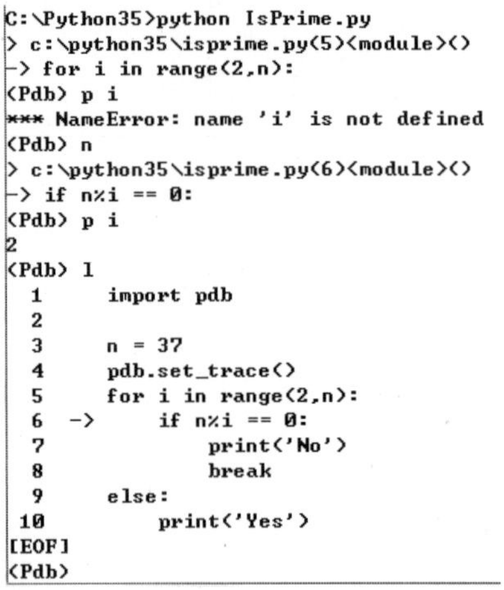

(3) 使用命令行调试程序。

在命令行提示符下执行"python -m pdb 脚本文件名"，可以直接进入调试环境，即使程序中并没有设置任何断点，也没有使用 pdb 的任何功能；当调试结束或程序正常结束以后，pdb 将重启该程序。例如，把上面的程序 IsPrime.py 中 pdb 模块的导入和断点插入函数都删除，然后在命令提示符环境中使用调试模式运行，如图 7-9 所示。

```
C:\Python35>python -m pdb IsPrime.py
> c:\python35\isprime.py(1)<module>()
-> n = 37
(Pdb) n
> c:\python35\isprime.py(2)<module>()
-> for i in range(2,n):
(Pdb) w
  c:\python35\lib\bdb.py(431)run()
-> exec(cmd, globals, locals)
  <string>(1)<module>()
> c:\python35\isprime.py(2)<module>()
-> for i in range(2,n):
(Pdb) n
> c:\python35\isprime.py(3)<module>()
-> if n%i == 0:
(Pdb) p i
2
(Pdb) p n
37
(Pdb)
```

图 7-9　使用命令行调试程序

阶段性寄语

　　看过《笑傲江湖》的朋友应该记得,华山派分为气宗和剑宗的事本来令狐冲以及其他华山气宗弟子是不知道的,后来剑宗打上门来并且重创气宗,这严重打击了华山弟子的自尊心和信心,导致气宗弟子们对自己练习的功夫产生了怀疑。虽然随着小说情节的推进,我们最终知道岳不群是个伪君子并且很鄙视他,但他在华山顶上对弟子们分析气宗和剑宗的一段话还是非常有道理的,大概意思是这样的:如果两个人同时学习武术,其中一个学习剑宗而另一个学习气宗,5年之内剑宗弟子远胜气宗弟子,10年之后不分伯仲,15年之后气宗弟子远胜剑宗弟子。内功的修炼就是这样的,修炼越久内功越深厚,功夫就越厉害(懂武术的朋友不要抬杠啊,咱不考虑那些练错甚至走火入魔的情况)。本书前7章可以看作 Python 内功,各位读者朋友应该反复修炼,必须达到"内三合"和"融会贯通"的境界。

　　看到这里读者可能会提出疑问,令狐冲后来没有了内功,机缘巧合得到风清扬传授独孤九剑,仅凭剑法也战胜了很多高手,甚至包括江南四友和魔教前任教主任我行,这又怎么解释呢?如果有这样的疑问那说明你看书或者电影不够仔细啊。令狐冲和江南四友比试时,向问天先忽悠(据说忽悠的英文翻译是 fool you)江南四友说不许使用内功,而令狐冲和任我行比试时,江南四友又施展激将法反复和任我行强调只比剑法不比内功,这样令狐冲才取胜的。如果允许使用内功,在绝对的实力面前,任何花哨的招式都是虚妄,最后任我行只是大声喊了一嗓子就把令狐冲和江南四友震晕了。

　　当然,只练内功不练招式也是不行的,再强的内功也必须要通过一定的外在形式表现出来,拳脚也好,器械也罢,神兵利器更佳。对于 Python 而言,利用各种标准库和扩展库开发各领域的应用程序就是表现 Python 内功的重要途径和形式,这也是本书接下来要重点介绍的内容。除了本书介绍的应用领域之外,Python 能做的还有很多很多,但是在一本书里展示 Python 的所有应用领域实在是不现实的,我只能控制住体内的洪荒之力,也给读者朋友一些空间去慢慢挖掘 Python 的强大功能。

　　另外,在阅读后面章节时除了 Python 内功和各种标准库与扩展库,你还需要另一种内功,那就是相关领域的专业知识。

第 8 章 数据库应用开发

毫无疑问,数据库技术的发展为各行各业都带来了很大的方便,数据库不仅支持各类数据的长期保存,更重要的是支持各种跨平台、跨地域的数据查询、共享以及修改,极大方便了人类生活和工作。电子邮箱、金融行业、聊天系统、各类网站、办公自动化系统、各种管理信息系统以及论坛、社区等,都少不了数据库技术的支持。另外,近些年来大数据相关技术的流行在一定程度上也促使了 NoSQL 数据库的快速发展。本书主要介绍 SQLite、Access、MySQL、MS SQL Server 等几种关系型数据库的 Python 接口,并通过几个示例来演示数据的增、删、改、查等操作。最后以 MongoDB 为例介绍了 Python 对 NoSQL 数据库的访问和操作。虽然不同数据库系统支持的 SQL 语法大致相同,但还是存在一定的差异的,读者可以根据自己使用的数据库系统查阅相应的 SQL 语法。

8.1 使用 Python 操作 SQLite 数据库

SQLite 是内嵌在 Python 中的轻量级、基于磁盘文件的数据库管理系统,不需要安装和配置服务器,支持使用 SQL 语句来访问数据库。该数据库使用 C 语言开发,支持大多数 SQL91 标准,支持原子的、一致的、独立的和持久的事务,不支持外键限制;通过数据库级的独占性和共享锁定来实现独立事务,当多个线程同时访问同一个数据库并试图写入数据时,每一时刻只有一个线程可以写入数据。

SQLite 支持最大 140TB 大小的单个数据库,每个数据库完全存储在单个磁盘文件中,以 B+树数据结构的形式存储,一个数据库就是一个文件,通过直接复制数据库文件就可以实现数据库的备份。如果需要使用可视化管理工具,可以下载并使用 SQLiteManager、SQLite Database Browser 或其他类似工具。

访问和操作 SQLite 数据时,需要首先导入 sqlite3 模块,然后创建一个与数据库关联的 Connection 对象,例如:

```
import sqlite3                              #导入模块
conn=sqlite3.connect('example.db')          #连接数据库
```

成功创建 Connection 对象以后,再创建一个 Cursor 对象,并且调用 Cursor 对象的 execute()方法来执行 SQL 语句创建数据表以及查询、插入、修改或删除数据库中的数据,例如:

```
c=conn.cursor()
#创建表
c.execute('''CREATE TABLE stocks(date text, trans text, symbol text, qty real,
price real)''')
#插入一条记录
c.execute("INSERT INTO stocks VALUES('2016-01-05','BUY','RHAT',100,35.14)")
#提交当前事务,保存数据
conn.commit()
#关闭数据库连接
conn.close()
```

如果需要查询表中内容,那么重新创建 Connection 对象和 Cursor 对象之后,可以使用下面的代码来查询。

```
for row in c.execute('SELECT * FROM stocks ORDER BY price'):
    print(row)
```

接下来重点介绍一下 sqlite3 模块中的 Connection、Cursor、Row 等对象。

8.1.1 Connection 对象

Connection 是 sqlite3 模块中最基本也是最重要的一个类,其主要方法如表 8-1 所示。

表 8-1 Connection 对象的主要方法

方 法	说 明
execute(sql[, parameters])	执行一条 SQL 语句
executemany(sql[, parameters])	执行多条 SQL 语句
cursor()	返回连接的游标
commit()	提交当前事务,如果不提交,那么自上次调用 commit()方法之后的所有修改都不会真正保存到数据库中
rollback()	撤销当前事务,将数据库恢复至上次调用 commit()方法后的状态
close()	关闭数据库连接
create_function(name, num_params, func)	创建可在 SQL 语句中调用的函数,其中 name 为函数名,num_params 表示该函数可以接收的参数个数,func 表示 Python 可调用对象

Connection 对象的其他几个函数都比较容易理解,下面的代码演示了如何在 sqlite3 连接中创建并调用自定义函数:

```
import sqlite3
import hashlib
```

```python
#自定义函数
def md5sum(t):
    return hashlib.md5(t).hexdigest()

#在内存中创建临时数据库
conn=sqlite3.connect(":memory:")
#创建可在SQL语句中调用的函数
conn.create_function("md5", 1, md5sum)
cur=conn.cursor()
#在SQL语句中调用自定义函数
cur.execute("SELECT md5(?)", ["中国山东烟台".encode()])
print(cur.fetchone()[0])
```

8.1.2 Cursor 对象

游标 Cursor 也是 sqlite3 模块中比较重要的一个类，下面简单介绍一下 Cursor 对象的常用方法。

1. execute(sql[，parameters])

该方法用于执行一条 SQL 语句，下面的代码演示了用法，以及为 SQL 语句传递参数的两种方法，分别使用问号和命名变量作为占位符。

```python
import sqlite3

conn=sqlite3.connect(":memory:")
cur=conn.cursor()
cur.execute("CREATE TABLE people(name_last, age)")
who="Dong"
age=38
#使用问号作为占位符
cur.execute("INSERT INTO people VALUES(?, ?)",(who, age))
#使用命名变量作为占位符
cur.execute("SELECT * FROM people WHERE name_last=:who AND age=:age",
            {"who": who, "age": age})
print(cur.fetchone())
```

运行结果如图 8-1 所示。

```
======================= RESTART: C:\Python 3.5\
('Dong', 38)
```

图 8-1　运行结果（一）

2. executemany(sql,seq_of_parameters)

该方法用来对于所有给定参数执行同一个 SQL 语句,参数序列可以使用不同的方式产生,例如,下面的代码使用迭代来产生参数序列:

```python
import sqlite3

#自定义迭代器,按顺序生成小写字母
class IterChars:
    def __init__(self):
        self.count=ord('a')
    def __iter__(self):
        return self
    def __next__(self):
        if self.count>ord('z'):
            raise StopIteration
        self.count +=1
        return(chr(self.count-1),)


conn=sqlite3.connect(":memory:")
cur=conn.cursor()
cur.execute("CREATE TABLE characters(c)")
#创建迭代器对象
theIter=IterChars()
#插入记录,每次插入一个英文小写字母
cur.executemany("INSERT INTO characters(c)VALUES(?)", theIter)
#读取并显示所有记录
cur.execute("SELECT c FROM characters")
print(cur.fetchall())
```

下面的代码则使用了更为简洁的生成器来产生参数:

```python
import sqlite3
import string

#包含 yield 语句的函数可以用来创建生成器对象
def char_generator():
    for c in string.ascii_lowercase:
        yield(c,)


conn=sqlite3.connect(":memory:")
cur=conn.cursor()
cur.execute("CREATE TABLE characters(c)")
#使用生成器对象得到参数序列
cur.executemany("INSERT INTO characters(c)VALUES(?)", char_generator())
```

```
cur.execute("SELECT c FROM characters")
print(cur.fetchall())
```

下面的代码则使用直接创建的序列作为 SQL 语句的参数：

```
import sqlite3

persons=[
        ("Hugo", "Boss"),
        ("Calvin", "Klein")
        ]
conn=sqlite3.connect(":memory:")
#创建表
conn.execute("CREATE TABLE person(firstname, lastname)")
#插入数据
conn.executemany(" INSERT INTO person (firstname, lastname) VALUES (?, ?)", persons)
#显示数据
for row in conn.execute("SELECT firstname, lastname FROM person"):
                        print(row)
print("I just deleted", conn.execute("DELETE FROM person").rowcount, "rows")
```

运行结果如图 8-2 所示。

```
======================= RESTART: C:\Python 
('Hugo', 'Boss')
('Calvin', 'Klein')
I just deleted 2 rows
```

图 8-2　运行结果（二）

3．fetchone()、fetchmany(size=cursor.arraysize)、fetchall()

这 3 个方法用来读取数据。假设数据库通过下面的代码创建并插入数据：

```
import sqlite3

conn=sqlite3.connect("D:/addressBook.db")
cur=conn.cursor()                          #创建游标
cur.execute('''INSERT INTO addressList(name , sex , phon , QQ , address)VALUES
('王小丫'，'女'，'13888997011'，'66735'，'北京市')''')
cur.execute('''INSERT INTO addressList(name, sex, phon, QQ, address)VALUES('李莉
','女', '15808066055', '675797', '天津市')''')
cur.execute('''INSERT INTO addressList(name, sex, phon, QQ, address)VALUES('李星
草', '男', '15912108090', '3232099', '昆明市')''')
conn.commit()                              #提交事务,把数据写入数据库
conn.close()
```

则下面的代码演示了使用 fetchall()读取数据的方法：

```python
import sqlite3

conn=sqlite3.connect('D:/addressBook.db')
cur=conn.cursor()
cur.execute('SELECT * FROM addressList')
li=cur.fetchall()                              #返回所有查询结果
for line in li:
    for item in line:
        print(item, end=' ')
    print()
conn.close()
```

注意：相信很多读者有 MS SQL Server、MySQL 或其他数据库的基础，在编写 SQL 语句时要注意，大多数 SQL 语法适用于多种类型的关系数据库，但不同数据库系统对某些特定操作的实现还是略有不同的。因此，如果某个按照以往经验写出的 SQL 语句无法执行或者结果与想象的不一样，可能需要查阅相关资料并对 SQL 做出相应的修改。

8.1.3 Row 对象

假设数据以下面的方式创建并插入数据：

```python
conn=sqlite3.connect("D:\\test.db")
c=conn.cursor()
c.execute('''CREATE TABLE stocks(date text, trans text, symbol text, qty real, price real)''')
c.execute("""INSERT INTO stocks VALUES ('2016-01-05','BUY','RHAT',100,35.14)""")
conn.commit()
c.close()
```

那么，可以使用下面的方式来读取其中数据：

```python
conn.row_factory=sqlite3.Row
c=conn.cursor()
c.execute('SELECT * FROM stocks')
r=c.fetchone()
print(type(r))
print(tuple(r))
print(r[2])
print(r.keys())
print(r['qty'])
for field in r:
    print(field)
```

8.2 使用 Python 操作其他关系型数据库

除了使用标准库 sqlite3 操作 SQLite 数据库以外，Python 还可以借助于功能强大的扩展库来操作 Access、MS SQL Server、MySQL 等多种类型的数据库，下面就简单介绍其中几个。

8.2.1 操作 Access 数据库

需要首先安装 Python for Windows extensions，即 Pywin32。然后可以参考下面的步骤和方式来访问 Access 数据库。

1. 建立数据库连接

```
import win32com.client
conn=win32com.client.Dispatch(r'ADODB.Connection')
DSN='PROVIDER=Microsoft.Jet.OLEDB.4.0;DATA SOURCE=C:/MyDB.mdb;'
conn.Open(DSN)
```

2. 打开记录集

```
rs=win32com.client.Dispatch(r'ADODB.Recordset')
rs_name='MyRecordset'                          #表名
rs.Open('['+rs_name+']', conn, 1, 3)
```

3. 操作记录集

```
rs.AddNew()
rs.Fields.Item(1).Value='data'
rs.Update()
```

4. 操作数据

```
conn=win32com.client.Dispatch(r'ADODB.Connection')
DSN='PROVIDER=Microsoft.Jet.OLEDB.4.0;DATA SOURCE=C:/MyDB.mdb;'
sql_statement="INSERT INTO [Table_Name]([Field_1], [Field_2])VALUES('data1','data2')"
conn.Open(DSN)
conn.Execute(sql_statement)
conn.Close()
```

5. 遍历记录

```
rs.MoveFirst()
```

```
count=0
while 1:
    if rs.EOF:
        break
    else:
        count=count+1
    rs.MoveNext()
```

在操作 Access 数据库时,如果一个记录集是空的,那么将指针移到第一个记录将导致一个错误,因为此时 RecordCount 是无效的。解决的方法是:打开一个记录集之前,先将 Cursorlocation 设置为 3,然后再打开记录集,此时 RecordCount 将是有效的。

```
rs.Cursorlocation=3
rs.Open('SELECT * FROM [Table_Name]', conn)   #确保 conn 处于打开状态
rs.RecordCount
```

8.2.2 操作 MS SQL Server 数据库

可以使用 pywin32、pymssql 和 pyodbc 等多种不同的方式来访问 MS SQL Server 数据库。

先来了解一下 pywin32 模块访问 MS SQL Server 数据库的步骤,如果下面的代码不能正常执行,很可能你还需要使用命令 pip install adodbapi 安装 adodbapi 扩展库。

1. 添加引用

```
import adodbapi
adodbapi.adodbapi.verbose=False              #adds details to the sample printout
import adodbapi.ado_consts as adc
```

2. 创建连接

```
Cfg={'server':'192.168.29.86\\eclexpress','password':'xxxx','db':'pscitemp'}
constr=r"Provider=SQLOLEDB.1; Initial Catalog=%s; Data Source=%s; user ID=%s; Password=%s;"%(Cfg['db'], Cfg['server'], 'sa', Cfg['password'])
conn=adodbapi.connect(constr)
```

3. 执行 sql 语句

```
cur=conn.cursor()
sql='''SELECT * FROM softextBook WHERE title='{0}' AND remark3!='{1}''''.format(bookName,flag)
cur.execute(sql)
data=cur.fetchall()
cur.close()
```

4. 执行存储过程

```
#假设 proName 有 3 个参数,最后一个参数传了 null
ret=cur.callproc('procName',(parm1,parm2,None))
conn.commit()
```

5. 关闭连接

```
conn.close()
```

接下来再通过一个示例来简单了解一下使用 pymssql 模块访问 MS SQL Server 数据库的方法,如果下面的代码提示无法导入 pymssql 模块,那么你很可能需要到 http://www.lfd.uci.edu/~gohlke/pythonlibs 下载与已安装 Python 版本对应的 pymssql 的 whl 文件,然后使用 pip 命令进行安装。

```
import pymssql
conn=pymssql.connect(host='SQL01', user='user', password='password',
                     database='mydatabase')
cur=conn.cursor()
cur.execute('CREATE TABLE persons(id INT, name VARCHAR(100))')
cur.executemany("INSERT INTO persons VALUES(%d, xinos.king)",
                [(1, 'John Doe'),(2, 'Jane Doe')])
conn.commit()
cur.execute('SELECT * FROM persons WHERE salesrep=xinos.king', 'John Doe')
row=cur.fetchone()
while row:
    print("ID=%d, Name=xinos.king" %(row[0], row[1]))
    row=cur.fetchone()
cur.execute("SELECT * FROM persons WHERE salesrep LIKE 'J%'")
conn.close()
```

最后让我们一起看看如何使用 pyodbc 扩展库读取 MS SQL Server 2008 数据库中的信息,如果下面的代码提示无法导入 pyodbc,请登录 http://www.lfd.uci.edu/~gohlke/pythonlibs 下载相应的 whl 文件之后再安装。

```
import pyodbc

s='DRIVER={SQL Server};SERVER=.;DATABASE=Test;UID=sa;PWD=test.'
conn=pyodbc.connect(s)
cur=conn.cursor()
cur.execute('SELECT * FROM yonghubiao')
row=cur.fetchone()
while row:
    print(row)
    row=cur.fetchone()
conn.close()
```

拓展知识：SQL 注入式攻击与防范。数据库广泛应用于各种场合，例如论坛、电子邮箱、社区、办公系统、银行、证券、游戏等。由于 B/S 模式的 Web 系统安装和配置简单，对客户端要求低，所以得到了越来越广泛的应用。在开发 Web 应用时，一定要注意防范 SQL 注入式攻击。网页表单接收用户输入之后把用户输入与代码中的 SQL 语句连接成完整的 SQL 语句，然后再提交数据库执行，这是比较常见的用法。如果恶意用户精心构造一些特殊的输入从而产生畸形但合法的 SQL 语句，就有可能导致 SQL 注入式攻击，可能会对服务器造成非常大的伤害，轻则导致私密信息泄露，重则严重危害服务器安全。所有类型的数据库都存在被 SQL 注入式攻击的危险，其根源在于允许把用户的输入与已有的不完整 SQL 语句进行连接，而精髓在于巧妙地构造输入来改变 SQL 语句的执行，最有效的防范措施则是培养和提高程序员的安全意识并对用户输入进行严格过滤，或者直接使用参数化查询而不是将用户输入和 SQL 语句简单地进行连接。经过大牛和前辈们十多年的研究，已经形成了一套完整而成熟的 SQL 注入式攻击体系，已经把 SQL 语句的编写发展成为一门艺术，当然同时也研究出了有效的防范措施和技术。关于 SQL 注入式攻击更加详细完整的内容请查阅有关资料，这里只举个简单但是已经过时的例子。例如，用户登录页面的 SQL 语句一般程序员会写成

```
sqlLogin="SELECT COUNT(id) FROM users WHERE userID='"+userName+"' AND userPwd='"+userPassword+"'"
```

如果用户按照正常操作进行输入，这个代码是没有问题的，但是如果用户输入任意密码(如 abc)而输入 admin' or 1=1-- 作为用户名(其中--是 MS SQL Server 的单行注释符)，上面的代码就变成了

```
sqlLogin="SELECT COUNT(id) FROM users WHERE userID='admin' OR 1=1--AND userPwd='abc'"
```

这样一来，"--"后面的代码将不被执行，而前面的代码中因为有 1=1 这样的条件总是成立，所以可能在不知道真实密码的情况下以任意账号身份进行登录。当然，这样简单的攻击手段早已失效，大家就不要尝试了。

8.2.3 操作 MySQL 数据库

Python 访问 MySQL 数据库可以使用 MySQLDb 模块，该模块主要方法如下。

(1) commit()：提交事务。
(2) rollback()：回滚事务。
(3) callproc(self, procname, args)：用来执行存储过程，接收的参数为存储过程名和参数列表，返回值为受影响的行数。
(4) execute(self, query, args)：执行单条 SQL 语句，接收的参数为 SQL 语句本身和使用的参数列表，返回值为受影响的行数。
(5) executemany(self, query, args)：执行单条 SQL 语句，但是重复执行参数列表里的参数，返回值为受影响的行数。

（6）nextset(self)：移到下一个结果集。

（7）fetchall(self)：接收全部的返回结果行。

（8）fetchmany(self，size=None)：接收 size 条返回结果行，如果 size 的值大于返回的结果行的数量，则会返回 cursor.arraysize 条数据。

（9）fetchone(self)：返回一条结果行。

（10）scroll(self，value，mode='relative')：移动指针到某一行，如果 mode='relative'，则表示从当前所在行移动 value 条记录；如果 mode='absolute'，则表示从结果集的第一行移动 value 条记录。

使用该模块查询 MySQL 数据库记录的方法如下面的代码所演示：

```
import MySQLdb
try:
    conn=MySQLdb.connect(host='localhost',user='root',passwd='root',
                        db='test',port=3306)
    cur=conn.cursor()
    cur.execute('SELECT * FROM user')
    cur.close()
    conn.close()
except MySQLdb.Error as e:
    print("Mysql Error %d: %s" %(e.args[0], e.args[1]))
```

插入数据的用法如下面的代码所演示：

```
import MySQLdb
try:
    conn=MySQLdb.connect(host='localhost',user='root',passwd='root',port=
                        3306)
    cur=conn.cursor()
    cur.execute('CREATE DATABASE IF NOT EXISTS python')
    conn.select_db('python')
    cur.execute('CREATE TABLE test(id int,info varchar(20))')
    value=[1,'hi rollen']
    cur.execute('INSERT INTO test VALUES(%s,%s)',value)
    values=[]
    for i in range(20):
        values.append((i,'hi rollen'+str(i)))
    cur.executemany('INSERT INTO test VALUES(%s,%s)',values)
    cur.execute('UPDATE test SET info="I am rollen" WHERE id=3')
    conn.commit()
    cur.close()
    conn.close()
except MySQLdb.Error as e:
    print("MySQL Error %d: %s" %(e.args[0], e.args[1]))
```

8.3 操作 MongoDB 数据库

一项权威调查显示，在大数据时代软件开发人员必备的十项技能中 MongoDB 数据库名列第二，仅次于 HTML5。MongoDB 是一个基于分布式文件存储的文档数据库，可以说是非关系型(Not Only SQL，NoSQL)数据库中比较像关系型数据库的一个，具有免费、操作简单、面向文档存储、自动分片、可扩展性强、查询功能强大等特点，对大数据处理支持较好，旨在为 Web 应用提供可扩展的高性能数据存储解决方案。MongoDB 将数据存储为一个文档，数据结构由键值(key→value)对组成。MongoDB 文档类似于 JSON 对象。字段值可以包含其他文档、数组及文档数组。

MongoDB 数据库可以到官方网站 https://www.mongodb.org/downloads 下载，安装之后打开命令提示符环境并切换到 MongoDB 安装目录中的 server\3.2\bin 文件夹，然后执行命令 mongod --dbpath D:\data --journal --storageEngine=mmapv1 启动 MongoDB，当然需要首先在 D 盘新建文件夹 data，让刚才那个命令提示符环境始终处于运行状态，然后再打开一个命令提示符环境，执行 mongo 命令连接 MongoDB 数据库，如果连接成功的话，会显示一个">"符号作为提示符，之后就可以输入 MongoDB 命令了，例如，下面的命令可以打开或创建数据库 students：

```
>use students
```

下面的命令用来在数据库中插入数据：

```
>zhangsan={'name':'Zhangsan', 'age':18, 'sex':'male'}
>db.students.insert(zhangsan)
>lisi={'name':'Lisi', 'age':19, 'sex':'male'}
>db.students.insert(lisi)
```

下面的命令用来查询数据库中的记录：

```
>db.students.find()
```

下面的命令用来查看系统中所有数据库名称：

```
>show dbs
```

其他更多 MongoDB 命令请读者查阅相关资料。另外，Python 扩展库 pymongo 完美支持 MongoDB 数据的操作，可以使用 pip 命令进行安装。下面的代码演示了 pymongo 操作 MongoDB 数据库的一部分用法，算是抛砖引玉吧，更多的用法可以使用学习 Python 的利器 dir() 和 help() 来获得，或者查阅 MongoDB 官方文档。

```
>>>import pymongo                                    #导入模块
>>>client=pymongo.MongoClient('localhost', 27017)
                                                     #连接数据库,27017是默认端口
>>>db=client.students                                #获取数据库
>>>db.collection_names()                             #查看数据集合名称列表
```

```
['students', 'system.indexes']
>>>students=db.students                              #获取数据集合
>>>students.find()
<pymongo.cursor.Cursor object at 0x00000000030934A8>
>>>for item in students.find():                      #遍历数据
    print(item)
{'age': 18.0, 'sex': 'male', '_id': ObjectId('5722cbcfeadfb295b4a52e23'),
'name': 'Zhangsan'}
{'age': 19.0, 'sex': 'male', '_id': ObjectId('5722cc6eeadfb295b4a52e24'),
'name': 'Lisi'}
>>>wangwu={'name':'Wangwu', 'age':20, 'sex':'male'}
>>>students.insert(wangwu)                           #插入一条记录
ObjectId('5723137346bf3d1804b5f4cc')
>>>for item in students.find({'name':'Wangwu'}):     #指定查询条件
    print(item)
{'age': 20, '_id': ObjectId('5723137346bf3d1804b5f4cc'), 'sex': 'male', 'name': '
Wangwu'}
>>>students.find_one()                               #获取一条记录
{'age': 18.0, 'sex': 'male', '_id': ObjectId('5722cbcfeadfb295b4a52e23'), 'name
': 'Zhangsan'}
>>>students.find_one({'name':'Wangwu'})
{'age': 20, '_id': ObjectId('5723137346bf3d1804b5f4cc'), 'sex': 'male', 'name': '
Wangwu'}
>>>students.find().count()                           #记录总数
3
>>>students.remove({'name':'Wangwu'})                #删除一条记录
{'ok': 1, 'n': 1}
>>>for item in students.find():
    print(item)
{'name': 'Zhangsan', '_id': ObjectId('5722cbcfeadfb295b4a52e23'), 'sex': 'male',
'age': 18.0}
{'name': 'Lisi', '_id': ObjectId('5722cc6eeadfb295b4a52e24'), 'sex': 'male',
'age': 19.0}
>>>students.find().count()
2
>>>students.create_index([('name', pymongo.ASCENDING)])    #创建索引
'name_1'
>>>students.update({'name':'Zhangsan'},{'$set':{'age':25}})   #更新数据库
{'nModified': 1, 'ok': 1, 'updatedExisting': True, 'n': 1}
>>>students.update({'age':25},{'$set':{'sex':'Female'}})      #更新数据库
{'nModified': 1, 'ok': 1, 'updatedExisting': True, 'n': 1}
>>>students.remove()                                 #清空数据库
{'ok': 1, 'n': 2}
>>>students.find().count()
```

```
0
>>>Zhangsan={'name':'Zhangsan', 'age':20, 'sex':'Male'}
>>>Lisi={'name':'Lisi', 'age':21, 'sex':'Male'}
>>>Wangwu={'name':'Wangwu', 'age':22, 'sex':'Female'}
>>>students.insert_many([Zhangsan, Lisi, Wangwu])          #插入多条数据
<pymongo.results.InsertManyResult object at 0x0000000003762750>
>>>for item in students.find().sort('name',pymongo.ASCENDING):  #对查询结果排序
    print(item)
{'name': 'Lisi', '_id': ObjectId('57240d3f46bf3d118ce5bbe4'), 'sex': 'Male', 'age': 21}
{'name': 'Wangwu', '_id': ObjectId('57240d3f46bf3d118ce5bbe5'), 'sex': 'Female', 'age': 22}
{'name': 'Zhangsan', '_id': ObjectId('57240d3f46bf3d118ce5bbe3'), 'sex': 'Male', 'age': 20}
>>>for item in students.find().sort([('sex',pymongo.DESCENDING),
                                     ('name',pymongo.ASCENDING)]):
    print(item)
{'name': 'Lisi', '_id': ObjectId('57240d3f46bf3d118ce5bbe4'), 'sex': 'Male', 'age': 21}
{'name': 'Zhangsan', '_id': ObjectId('57240d3f46bf3d118ce5bbe3'), 'sex': 'Male', 'age': 20}
{'name': 'Wangwu', '_id': ObjectId('57240d3f46bf3d118ce5bbe5'), 'sex': 'Female', 'age': 22}
```

第 9 章 网络应用开发

Socket 是计算机之间进行网络通信的一套程序接口,最初由 Berkeley 大学研发,目前已经成为网络编程的标准,可以实现跨平台的数据传输。Socket 相当于在发送端和接收端之间建立了一个管道来实现数据和命令的相互传递。Python 标准库 socket 对 Socket 进行了封装,支持 Socket 接口的访问,大幅度简化了程序的开发步骤,提高了开发效率。除此之外,Python 还提供了 urllib 等大量模块可以对网页内容进行读取和处理,在此基础上结合多线程编程以及其他有关模块可以快速开发网页爬虫之类的应用。可以使用 Python 语言编写 CGI 程序,也可以把 Python 代码嵌入到网页中运行,而借助于 web2py、django、Flask 或其他框架,则可以快速开发网站应用。

9.1 计算机网络基础知识

为了更好地理解本章后面的内容,首先简要介绍一下计算机网络的基本概念,如果读者确实对网络编程感兴趣可以参考《计算机网络》、《网络应用开发实践》、《网站设计与开发》之类的书籍了解更详细的知识。

1. 网络体系结构

目前较为主流的网络体系结构是 ISO/OSI 参考模型和 TCP/IP 协议族。这两种体系结构都采用了分层设计和实现的方式,ISO/OSI 参考模型从上而下划分为应用层、表示层、会话层、传输层、网络层、数据链路层和物理层,而 TCP/IP 则将网络划分为应用层、传输层、网络层和链路层。分层设计的好处是,各层可以独立设计和实现,只要保证相邻层之间的调用规范和接口不变,就可以方便、灵活地改变各层的内部实现以进行优化或完成其他需求。

2. 网络协议

网络协议是计算机网络中为了进行数据交换而建立的规则、标准或约定的集合,语法、语义和时序是网络协议的三要素。简单地讲,可以这样理解,语义表示要做什么,语法表示要怎么做,时序规定了各种事件出现的顺序。语法和语义相对来说比较容易理解,可能有读者在想为啥要严格规定各类事件的时间和顺序。试想,假设早上 8 点 A 和 B 两个同事上班时在公司门口偶遇,A 问 B"吃了吗",B 没做任何回答就走了(如何计算 A 的心理阴影面积),中午 12 点下班时两人在公司门口再次偶遇,B 对 A 说"吃了",我们可以想象到 A 看 B 的眼神会是什么样的。

(1) 语法:语法规定了用户数据与控制信息的结构与格式。

(2) 语义:语义用来解释控制信息每个部分的含义,规定了需要发出何种控制信息,以及需要完成的动作和做出什么样的响应。

(3) 时序:时序是对事件发生顺序的详细说明,也可称为"同步"。

3. 应用层协议

应用层协议直接与最终用户进行交互,用来确定运行在不同终端系统上的应用进程之间如何传递报文。下面简单列出了几种常见的应用层协议。

(1) DNS:域名系统(Domain Name System),用来实现域名与 IP 地址的转换,运行于 UDP 之上,默认使用 53 号端口。

(2) FTP:文件传输协议(File Transfer Protocol),可以通过网络在不同平台之间实现文件的传输,是一种基于 TCP 的明文传输协议,默认工作在 21 号端口。

(3) HTTP:超文本传输协议(HyperText Transfer Protocol),运行于 TCP 之上,默认使用 80 号端口。

(4) SMTP:简单邮件传输协议(Simple Mail Transfer Protocol),建立在 TCP 的基础上,使用明文传递邮件和发送命令,默认使用 25 号端口。

(5) TELNET:远程登录协议,运行于 TCP 之上,默认使用 23 号端口。

4. 传输层协议

在传输层主要运行着传输控制协议(Transmission Control Protocol,TCP)和用户数据报协议(User Datagram Protocol,UDP)两个协议,其中 TCP 是面向连接的、具有质量保证的可靠传输协议,但开销较大;UDP 是尽最大能力传输的无连接协议,开销小,常用于视频在线点播(Video On Demand,VOD)之类的应用。TCP 和 UDP 本身并没有优劣之分,仅仅是适用场合有所不同。在传输层,使用端口号来标识和区分同一台计算机上运行的多个应用层进程,每当创建一个应用层网络进程时系统就会自动分配一个端口号与之关联,是实现网络上端到端通信的重要基础。例如,MS SQL Server 默认占用 1433 端口,远程桌面连接默认占用 3389 端口,HTTP 默认使用 80 端口,MySQL 使用 3306 端口,MongoDB 使用 27017 端口,大多数情况下 IRC 服务器使用 6667 端口,IMAP 使用 143 端口,Oracle 使用 1521、1158、8080、210 等几个端口,等等。

5. IP 地址

IP 运行于网络层,是网络互连的重要基础。IP 地址(32 位或 128 位二进制数)用来标识网络上的主机,在公开网络上或同一个局域网内部,每台主机都必须使用不同的 IP 地址;而由于网络地址转换(Network Address Translation,NAT)和代理服务器等技术的广泛应用,不同内网之间的主机可以使用相同的 IP 地址。IP 地址与端口号共同来标识网络上特定主机上的特定应用进程,俗称 Socket。

6. MAC 地址

MAC 地址也称为网卡物理地址,是一个 48 位的二进制数,用来标识不同的网卡物理地址。本机的 IP 地址和 MAC 地址可以在命令提示符窗口中使用 ipconfig/all 命令查看,如图 9-1 所示。

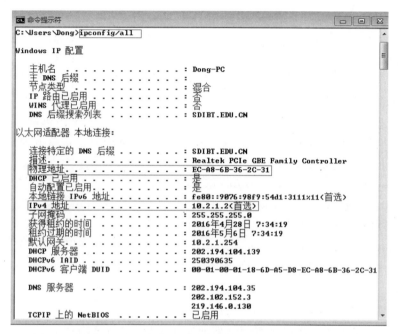

图 9-1 使用 ipconfig 命令查看本机的 IP 地址和网卡的物理地址

拓展知识:如果想知道某个 IP 地址的详细信息,如国家、城市、经纬度等信息,可以使用 Python 扩展库 pygeoip 配合数据库 GeoLiteCity.dat 来获取这些信息。其中,pygeoip 可以使用 pip 工具安装,GeoLiteCity.dat 数据库可以从网上下载。当然,信息的准确程度主要取决于下载的数据库版本。

```
>>>import pygeoip
>>>gi=pygeoip.GeoIP('GeoLiteCity.dat')
>>>gi.record_by_name('221.0.95.247')
{'longitude': 116.99720000000002, 'area_code': 0, 'region_code': '25', 'country_name': 'China', 'dma_code': 0, 'postal_code': None, 'continent': 'AS', 'city': 'Jinan', 'time_zone': 'Asia/Shanghai', 'country_code': 'CN', 'country_code3': 'CHN', 'metro_code': None, 'latitude': 36.66829999999999}
```

9.2 Socket 编程

远程管理软件和黑客软件大多依赖于 Socket 来实现特定功能,前几年流行的端口反弹更是把这项技术发挥到了极致。

如前所述，UDP 和 TCP 是网络体系结构的传输层运行的两大重要协议，其中，TCP 适用于对效率要求相对低而对准确性要求相对高的场合，如文件传输、电子邮件等；而 UDP 适用于对效率要求相对高，对准确性要求相对低的场合，如视频在线点播、网络语音通话等。在 Python 中，主要使用 socket 模块来支持 TCP 和 UDP 编程。

9.2.1 UDP 编程

在很多年以前普通家庭还没有手机、电话、传呼机的时候，主要靠信件往来联系，发信人填写好收信人地址然后把信件邮寄出去就可以了，但是没法保证对方一定能收到这封信（例如对方换了地址），也不能保证不同时间的几封信按照发出的顺序到达目的地。UDP 的工作过程就类似于邮寄普通信件，它属于无连接协议，在 UDP 编程时不需要首先建立连接，而是直接向接收方发送信息。UDP 也不提供应答和重传机制，无法保证数据一定能够到达目的地。UDP 最大的优点是效率高，其首部中只包含双方地址与校验和等很少的字段，额外开销很小。UDP 编程经常用到的 socket 模块方法如下。

（1）socket([family[, type[, proto]]])：创建一个 Socket 对象，其中 family 为 socket.AF_INET 表示 IPv4，socket.AF_INET6 表示 IPv6；type 为 SOCK_STREAM 表示使用 TCP，SOCK_DGRAM 表示使用 UDP。

（2）sendto(string, address)：把 string 指定的内容发送给 address 指定的地址，其中 address 是一个包含接收方主机 IP 地址和应用进程端口号的元组，格式为（IP 地址，端口号）。

（3）recvfrom(bufsize[, flags])：接收数据。

下面通过一个示例来简单了解一下如何使用 UDP 进行网络通信。

例 9-1 UDP 通信程序。

发送端发送一个字符串，假设接收端在本机 5000 端口进行监听，并显示接收的内容，如果收到字符串'bye'（忽略大小写）则结束监听。

接收端代码：

```python
import socket
#使用 IPv4 协议,使用 UDP 传输数据
s=socket.socket(socket.AF_INET, socket.SOCK_DGRAM)
#绑定端口和端口号,空字符串表示本机任何可用 IP 地址
s.bind(('', 5000))
while True:
    data, addr=s.recvfrom(1024)
    #显示接收到的内容
    print('received message:{0} from PORT {1} on {2}'.format(data.decode(),
                                                    addr[1], addr[0]))
    if data.decode().lower()=='bye':
        break
s.close()
```

发送端代码：

```
import socket
import sys
s=socket.socket(socket.AF_INET, socket.SOCK_DGRAM)
#假设192.168.0.103是接收端机器的IP地址
s.sendto(sys.argv[1].encode(),("192.168.0.103",5000))
s.close()
```

将上面的代码分别保存为 receiver.py 和 sender.py,然后首先启动一个命令提示符环境并运行接收端程序,这时接收端程序处于阻塞状态,接下来再启动一个新的命令提示符环境并运行发送端程序,此时会看到接收端程序继续运行并显示接收到的内容以及发送端程序所在计算机 IP 地址和占用的端口号。当发送端发送字符串'bye'后,接收端程序结束,此后再次运行发送端程序时接收端没有任何反应,但发送端程序也并不报错。这正是 UDP 的特点,即"尽最大努力传输",并不保证非常好的服务质量。运行过程如图 9-2 所示。

图 9-2　UDP 通信程序运行结果

拓展知识:使用 Python 查看本机的 IP 地址与网卡的物理地址。在上面的发送端程序中假设接收端主机的 IP 地址为 192.168.0.103,可能与你的计算机配置并不一样。可以在命令提示符环境中使用命令 ipconfig/all 查看本机的 IP 地址,如图 9-1 所示,然后对发送端代码中的 IP 地址进行相应修改。如果对命令提示符不熟悉,也可以使用下面的 Python 代码来获取本机的 IP 地址和网卡的物理地址。

```
import socket
```

```
import uuid

ip=socket.gethostbyname(socket.gethostname())    #本机的 IP 地址
node=uuid.getnode()
macHex=uuid.UUID(int=node).hex[-12:]
mac=[]
for i in range(len(macHex))[::2]:
    mac.append(macHex[i:i+2])
mac=':'.join(mac)                                #网卡的物理地址
print('IP:', ip)
print('MAC:', mac)
```

拓展知识：发送数据时，如果目标 IP 地址中最后一组数字是 255，表示广播地址，也就是说局域网内的所有主机都会收到信息。在本书第 16 章介绍的课堂教学管理系统中服务器自动发现功能就用到了 UDP 广播技术。

9.2.2 TCP 编程

TCP 一般用于要求可靠数据传输的场合。编写 TCP 程序时需要用到的 socket 模块方法主要如下。

(1) connect(address)：连接远程计算机。
(2) send(bytes[,flags])：发送数据。
(3) recv(bufsize[,flags])：接收数据。
(4) bind(address)：绑定地址。
(5) listen(backlog)：开始监听，等待客户端连接。
(6) accept()：响应客户端的请求，接收一个连接。

下面通过一个示例来演示如何使用 TCP 进行通信，在本书第 16 章介绍的课堂教学管理系统中还演示了如何通过 TCP 来传送文件，以及如何进行远程屏幕截图。

例 9-2 会聊天的小机器人。

使用 TCP 进行通信需要首先在客户端和服务端之间建立连接，并且要在通信结束后关闭连接以释放资源。TCP 能够提供比 UDP 更好的服务质量（Quality of Service，QoS），通信可靠性有本质上的提高。下面的代码简单模拟了机器人聊天软件原理，服务端提前建立好字典，然后根据接收到的内容自动回复。当然，这个程序对客户端的信息是进行严格匹配，大家可以尝试结合 5.1.3 节介绍的分词功能，设计一个合适的模糊匹配算法，就可以实现聊天机器人了。

服务端代码：

```
import socket

words={'how are you?':'Fine,thank you.', 'how old are you?':'38',
       'what is your name?':'Dong FuGuo', 'what''s your name?':'Dong FuGuo',
       'where do you work?':'SDIBT', 'bye':'Bye'}
```

```python
HOST=''
PORT=50007
s=socket.socket(socket.AF_INET, socket.SOCK_STREAM)
#绑定socket
s.bind((HOST, PORT))
#开始监听一个客户端连接
s.listen(1)
print('Listening at port:',PORT)
conn, addr=s.accept()
print('Connected by', addr)
while True:
    data=conn.recv(1024)
    data=data.decode()
    if not data:
        break
    print('Received message:', data)
    conn.sendall(words.get(data, 'Nothing').encode())
conn.close()
s.close()
```

客户端代码：

```python
import socket, sys

#服务端主机的IP地址和端口号
HOST='127.0.0.1'
PORT=50007
s=socket.socket(socket.AF_INET, socket.SOCK_STREAM)
try:
    #连接服务器
    s.connect((HOST, PORT))
except Exception as e:
    print('Server not found or not open')
    sys.exit()
while True:
    c=input('Input the content you want to send:')
    #发送数据
    s.sendall(c.encode())
    #从服务端接收数据
    data=s.recv(1024)
    data=data.decode()
    print('Received:', data)
    if c.lower()=='bye':
        break
#关闭连接
```

```
s.close()
```

将上面的代码分别保存为 server.py 和 client.py 文件,然后启动一个命令提示符环境并运行服务端程序,服务端开始监听;启动一个新的命令提示符环境并运行客户端程序,服务端提示连接已建立;在客户端输入要发送的信息后,服务端会根据提前建立的字典来自动回复。服务端每次都在固定的端口进行监听,而客户端每次建立连接时可能会使用不同的端口。如果服务端程序没有运行,那么客户端就无法建立连接,当然也无法发送任何信息,这正是 TCP 区别于 UDP 的地方。运行过程如图 9-3 所示。

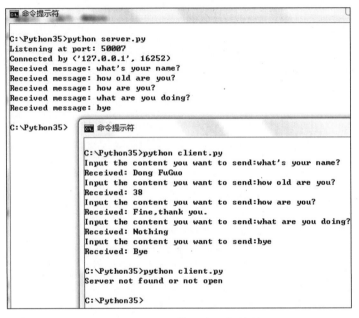

图 9-3 TCP 通信程序运行结果

拓展知识:Python 标准库 socket 除了支持 UDP 和 TCP 编程之外,还提供了用来获取本地主机名的 gethostname()、根据主机名获取 IP 地址的 gethostbyname()、根据 IP 地址获取主机名的 gethostbyaddr()、根据端口号获取对应服务名称的 getservbyport()、根据服务名称获取对应端口号的 getservbyname()等方法。

9.2.3 网络嗅探器

嗅探器程序可以检测本机所在局域网内的网络流量和数据包收发情况,对于网络管理具有重要作用,也属于系统运维内容之一。为了实现网络流量嗅探,需要将网卡设置为混杂模式,并且运行嗅探器程序的用户账号需要拥有系统管理员权限。

例 9-3 网络嗅探器程序。

下面的代码运行 60s,然后输出本机所在局域网内非本机发出的数据包,并统计不同主机发出的数据包数量。关于多线程的知识请参考第 10 章。

```
import socket
```

```python
import threading
import time

activeDegree=dict()
flag=1
def main():
    global activeDegree
    global flag
    #获取本机IP地址
    HOST=socket.gethostbyname(socket.gethostname())
    #创建原始套接字,适用于Windows平台
    #对于其他操作系统,要把socket.IPPROTO_IP替换为socket.IPPROTO_ICMP
    s=socket.socket(socket.AF_INET, socket.SOCK_RAW, socket.IPPROTO_IP)
    s.bind((HOST, 0))
    #设置在捕获的数据包中含有IP包头
    s.setsockopt(socket.IPPROTO_IP, socket.IP_HDRINCL, 1)
    #启用混杂模式,捕捉所有数据包
    s.ioctl(socket.SIO_RCVALL, socket.RCVALL_ON)
    #开始捕捉数据包
    while flag:
        c=s.recvfrom(65565)
        host=c[1][0]
        activeDegree[host]=activeDegree.get(host, 0)+1
        #假设本机IP地址为10.2.1.8
        if c[1][0]!='10.2.1.8':
            print(c)
    #关闭混杂模式
    s.ioctl(socket.SIO_RCVALL, socket.RCVALL_OFF)
    s.close()
t=threading.Thread(target=main)
t.start()
time.sleep(60)
flag=0
t.join()
for item in activeDegree.items():
    print(item)
```

拓展知识：sniffer pro是NAI公司出品的一款一流的便携式网管和应用故障诊断分析软件,拥有强大的网络抓包和协议分析能力,软件能够完美支持全系统Windows平台,性能优越,是网络管理员必备的一款网络协议分析软件。

拓展知识：scapy是一款功能非常强大的交互式包处理程序,可以伪造或解码很多种网络协议的数据包,可以发送和捕获数据包,可以对请求数据包和回复数据包进行匹配,可以处理扫描、路由跟踪、探测、单元测试、攻击、网络发现等任务,还具有很多其他工

具所不具有的功能。如果使用 scapy，可能需要暂时切换到 Linux＋Python 2.7.x 环境（这也是很多涉及软件安全、系统安全和网络安全的 Python 扩展库的推荐运行平台，最起码目前是这样的），因为在 Windows 平台上搭建 scapy 环境实在是有点麻烦。如果对 Linux 不太熟悉的话，也不用过于担心，配置的过程并不是特别复杂。以 Ubuntu 为例，使用 pip install scapy 即可安装，然后使用 vim 编写下面的 Python 代码并保存为 sniff.py 文件，然后执行 sudo python sniff.py 即可，最后启动 9.2.6 节中的 FTP 程序，可以发现 sniff.py 程序成功地捕获了我们感兴趣的信息。当然，如果需要编写对特定程序网络数据包的收发情况嗅探程序，需要对目标程序的通信协议有一定的了解才行。另外，dpkt 也是一款不错的网络数据包分析工具，大家可以查阅相关资料。

```python
from scapy.all import *
#回调函数，每个捕获的数据包都用这个函数处理
def packet_callback(packet):
    if packet[TCP].payload:
        temp_packet=str(packet[TCP].payload).lower()
        if 'zhangsan' in temp_packet:
            print '%s'%packet[TCP].payload
#开始嗅探与 TCP 端口 10600 有关的数据包，并且不在内存中存储
sniff(filter='tcp port 10600', prn=packet_callback, store=0)
```

9.2.4　多进程端口扫描器

说到扫描器，就不得不提到一个经典的扫描软件 xscan，相信热爱黑客和网络安全技术的读者都用过这个软件，这个小软件能够扫描的内容包括远程服务类型、操作系统类型及版本、各种弱口令漏洞、后门、应用服务漏洞、网络设备漏洞、拒绝服务漏洞 20 多个大类。

一般而言，绝大多数成功的网络攻击都是以端口扫描开始的。在网络安全和黑客领域，端口扫描是经常用到的技术，可以探测指定主机上是否开放了特定端口，进一步判断主机上是否运行某些重要的网络服务，最终判断是否存在潜在的安全漏洞，从一定意义上讲也属于系统运维的范畴。

例 9-4　端口扫描器程序。

下面代码模拟了端口扫描器的工作原理，并采用多进程技术提高扫描速度，关于多进程编程请参考第 10 章。

```python
import socket, sys
import multiprocessing

def ports(ports_service):
    #获取常用端口对应的服务名称
    for port in list(range(1,100))+[143, 145, 113, 443, 445, 3389, 8080]:
        try:
            ports_service[port]=socket.getservbyport(port)
```

```python
        except socket.error:
            pass

def ports_scan(host, ports_service):
    ports_open=[]
    try:
        sock=socket.socket(socket.AF_INET, socket.SOCK_STREAM)
        #超时时间的不同会影响扫描结果的精确度
        sock.settimeout(0.01)
    except socket.error:
        print('socket creation error')
        sys.exit()
    for port in ports_service:
        try:
            #尝试连接指定端口
            sock.connect((host,port))
            #记录打开的端口
            ports_open.append(port)
            sock.close()
        except socket.error:
            pass
    return ports_open

if __name__=='__main__':
    m=multiprocessing.Manager()
    ports_service=dict()
    results=dict()
    ports(ports_service)
    #创建进程池,允许最多8个进程同时运行
    pool=multiprocessing.Pool(processes=8)
    net='10.2.1.'
    for host_number in map(str, range(8,10)):
        host=net+host_number
        #创建一个新进程,同时记录其运行结果
        results[host]=pool.apply_async(ports_scan,(host, ports_service))
        print('starting '+host+'…')
    #关闭进程池,close()必须在join()之前调用
    pool.close()
    #等待进程池中的进程全部执行结束
    pool.join()

    #打印输出结果
    for host in results:
        print('=' * 30)
```

```
        print(host, '.' * 10)
        for port in results[host].get():
            print(port, ':', ports_service[port])
```

拓展知识：运行例9-4的代码会发现，虽然扫描效果不错，但是速度非常慢，远不如xscan快。究其原因，是上面的代码使用TCP进行探测，而TCP需要首先通过"三次握手"建立连接，通信后还需要断开连接，这些操作对速度的影响非常大。为了追求更快的速度，可以考虑改用UDP进行探测。当发送一个UDP数据包到网络上主机的某个关闭的UDP端口时，目标主机通常会返回一个ICMP包指示目标端口不可达，这就意味着目标主机是存活的。可以使用UDP对整个子网内所有主机发送信息，然后等待这些主机的ICMP响应，而这个过程的开销几乎可以忽略不计。

拓展知识：Python扩展库netaddr提供了大量可以处理网络地址的类和对象。例如，netaddr.valid_ipv4(addr)可以判断addr是否为合法的IPv4地址，netaddr.IPNetwork('10.2.1.0/24')和netaddr.IPRange('10.2.1.0','10.2.1.255')都可以用来生成包含介于10.2.1.0到10.2.1.255之间IP地址的迭代对象。

拓展知识：在例9-4的代码中是使用IP地址来表示目标主机的，但是很多网站为了防止黑客攻击或者进行负载均衡，会经常变换主机，这样同一个域名在不同时间可能会对应不同的IP地址，在这种情况下可以通过socket模块的gethostbyname()函数来实时获取目标主机的IP地址。下面的代码连续不间断地跟踪指定目标主机的IP地址变化情况。

```
from time import sleep
from socket import gethostbyname
from datetime import datetime

def get_ipAddresses(url):
    ipAddresses=[0]
    while True:
        sleep(0.5)                                      #暂停0.5s
        ip=gethostbyname(url)
        if ip!=ipAddresses[-1]:                         #目标主机IP地址发生变化
            ipAddresses.append(ip)
            print(str(datetime.now())[:19]+'===>'+ip)
get_ipAddresses(r'www.microsoft.com')
```

拓展知识：Nmap是一款非常棒的网络扫描工具，首先下载并安装Nmap工具，把安装路径添加到系统Path环境变量，然后使用pip安装python-nmap，就可以使用了。例如下面的代码：

```
import socket
import nmap
```

```python
nmScan=nmap.PortScanner()                                #创建端口扫描对象
ip=socket.gethostbyname('www.microsort.com')             #获取目标主机的IP地址
nmScan.scan(ip,'80')                                     #扫描指定端口
print(nmScan[ip]['tcp'][80]['state'])                    #查看端口状态
```

9.2.5 代理服务器端口映射功能的实现

端口映射是网络地址转换（Network Address Translation，NAT）技术的实现方法之一，也是代理服务器的必备功能之一，其基本原理是在两个端口之间进行消息转发，可以用来实现内网和外网之间的通信，有效利用有限的IP地址，并且可以进行必要的过滤来保护内网的安全。

(1) 模拟服务端代码（sockMiddle_server.py）如下：

```python
import sys
import socket
import threading

#回复消息,原样返回
def replyMessage(conn):
    while True:
        data=conn.recv(1024)
        conn.send(data)
        if data.decode().lower()=='bye':
            break
    conn.close()

def main():
    sockScr=socket.socket(socket.AF_INET, socket.SOCK_STREAM)
    sockScr.bind(('', port))
    sockScr.listen(200)
    while True:
        try:
            conn, addr=sockScr.accept()
            #只允许特定主机访问本服务器
            if addr[0] !=onlyYou:
                conn.close()
                continue
            #创建并启动线程
            t=threading.Thread(target=replyMessage, args=(conn,))
            t.start()
        except:
            print('error')

if __name__ =='__main__':
```

```python
    try:
        #获取命令行参数,port 为服务器监听端口
        #只允许 IP 地址为 onlyYou 的主机访问
        port=int(sys.argv[1])
        onlyYou=sys.argv[2]
        main()
    except:
        print('Must give me a number as port')
```

(2) 模拟代理服务器代码(sockMiddle.py)如下:

```python
import sys
import socket
import threading

def middle(conn, addr):
    #面向服务器的 Socket
    sockDst=socket.socket(socket.AF_INET, socket.SOCK_STREAM)
    sockDst.connect((ipServer,portServer))
    while True:
        data=conn.recv(1024).decode()
        print('收到客户端消息:'+data)
        if data =='不要发给服务器':
            conn.send('该消息已被代理服务器过滤'.encode())
            print('该消息已过滤')
        elif data.lower()=='bye':
            print(str(addr)+'客户端关闭连接')
            break
        else:
            sockDst.send(data.encode())
            print('已转发服务器')
            data_fromServer=sockDst.recv(1024).decode()
            print('收到服务器回复的消息:'+data_fromServer)
            if data_fromServer =='不要发给客户端':
                conn.send('该消息已被代理服务器修改'.encode())
                print('消息已被篡改')
            else:
                conn.send(b'Server reply:'+data_fromServer.encode())
                print('已转发服务器消息给客户端')

    conn.close()
    sockDst.close()

def main():
    sockScr=socket.socket(socket.AF_INET, socket.SOCK_STREAM)
```

```python
        sockScr.bind(('', portScr))
        sockScr.listen(200)
        print('代理已启动')
        while True:
            try:
                conn, addr=sockScr.accept()
                t=threading.Thread(target=middle, args=(conn, addr))
                t.start()
                print('新客户:'+str(addr))
            except:
                pass

if __name__ =='__main__':
    try:
        #(本机IP地址,portScr)<==>(ipServer,portServer)
        #代理服务器监听端口
        portScr=int(sys.argv[1])
        #服务器IP地址与端口号
        ipServer=sys.argv[2]
        portServer=int(sys.argv[3])
        main()
    except:
        print('Sth error')
```

(3) 模拟客户端代码(sockMiddle_client.py)如下:

```python
import sys
import socket

def main():
    sock=socket.socket(socket.AF_INET, socket.SOCK_STREAM)
    sock.connect((ip, port))
    while True:
        data=input('What do you want to ask:')
        sock.send(data.encode())
        print(sock.recv(1024).decode())
        if data.lower()=='bye':
            break
    sock.close()

if __name__ =='__main__':
    try:
        #代理服务器的IP地址和端口号
        ip=sys.argv[1]
        port=int(sys.argv[2])
```

```
        main()
except:
        print('Sth error')
```

代码编写完成后,启动3个命令提示符窗口,分别执行以下3个命令启动服务器、代理服务器和客户端,然后就可以在客户端发送消息了。

```
python sockMiddle_server.py 10000 10.2.1.2
python sockMiddle.py 30800 10.2.1.2 10000
python sockMiddle_client.py 10.2.1.2 30800
```

可以看出,代理服务器代码能够对客户端和服务端之间的通信内容进行记录,也能够修改双方的通信内容,这样实际上是存在潜在危险的。只要代理服务器想这样做,客户在网络上的通信基本上就没有什么隐私可言了。因此,如果涉及金钱交易最好不要使用代理服务器,也尽量不要使用外面免费的 Wi-Fi 网络,因为我们没法保证那些代理服务器和路由器除了提供正常通信功能之外还会做些什么。

9.2.6 自己编写 FTP 通信软件

看到这一节的标题,大家的第一反应大概可以分为3类:①不是已经有很成熟的FTP程序了吗,还有必要自己编写吗?②太高深了吧,真的可以自己编写 FTP 服务器和客户端吗?③太难,我选择直接跳过这一节内容。

现有的 FTP 程序确实已经非常成熟、稳定了,那么为什么还要自己编写 FTP 程序呢?首先,可以更加熟悉 Python 代码的编写。其次,并不是时时刻刻都可以使用现有的 FTP 程序的,比如目标主机上没有安装 FTP 服务或者防火墙拦截了外部的 FTP 访问。比如说,我们"一不小心"进入了一台不属于自己的计算机,并且想把自己计算机上的一个文件传送到对方计算机上,而对方计算机的防护机制不允许直接上传文件,却恰好安装了 Python 解释器(肯定有读者会问怎么这么巧,但事实是绝大多数版本的 Linux 系统都默认安装了 Python 解释器),这样我们就可以编写两段代码,把自己的计算机当做服务器,从目标主机上把文件下载过去就可以了。关键是一切都在自己的掌握之中,感觉确实不错。

凡事预则立,不预则废。日本近代作家夏目漱石的作品《心》中有句名言"没有天生的武士,也没人生来平凡,是我们自己成就了自己",也表达了相同的意思。我们确实可以自己开发 FTP 程序,看完下面的代码就会发现,自己编写 FTP 程序还真不是什么难事,所以,既然走过路过就千万不要错过,让我们一起看看如何编写 FTP 服务器和客户端代码。

(1) 服务端代码(ftpServer.py)为

```
import socket
import threading
import os
import struct
```

```python
#用户账号、密码、主目录
#也可以把这些信息存放到数据库中
users={'zhangsan':{'pwd':'zhangsan1234', 'home':r'c:\python 3.5'},
       'lisi':{'pwd':'lisi567', 'home':'c:\\'}}

def server(conn,addr, home):
    print('新客户端:'+str(addr))
    #进入当前用户主目录
    os.chdir(home)
    while True:
        data=conn.recv(100).decode().lower()
        #显示客户端输入的每一条命令
        print(data)
        #客户端退出
        if data in('quit', 'q'):
            break
        #查看当前文件夹的文件列表
        elif data in('list', 'ls', 'dir'):
            files=str(os.listdir(os.getcwd()))
            files=files.encode()
            #先发送字节串大小,再发送字节串
            conn.send(struct.pack('I', len(files)))
            conn.send(files)
        #切换至上一级目录
        elif ''.join(data.split())=='cd..':
            cwd=os.getcwd()
            newCwd=cwd[:cwd.rindex('\\')]
            #考虑根目录的情况
            if newCwd[-1] ==':':
                newCwd +='\\'
            #限定用户主目录
            if newCwd.lower().startswith(home):
                os.chdir(newCwd)
                conn.send(b'ok')
            else:
                conn.send(b'error')
        #查看当前目录
        elif data in('cwd', 'cd'):
            conn.send(str(os.getcwd()).encode())
        elif data.startswith('cd '):
            #指定最大分隔次数,考虑目标文件夹带有空格的情况
            #只允许使用相对路径进行跳转
            data=data.split(maxsplit=1)
            if len(data)==2 and  os.path.isdir(data[1])\
```

```python
                        and data[1]!=os.path.abspath(data[1]):
                        os.chdir(data[1])
                        conn.send(b'ok')
                    else:
                        conn.send(b'error')
                #下载文件
                elif data.startswith('get '):
                    data=data.split(maxsplit=1)
                    #检查文件是否存在
                    if len(data)==2 and os.path.isfile(data[1]):
                        conn.send(b'ok')
                        fp=open(data[1], 'rb')
                        while True:
                            content=fp.read(4096)
                            #发送文件结束
                            if not content:
                                conn.send(b'overxxxx')
                                break
                            #发送文件内容
                            conn.send(content)
                            if conn.recv(10)==b'ok':
                                continue
                        fp.close()
                    else:
                        conn.send(b'no')
                #无效命令
                else:
                    pass

        conn.close()
        print(str(addr)+'关闭连接')


#创建Socket,监听本地端口,等待客户端连接
sock=socket.socket(socket.AF_INET, socket.SOCK_STREAM)
sock.bind(('', 10600))
sock.listen(5)
while True:
    conn, addr=sock.accept()
    #验证客户端输入的用户名和密码是否正确
    userId, userPwd=conn.recv(1024).decode().split(',')
    if userId in users and users[userId]['pwd']==userPwd:
        conn.send(b'ok')
        #为每个客户端连接创建并启动一个线程
        #参数为连接、客户端地址、客户主目录
```

```
            home=users[userId]['home']
            t=threading.Thread(target=server, args=(conn,addr,home))
            t.daemon=True
            t.start()
        else:
            conn.send(b'error')
```

(2) 客户端代码(ftpClient.py)为

```python
import socket
import sys
import re
import struct
import getpass

def main(serverIP):
    sock=socket.socket(socket.AF_INET, socket.SOCK_STREAM)
    sock.connect((serverIP, 10600))
    userId=input('请输入用户名:')
    #使用getpass模块的getpass()方法获取密码,不回显
    userPwd=getpass.getpass('请输入密码:')
    message=userId+','+userPwd
    sock.send(message.encode())
    login=sock.recv(100)
    #验证是否登录成功
    if login ==b'error':
        print('用户名或密码错误')
        return
    #整数编码大小
    intSize=struct.calcsize('I')
    while True:
        #接收客户端命令,其中##>是提示符
        command=input('##>').lower().strip()
        #没有输入任何有效字符,提前进入下一次循环,等待用户继续输入
        if not command:
            continue
        #向服务端发送命令
        command=' '.join(command.split())
        sock.send(command.encode())
        #退出
        if command in('quit', 'q'):
            break
        #查看文件列表
        elif command in('list', 'ls', 'dir'):
            #先接收字节串大小,再根据情况接收合适数量的字节串
```

```python
                loc_size=struct.unpack('I', sock.recv(intSize))[0]
                files=eval(sock.recv(loc_size).decode())
                for item in files:
                    print(item)
            #切换至上一级目录
            elif ''.join(command.split())=='cd..':
                print(sock.recv(100).decode())
            #查看当前工作目录
            elif command in('cwd', 'cd'):
                print(sock.recv(1024).decode())
            #切换至子文件夹
            elif command.startswith('cd '):
                print(sock.recv(100).decode())
            #从服务器下载文件
            elif command.startswith('get '):
                isFileExist=sock.recv(20)
                #文件不存在
                if isFileExist !=b'ok':
                    print('error')
                #文件存在,开始下载
                else:
                    print('downloading.', end='')
                    fp=open(command.split()[1], 'wb')
                    while True:
                        #显示进度
                        print('.', end='')
                        data=sock.recv(4096)
                        if data ==b'overxxxx':
                            break
                        fp.write(data)
                        sock.send(b'ok')
                    fp.close()
                    print('ok')
            
            #无效命令
            else:
                print('无效命令')
    sock.close()

if __name__ =='__main__':
    if len(sys.argv)!=2:
        print('Usage:{0} serverIPAddress'.format(sys.argv[0]))
        exit()
    serverIP=sys.argv[1]
```

```
#使用正则表达式判断服务器地址是否为合法的IP地址
if re.match(r'^\d{1,3}.\d{1,3}.\d{1,3}.\d{1,3}$', serverIP):
    main(serverIP)
else:
    print('服务器地址不合法')
    exit()
```

拓展知识：Python 标准库 ftplib 提供了 FTP 客户端的主要功能。下面的代码可以用来测试目标主机上的 FTP 服务是否允许匿名登录，稍加改写后也可以用来暴力破解 FTP 服务器的账号和密码，请参考第 6 章破解压缩文件密码和第 14 章破解 MD5 值有关的内容。

```
>>>import ftplib
>>>ftp=ftplib.FTP('127.0.0.1')
>>>try:
    ftp.login('anonymous','1234@567.8')
    print('anonymous login allowed.')
except:
    print('anonymous login denied.')
```

9.3 域名解析与网页爬虫

9.3.1 网页内容读取与域名分析

Python 3.x 标准库 urllib 提供了 urllib.request、urllib.response、urllib.parse 和 urllib.error 4 个模块，很好地支持了网页内容读取功能。

下面的代码演示了如何读取并显示指定网页的内容。

```
>>>import urllib.request
>>>fp=urllib.request.urlopen(r'http://www.python.org')
>>>print(fp.read(100))
>>>print(fp.read(100).decode())
>>>fp.close()
```

下面的代码演示了如何使用 GET 方法读取并显示指定 URL 的内容。

```
>>>import urllib.request
>>>import urllib.parse
>>>params=urllib.parse.urlencode({'spam': 1, 'eggs': 2, 'bacon': 0})
>>>url="http://www.musi-cal.com/cgi-bin/query?%s" %params
>>>with urllib.request.urlopen(url) as f:
    print(f.read().decode('utf-8'))
```

下面的代码演示了如何使用 POST 方法提交参数并读取指定页面内容。

```
>>>import urllib.request
>>>import urllib.parse
>>>data=urllib.parse.urlencode({'spam': 1, 'eggs': 2, 'bacon': 0})
>>>data=data.encode('ascii')
>>>with urllib.request.urlopen("http://requestb.in/xrbl82xr", data) as f:
    print(f.read().decode('utf-8'))
```

下面的代码演示了如何使用 HTTP 代理访问指定页面。

```
>>>import urllib.request
>>>proxies={'http': 'http://proxy.example.com:8080/'}
>>>opener=urllib.request.FancyURLopener(proxies)
>>>with opener.open("http://www.python.org") as f:
    f.read().decode('utf-8')
```

另外，Python 标准库 webbrowser 支持使用已安装的浏览器直接打开网页。可以在命令提示符环境中执行下面的命令：

```
python -m webbrowser -t "http://www.python.org"
```

也可以在 IDLE 或者 Python 程序中使用下面的代码调用浏览器打开指定网页：

```
import webbrowser
webbrowser.open('http://www.python.org')
```

最后，标准库 urllib.parse 提供了域名解析的功能，支持 URL 的拆分与合并以及相对地址到绝对地址的转换。

```
>>>from urllib.parse import urlparse
>>>o=urlparse('http://www.cwi.nl:80/%7Eguido/Python.html')
>>>o.port
80
>>>o.hostname
'www.cwi.nl'
>>>urlparse('//www.cwi.nl:80/%7Eguido/Python.html')
ParseResult(scheme='', netloc='www.cwi.nl:80', path='/%7Eguido/Python.html', params='', query='', fragment='')
>>>urlparse('www.cwi.nl/%7Eguido/Python.html')
ParseResult(scheme='', netloc='', path='www.cwi.nl/%7Eguido/Python.html', params='', query='', fragment='')
>>>from urllib.parse import urljoin
>>>urljoin('http://www.cwi.nl/%7Eguido/Python.html', 'FAQ.html')
'http://www.cwi.nl/%7Eguido/FAQ.html'
>>>urljoin('http://www.cwi.nl/%7Eguido/Python.html', '//www.python.org/%7Eguido')
'http://www.python.org/%7Eguido'
>>>from urllib.parse import urlsplit
```

```
>>>url=r'https://docs.python.org/3/library/urllib.parse.html'
>>>r1=urlsplit(url)
>>>r1.hostname
'docs.python.org'
>>>r1.geturl()
'https://docs.python.org/3/library/urllib.parse.html'
>>>r1.netloc
'docs.python.org'
>>>r1.scheme
'https'
```

拓展知识：如果你仍然不舍得放弃 Python 2.x，或许可以试试 Python 扩展库 machanize，这也是一款不错的网页内容读取工具。

```
>>>import mechanize
>>>browser=mechanize.Browser()
>>>page=browser.open(r'http://www.python.org')
>>>source_code=page.read()
>>>print source_code
```

9.3.2 网页爬虫

网页爬虫常用来在互联网上爬取感兴趣的页面或文件，结合数据处理与分析技术可以得到更深层次的信息。下面的代码实现了网页爬虫，可以抓取指定网页中的所有链接，并且可以指定关键字和抓取深度。

例 9-5 网页爬虫程序。

```
import sys
import multiprocessing
import re
import os
import urllib.request as lib

def craw_links(url, depth, keywords, processed):
    '''url:the url to craw
     depth:the current depth to craw
     keywords:the tuple of keywords to focus
     pool:process pool
    '''
    contents=[]
    if url.startswith(('http://', 'https://')):
        if url not in processed:
            #mark this url as processed
            processed.append(url)
```

```
        else:
            #avoid processing the same url again
            return
        print('Crawing '+url+'…')
        fp=lib.urlopen(url)
        #Python3 returns bytes, so need to decode
        contents=fp.read()
        contents_decoded=contents.decode('UTF-8')
        fp.close()
        pattern='|'.join(keywords)
        #if this page contains certain keywords, save it to a file
        flag=False
        if pattern:
            searched=re.search(pattern, contents_decoded)
        else:
            #if the keywords to filter is not given, save current page
            flag=True
        if flag or searched:
            with open('craw\\'+url.replace(':','_').replace('/','_'), 'wb')
                as fp:
                    fp.write(contents)
        #find all the links in the current page
        links=re.findall('href="(.*?)"', contents_decoded)
        #craw all links in the current page
        for link in links:
            #consider the relative path
            if not link.startswith(('http://','https://')):
                try:
                    index=url.rindex('/')
                    link=url[0:index+1]+link
                except:
                    pass
            if depth>0 and link.endswith(('.htm','.html')):
                craw_links(link, depth-1, keywords, processed)

if __name__=='__main__':
    processed=[]
    keywords = ('datetime','KeyWord2')
    if not os.path.exists('craw') or not os.path.isdir('craw'):
        os.mkdir('craw')
    craw_links(r'https://docs.python.org/3/library/index.html', 1, keywords,
    processed)
```

9.3.3 scrapy 框架

scrapy(注意,不是 scapy)是一个非常好用的 Web 爬虫框架,非常适合抓取 Web 站点从网页中提取结构化的数据,并且支持自定义需求。在使用 scrapy 爬取网页数据时,除了熟悉 HTML 标签,还需要了解目标网页的数据组织结构,确定要爬取什么信息,这样才能有针对性地编写爬虫程序。

使用 pip 命令安装好 scrapy 之后,在命令提示符环境中执行下面的命令创建一个项目 MyCraw:

scrapy startproject MyCraw

然后编写 Python 程序 MyCraw\MyCraw\spiders\MySpider.py,用于爬取指定页面的内容,把网页内容和图片分别保存为文件,代码如下:

```python
import os
import urllib.request
import scrapy

class MySpider(scrapy.spiders.Spider):
    #爬虫的名字,每个爬虫必须有不同的名字
    name='mySpider'
    allowed_domains=['www.sdibt.edu.cn']
    #要爬取的起始页面,必须是列表,可以包含多个 URL
    start_urls=['http://www.sdibt.edu.cn/info/1026/11238.htm']

    #对每个要爬取的页面,会自动调用下面这个方法
    def parse(self, response):
        self.downloadWebpage(response)
        self.downloadImages(response)

        #检查页面中的超链接,并继续爬取
        hxs=scrapy.Selector(response)
        sites=hxs.xpath('//ul/li')
        for site in sites:
            link=site.xpath('a/@href').extract()[0]
            if link =='#':
                continue
            #把相对地址转换成绝对地址
            elif link.startswith('..'):
                next_url=os.path.dirname(response.url)
                next_url +='/'+link
            else:
                next_url=link
            #生成 Request 对象,并指定回调函数
            yield scrapy.Request(url=next_url, callback=self.parse_item)

    #回调函数,对起始页面中的每个超链接起作用
```

```python
    def parse_item(self, response):
        self.downloadWebpage(response)
        self.downloadImages(response)

    #下载当前页面中的所有图片
    def downloadImages(self, response):
        hxs=scrapy.Selector(response)
        images=hxs.xpath('//img/@src').extract()
        for image_url in images:
            imageFilename=image_url.split('/')[-1]
            if os.path.exists(imageFilename):
                continue
            #把相对地址转换成绝对地址
            if image_url.startswith('..'):
                image_url=os.path.dirname(response.url)+'/'+image_url
            #打开网页图片
            fp=urllib.request.urlopen(image_url)
            #创建本地图片文件
            with open(imageFilename,'wb') as f:
                f.write(fp.read())
            fp.close()

    #把网页内容保存为本地文件
    def downloadWebpage(self, response):
        filename=response.url.split('/')[-1]
        with open(filename, 'wb') as f:
            f.write(response.body)
```

最后在命令提示符环境中执行下面的命令启动爬虫程序开始爬取数据：

```
scrapy crawl mySpider
```

9.3.4 BeautifulSoup4

BeautifulSoup 是一个非常优秀的 Python 扩展库，可以用来从 HTML 或 XML 文件中提取我们感兴趣的数据，并且允许指定使用不同的解析器。由于 beautifulsoup3 已经不再继续维护，新的项目中应使用 beautifulsoup4，目前最新版本是 4.5.0，可以使用 pip install beautifulsoup4 直接进行安装，安装之后应使用 from bs4 import BeautifulSoup 导入并使用。下面我们就一起来简单看一下 BeautifulSoup4 的强大功能，更加详细完整的学习资料请参考 https://www.crummy.com/software/BeautifulSoup/bs4/doc/。

```
>>>BeautifulSoup('hello world!', 'lxml')          #自动添加和补全标签
<html><body><p>hello world!</p></body></html>
>>>html_doc="""
```

```
<html><head><title>The Dormouse's story</title></head>
<body>
<p class="title"><b>The Dormouse's story</b></p>

<p class="story">Once upon a time there were three little sisters; and their names were
<a href="http://example.com/elsie" class="sister" id="link1">Elsie</a>,
<a href="http://example.com/lacie" class="sister" id="link2">Lacie</a>and
<a href="http://example.com/tillie" class="sister" id="link3">Tillie</a>;
and they lived at the bottom of a well.</p>

<p class="story">…</p>
"""
>>>from bs4 import BeautifulSoup
>>>soup=BeautifulSoup(html_doc, 'html.parser')      #也可以使用 lxml 或其他解析器
>>>print(soup.prettify())                           #以优雅的方式显示出来
<html>
 <head>
  <title>
   The Dormouse's story
  </title>
 </head>
 <body>
  <p class="title">
   <b>
    The Dormouse's story
   </b>
  </p>
  <p class="story">
   Once upon a time there were three little sisters; and their names were
   <a class="sister" href="http://example.com/elsie" id="link1">
    Elsie
   </a>
   ,
   <a class="sister" href="http://example.com/lacie" id="link2">
    Lacie
   </a>
   and
   <a class="sister" href="http://example.com/tillie" id="link3">
    Tillie
   </a>
   ;
   and they lived at the bottom of a well.
  </p>
```

```
    <p class="story">
     ...
    </p>
 </body>
</html>
>>>soup.title                                          #访问特定的标签
<title>The Dormouse's story</title>
>>>soup.title.name                                     #标签名字
'title'
>>>soup.title.text                                     #标签文本
"The Dormouse's story"
>>>soup.title.string
"The Dormouse's story"
>>>soup.title.parent                                   #上一级标签
<head><title>The Dormouse's story</title></head>
>>>soup.head
<head><title>The Dormouse's story</title></head>
>>>soup.b
<b>The Dormouse's story</b>
>>>soup.body.b
<b>The Dormouse's story</b>
>>>soup.name                                           #把整个BeautifulSoup对象看作标签对象
'[document]'
>>>soup.body
<body>
<p class="title"><b>The Dormouse's story</b></p>
<p class="story">Once upon a time there were three little sisters; and their names were
<a class="sister" href="http://example.com/elsie" id="link1">Elsie</a>,
<a class="sister" href="http://example.com/lacie" id="link2">Lacie</a> and
<a class="sister" href="http://example.com/tillie" id="link3">Tillie</a>;
and they lived at the bottom of a well.</p>
<p class="story">...</p>
</body>
>>>soup.p
<p class="title"><b>The Dormouse's story</b></p>
>>>soup.p['class']                                     #标签属性
['title']
>>>soup.p.get('class')                                 #也可以这样查看标签属性
['title']
>>>soup.p.text
"The Dormouse's story"
>>>soup.p.contents
[<b>The Dormouse's story</b>]
```

```
>>>soup.a
<a class="sister" href="http://example.com/elsie" id="link1">Elsie</a>
>>>soup.a.attrs                                  #查看标签所有属性
{'class': ['sister'], 'href': 'http://example.com/elsie', 'id': 'link1'}
>>>soup.find_all('a')                            #查找所有<a>标签
[<a class="sister" href="http://example.com/elsie" id="link1">Elsie</a>, <a class="sister" href="http://example.com/lacie" id="link2">Lacie</a>, <a class="sister" href="http://example.com/tillie" id="link3">Tillie</a>]
>>>soup.find_all(['a', 'b'])                     #同时查找<a>和<b>标签
[<b>The Dormouse's story</b>, <a class="sister" href="http://example.com/elsie" id="link1">Elsie</a>, <a class="sister" href="http://example.com/lacie" id="link2">Lacie</a>, <a class="sister" href="http://example.com/tillie" id="link3">Tillie</a>]
>>>import re
>>>soup.find_all(href=re.compile("elsie"))       #查找href包含特定关键字的标签
[<a class="sister" href="http://example.com/elsie" id="link1">Elsie</a>]
>>>soup.find(id='link3')
<a class="sister" href="http://example.com/tillie" id="link3">Tillie</a>
>>>soup.find_all('a', id='link3')
[<a class="sister" href="http://example.com/tillie" id="link3">Tillie</a>]
>>>for link in soup.find_all('a'):
    print(link.text,':',link.get('href'))

Elsie : http://example.com/elsie
Lacie : http://example.com/lacie
Tillie : http://example.com/tillie
>>>print(soup.get_text())                        #返回所有文本
The Dormouse's story
The Dormouse's story
Once upon a time there were three little sisters; and their names were
Elsie,
Lacie and
Tillie;
and they lived at the bottom of a well.
...
>>>soup.a['id']='test_link1'                     #修改标签属性的值
>>>soup.a
<a class="sister" href="http://example.com/elsie" id="test_link1">Elsie</a>
>>>soup.a.string.replace_with('test_Elsie')      #修改标签文本
'Elsie'
>>>soup.a.string
'test_Elsie'
>>>print(soup.prettify())
<html>
```

```
  <head>
    <title>
      The Dormouse's story
    </title>
  </head>
  <body>
    <p class="title">
      <b>
        The Dormouse's story
      </b>
    </p>
    <p class="story">
      Once upon a time there were three little sisters; and their names were
      <a class="sister" href="http://example.com/elsie" id="test_link1">
        test_Elsie
      </a>
      ,
      <a class="sister" href="http://example.com/lacie" id="link2">
        Lacie
      </a>
      and
      <a class="sister" href="http://example.com/tillie" id="link3">
        Tillie
      </a>
      ;
and they lived at the bottom of a well.
    </p>
    <p class="story">
      ...
    </p>
  </body>
</html>
>>>for child in soup.body.children:                    #遍历直接子标签
    print(child)


<p class="title"><b>The Dormouse's story</b></p>
<p class="story">Once upon a time there were three little sisters; and their names were
<a class="sister" href="http://example.com/elsie" id="test_link1">test_Elsie
</a>,
<a class="sister" href="http://example.com/lacie" id="link2">Lacie</a>and
<a class="sister" href="http://example.com/tillie" id="link3">Tillie</a>;
and they lived at the bottom of a well.</p>
<p class="story">…</p>
```

```
>>>for string in soup.strings:                    #遍历所有文本,结果略
    print(string)
>>>test_doc='<html><head></head><body><p></p><p></p></body></heml>'
>>>s=BeautifulSoup(test_doc, 'lxml')
>>>for child in s.html.children:                  #遍历直接子标签
    print(child)

<head></head>
<body><p></p><p></p></body>
>>>for child in s.html.descendants:               #遍历子孙标签
    print(child)

<head></head>
<body><p></p><p></p></body>
<p></p>
<p></p>
```

9.4 网站开发

9.4.1 使用 IIS 运行 Python CGI 程序

在 Windows 平台上,为了使 IIS 能够运行 Python 程序,需要完成以下几步设置(以 Win 7 旗舰版为例)。

(1)依次打开"开始"→"控制面板"→"管理工具"→"Internet 信息服务(IIS)管理器",右击"网站",新建网站并填写网站基本信息,如图 9-4 所示。

图 9-4　在 IIS 中创建网站

（2）选择刚创建的 Python 网站，在右侧窗口中选择"处理程序映射"，然后在打开的窗口中右侧单击"添加脚本映射"，在弹出的窗口中填写信息，如图 9-5 所示。

图 9-5　配置 IIS 的程序映射

（3）编写 Python 程序文件 index.py，并放置到刚刚创建的网站根目录中。代码如下：

```
print('Status: 200 OK')
print('Content-type: text/html')
print('')
header='''<head runat="server">
    <title></title>
    <script language="javascript" type="text/javascript">
// <![CDATA[

        function Button2_onclick(){
            alert(Text1.value);
        }

// ]]>
    </script>
</head>'''
print(header)
print('<h1>This is a header</h1>')
print('<input id="Text1" type="text" />')
button='''<input id="Button2" type="button" value="button"
        onclick="return Button2_onclick()"/>'''
print(button)
print('<p>note:this is only a test.</p>')
```

（4）选择上面创建的 Python 网站，单击右侧窗口中的"默认文档"，然后添加 index.py。

（5）打开浏览器并输入刚才配置的网址，浏览创建的网站，如图 9-6 所示。在网页中的文本框内输入内容以后，单击右侧的按钮，可以弹出网页信息提示输入的内容。

图 9-6　网站运行效果（一）

9.4.2　Python 在 ASP.NET 中的应用

IronPython 插件使得可以在 ASP.NET 网站中使用 Python 代码，不过目前只支持 Python 2.7.x，暂时还不支持 Python 3.x。下面以 Visual Sutdio 2008 和 Visual Studio 2010 两个版本为例介绍如何在 ASP.NET/C♯ 网站中使用 Python 代码，Visual Studio 2015 中已经内置支持 Python，不再赘述。

1. C♯ 2008 结合 Python 开发 ASP.NET 网站

（1）创建 Web Site，并添加 IronPython.dll 和 Micrsoft.Scripting.dll 两个引用。

（2）编写 Python 程序文件 demo.py，并放置到网站根目录中，内容如下：

```
def demo(a):
    return a.split(',')
```

（3）在默认页面文件 Default.aspx.cs 中添加命名空间 IronPython.Hosting 和 Microsoft.Scripting.Hosting。

（4）在默认 Web 页面 Default.aspx 中放置一个按钮和下拉列表框，然后在按钮的单击事件处理函数中添加如下代码：

```
var engine=Python.CreateEngine();
var scope=engine.CreateScope();
var source= engine.CreateScriptSourceFromFile(Server.MapPath("~")+"\\demo.py");
source.Execute(scope);
#获取 Python 程序文件中的函数
var demo=scope.GetVariable<Func<object, object>>("demo");
#调用 Python 程序文件中的函数
DropDownList1.DataSource=demo("a,b,c,d,e");
DropDownList1.DataBind();
```

（5）运行网站，初始时下拉列表框中没有任何可选项，单击按钮为下拉列表框添加几个英文字母的可选项,如图9-7所示。

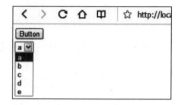

图9-7 网站运行效果(二)

2. C♯ 2010 结合 Python 开发 ASP.NET 网站

Visual Studio 2010 的 C♯ 4.0 提供了一个新的关键字 dynamic,大幅度方便了 C♯ 与 Python 语言的混合编程。

在 Visual Studio 2010 中创建 Web Site,添加按钮和下拉列表框,添加引用(选择 ASP.NET 4.0 版本,在 IronPython 的安装目录下)IronPython.dll 和 Micrsoft.Scripting.dll,添加命名空间 IronPython.Hosting 和 Microsoft.Scripting.Hosting,编写并添加 demo.py 文件,与 Visual Studio 2008 中的操作步骤一样,然后在按钮的单击事件处理函数中添加如下代码:

```
ScriptRuntime pyRuntime=Python.CreateRuntime();
dynamic obj=pyRuntime.UseFile(Server.MapPath("~")+"\\demo.py");
#调用 Python 程序文件中的函数
DropDownList1.DataSource=obj.Split("a,b,c,d,e");
DropDownList1.DataBind();
```

运行网站,单击按钮,运行效果与图9-7相同。

3. 在 ASP.NET 网站中使用 C♯ 和 Python 混合处理数据

以 Visual Studio 2010 为例,假设有 Access 数据库 db5.mdb,表名为 test,第一个字段为整数类型。创建 ASP.NET 新页面,放置一个按钮,在其 Click 事件中编写代码:

```
string connectionString=@"Provider=Microsoft.Jet.OLEDB.4.0;Data Source="+
Server.MapPath("~")+@"\App_Data\db5.mdb";
OleDbConnection conn=new OleDbConnection();
conn.ConnectionString=connectionString;
string sql="SELECT * FROM test";
OleDbDataAdapter adapter=new OleDbDataAdapter(sql, conn);
DataSet ds=new DataSet();
adapter.Fill(ds);
ScriptRuntime pyRuntime=Python.CreateRuntime();
dynamic obj=pyRuntime.UseFile(Server.MapPath("~")+"\\demo.py");
Button1.Text=obj.demo1(ds.Tables[0].Rows).ToString();
```

上面代码从 Access 数据库 db5.mdb 中 test 表查询所有数据,将返回的所有数据行传递给 Python 程序中的函数 demo1()。编写 demo.py 文件并放在网站根目录中，代码如下：

```
def demo1(v):
    return v[0][0]+v[1][0]
```

运行网站并单击按钮后按钮文字变为 test 数据表中前两列第一个字段数字之和。

9.4.3 Flask 框架简单应用

Flask 是 Python 社区比较主流的 Web 框架之一，可以使用 pip 工具安装 Flask 框架及其扩展包。

例 9-6 使用 Flask 框架编写网站程序。

```
from flask import Flask

app=Flask(__name__)
@app.route("/")
def hello():
    return "Hello World!"

if __name__=="__main__":
    app.run()
```

将代码保存为 flask_test.py 并运行，在 IDLE 中显示网站已启动，如图 9-8 所示。

使用浏览器打开网址 http://127.0.0.1:5000，显示文本"Hello World!"，如图 9-9 所示。

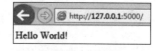

图 9-8　网站启动界面（三）　　　图 9-9　网站运行效果（四）

例 9-7 Python＋Flask＋flask-email 发送带附件的电子邮件。

运行下面的程序之前，需要使用 pip install flask-mail 安装电子邮件扩展包。

```
import os.path
from flask import Flask
from flask.ext.mail import Mail, Message

app=Flask(__name__)
#以 126 免费邮箱为例
app.config['MAIL_SERVER']='smtp.126.com'
app.config['MAIL_PORT']=25
app.config['MAIL_USE_TLS']=True
#如果电子邮箱地址是 abcd@126.com,那么应填写 abcd
```

```python
app.config['MAIL_USERNAME']='your own username of your email'
app.config['MAIL_PASSWORD']='your own password of the username'

def sendEmail(From, To, Subject, Body, Html, Attachments):
    '''To:must be a list'''
    msg=Message(Subject, sender=From, recipients=To)
    msg.body=Body
    msg.html=Html
    for f in Attachments:
        with app.open_resource(f) as fp:
            msg.attach(filename=os.path.basename(f), data=fp.read(),
                    content_type='application/octet-stream')
    mail=Mail(app)
    with app.app_context():
        mail.send(msg)

if __name__=='__main__':
    #From填写的电子邮箱地址必须与前面配置的相同
    From='<your email address>'
    #这是我本人的QQ邮箱,大家测试的时候一定要修改一下啊
    To=['<306467355@qq.com>']
    Subject='hello world'
    Body='Only a test.'
    Html='<h1>test test test.</h1>'
    Attachments = ['c:\\python35\\python.exe']
    sendEmail(From, To, Subject, Body, Html, Attachments)
```

9.4.4 django 框架简单应用

django 是非常成熟的 Web 框架,采用经典的 MVC 模式,支持使用 pip 工具安装。安装成功之后,在 Python 安装目录中的 scripts 文件夹中有个文件 django-admin.py,这是用来创建和管理 Web 应用的重要程序。

例 9-8 创建第一个 django Web 应用。

在命令提示符环境中使用下面的命令创建一个 Web 应用 first_django:

```
python django-admin.py startproject first_django
```

如果这个命令执行成功(一般都会执行成功,如果不成功的话,很可能是 django 安装不成功或者 Web 应用的名字有冲突),会在当前文件夹中创建一个子文件夹 first_django。

在 first_django\first_django 文件夹中创建一个 Python 程序 view.py(当然也可以不叫这个名字,只要和后面代码中一致就可以了),内容如下:

```python
from django.http import HttpResponse
```

```python
def hello(request):
    return HttpResponse('This is my first django application!')

def greeting(request):
    return HttpResponse('Hello everyone!')
```

然后,打开文件 first_django\first_django\urls.py,修改其中的代码如下:

```python
from django.conf.urls import url
from first_django.view import hello, greeting
                                            #这里 view 对应刚才创建的 view.py 文件


urlpatterns=[                               #建立 URL 和函数的对应关系
    url(r'^hello/', hello),                 #第一个参数是正则表达式
    url(r'^greeting/', greeting),           #第二个参数是 view.py 中的函数
]
```

好了,一个简单的网站已经建好了,接下来到命令提示符环境中执行下面的命令启动这个网站:

```
python manage.py runserver 127.0.0.1:8000
```

在上面的命令中,127.0.0.1 表示本机 IP 地址,当然读者可以修改为自己计算机上配置的 IP 地址,8000 表示这个网站使用的端口,也可以修改为其他的端口,如果端口没有冲突的话,会出现下面的图 9-10 中的提示,表示网站启动成功了。最后可以使用浏览器打开网址 http://127.0.0.1:8000/greeting/ 和 http://127.0.0.1:8000/hello/ 查看效果。

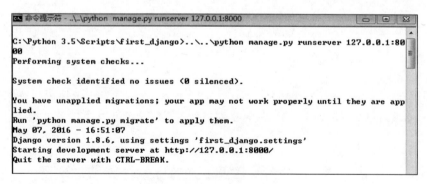

图 9-10 启动 django 网站

小提示:当修改网站中的文件时,不需要手工重新启动网站,django 会自动检查并重启服务。另外,当用户通过浏览器访问页面时,命令提示符窗口会提示相应的日志信息。

例 9-9 素数判断。

按照上例的步骤,首先创建网站 django_IsPrime,然后编写 Python 程序文件 django_

IsPrime\django_IsPrime\view.py,其中代码如下:

```python
from django.http import HttpResponse, Http404
from math import sqrt

def isPrime(request, number):                    #number 表示接收到的参数
    try:
        number=int(number)
    except:
        raise Http404
    if number ==2:                               #2 是最小的素数
        flag=' is '
    else:
        for i in range(2, int(sqrt(number)+2)):  #判断 number 是否为素数
            if number%i ==0:
                flag=' is not '
                break
            else:
                flag=' is '
    txt=str(number)+flag+' a Prime'
    return HttpResponse('<h1>'+txt+'</h1>')
```

然后修改 django_IsPrime\django_IsPrime\urls.py 文件中的代码为

```python
from django.conf.urls import include, url
from django.contrib import admin
from django_IsPrime.view import isPrime

urlpatterns=[                                    #URL 与函数之间的对应关系
    url(r'^admin/', include(admin.site.urls)),
    url(r'^isPrime/(\d+)/$', isPrime),           #正则表达式,提取数字参数
]
```

然后启动网站,在浏览器中输入 http://127.0.0.1:8000/isPrime/91,按 Enter 键后查看结果,如图 9-11 所示,其中 91 表示要判断的整数,读者可以改成其他数字进行检验。

图 9-11 素数判断

例 9-10 使用网页模板。

使用网页模板可以避免重复设计网页布局带来的工作量,有效利用已有代码,保持风格的一致性。按前面的步骤,首先创建网站 django_template,然后创建文件夹 django_ template\templates,并创建网页模板文件 django_template\templates\greeting.html,其中内容为

```html
<html>
    <body>
        <h1>Good {{morning_afternoon_evening}}, I am {{name}}.</h1>
    </body>
</html>
```

其中,两对大括号{{}}中的内容为变量,也就是运行时要替换的内容。接下来创建 Python 程序文件 django_template\django_template\view.py,其中代码为

```python
from django.http import HttpResponse, Http404
import datetime
from django import template
from django.conf import settings
import os
import os.path
from random import choice

#settings.configure()
names = ('Zhang san', 'Li si', 'Wang wu', 'Ma liu')

def greeting(request):
    templateFile=os.path.join(os.path.split(os.path.dirname(__file__))[0],'
                              templates')+'\\greeting.html'
    with open(templateFile) as fp:           #打开网页模板文件,创建模板
        t=template.Template(fp.read())

    current_name=choice(names)                #随机选择一个问候人
    h=datetime.datetime.now().hour            #当前时间
    if 0<=h<12:                               #上午
        mae='Morning'
    elif 12<=h<18:                            #下午
        mae='Afternoon'
    else:                                     #晚上
        mae='Evening'
    con=template.Context({'name':current_name, 'morning_afternoon_evening':
    mae})
    html=t.render(con)                        #渲染模板
    return HttpResponse(html)
```

最后,修改 django_template\django_template\urls.py 文件如下:

```python
from django.conf.urls import include, url
from django.contrib import admin
```

```
from django_template.view import greeting

urlpatterns=[
    url(r'^admin/', include(admin.site.urls)),
    url(r'^greeting/$', greeting),
]
```

好了,网站设计完了,把上面所有文件都保持到正确的文件夹中,启动网站,然后使用浏览器访问 http://127.0.0.1:8000/greeting/,多刷新几次,会看到网页上的问候人会有所变化,并且问候的时间也会根据当前时间进行相应的变化,某时刻的运行效果如图 9-12 所示。

图 9-12 使用网页模板

第 10 章 多线程与多进程

日常生活中我们每天都在不知不觉地使用多线程和多进程的技术来安排繁忙的事务。例如，早上 6:20 起床后，先把牛奶热上（启动煮牛奶进程），把鸡蛋煮上（启动煮鸡蛋进程），烧上一壶水（启动烧水进程），然后开始洗漱并等待牛奶或鸡蛋煮好的信号（进程同步），当收到相应的信号时关闭相应的电源或天然气阀门，洗漱结束后洗个青菜开始炒，其间如果已到达 6:50 则叫孩子起床洗漱吃饭（又启动一个进程）……如果这些任务逐个进行，完成一个再开始下一个，时间根本不够。而由于多个任务同时进行，大大提高了整体效率。

硬件技术的发展速度已经超越了"摩尔定律"的限制，早期的多核、多处理器这些高大上的技术已经走进了普通家庭，再加上内存、主频、硬盘等各种硬件配置的飞速提高，大幅度提高了普通 PC 的运算速度和数据处理能力，甚至普通手机也具有了四核以上的处理器、4GB 以上的运行内存以及 32GB 以上的存储卡。在多核、多处理器平台上，在任意时刻每个核可以运行一个线程，多个线程同时运行并相互协作，从而达到高速处理任务的目的。然而，即使是高端服务器或工作站甚至集群系统，处理器和核的数量总是有限的，如果线程的数量多于核的数量，就必然需要进行调度。在调度时，处理器为每个线程分配一个很短的时间片，所有线程根据具体的调度算法轮流获得该时间片。当时间片用完以后，即使该线程还没有执行完也要退出处理器并等待下次调度（也就是说，一个线程可能需要被调度并执行很多次才会结束，图 10-1 中椭圆内的上下文开关次数反映了线程被调度的次数），同时由操作系统按照优先级再选择一个线程进入 CPU 运行（多处理器的调度算法更加复杂，可以参考《操作系统》之类的书籍）。由于处理器中寄存器的数量有限，而不同的线程很可能需要使用到相同的一组寄存器来保存中间计算结果或当前运行状态。因此，在调度线程时必须要做好上下文保存和恢复工作，以保证该线程下次被调度进处理器后能够继续上次的工作。虽然这些工作并不需要 Python 程序员操心，但是我们必须清楚的一件事是，并不是使用的线程数量越多越好，如果线程太多，线程调度带来的开销可能会比线程实际执行的开销还大，这样使用多线程就失去本来的意义了。Python 多线程编程技术存在 GIL 问题，而使用多进程则有效地避免了这个问题，进一步提高了系统吞吐量。

图 10-1 WPS 进程中某个线程的属性

10.1 多线程编程

磁盘上的应用程序文件被打开并执行时就创建了一个进程，但进程是"懒惰"的，本身并不是可执行单元，从来不执行任何东西，主要是线程的容器。要使进程中的代码真正运行起来，必须拥有至少一个能够在这个环境中运行代码的执行单元，就是线程。线程是操作系统调度和分配处理器时间的基本单位，负责执行包含在进程地址空间中的代码。当一个进程被创建时，操作系统会自动为之建立一个线程，通常称为主线程。一个进程可以包含多个线程，主线程根据需要再动态创建其他子线程，操作系统为每个线程保存单独的寄存器环境和单独的堆栈，但是它们共享进程的地址空间、对象句柄、代码、数据和其他资源。线程总是在某个进程的上下文中被创建，并且会在这个进程空间中"终其一生"，不可以脱离进程而独立存在。虽然线程总是像大家闺秀一样被限制在进程的地址空间中，但是因为有其他小伙伴（线程）的存在所以一般来说并不会感到寂寞，同一个进程中的多个线程之间是允许进行交流的（数据共享、同步等）。另外，一般来说，除主线程外，其他线程的生命周期都小于其所属进程的生命周期。

标准库 threading 是 Python 支持多线程编程的重要模块，该模块是在底层模块"_thread"的基础上开发的更高层次的线程编程接口，提供了大量的方法和类来支持多线程编程，极大地方便了用户。threading 模块常用方法与类如表 10-1 所示。

表 10-1　threading 模块常用方法与类

方法/类	功 能 说 明
active_count()、activeCount()	返回当前处于 alive 状态的 Thread 对象数量
current_thread()、currentThread()	返回当前 Thread 对象
threading.get_ident()	返回当前线程的线程标识符。线程标识符是一个非负整数，并没有特殊含义，只是用来标识线程，该整数可能会被循环利用。Python 3.3 及以后版本支持该方法
threading.enumerate()	返回当前处于 alive 状态的所有 Thread 对象列表
threading.main_thread()	返回主线程对象，即启动 Python 解释器的线程对象。Python 3.4 及以后版本支持该方法
threading.stack_size([size])	返回创建线程时使用的栈的大小，如果指定 size 参数，则用来指定后续创建的线程使用的栈的大小，size 必须是 0（表示使用系统默认值）或大于 32K（K 表示 1024）的正整数
Thread	线程类，用于创建和管理线程
Event	事件类，用于线程同步
Condition	条件类，用于线程同步
Lock、RLock	锁类，用于线程同步
Semaphore	信号量类，用于线程同步
Timer	用于在指定时间之后调用一个函数的情况

下面的代码简单演示了该模块方法的用法：

```
>>>import threading
>>>threading.stack_size()                     #查看当前线程栈的大小
0
>>>threading.stack_size(64 * 1024)            #设置当前线程栈的大小
0
>>>threading.stack_size()
65536
>>>threading.active_count()                   #查看活动线程数量
2
>>>threading.current_thread()                 #返回当前线程对象
<_MainThread(MainThread, started 4852)>
>>>threading.enumerate()                      #枚举所有线程
[<Thread(SockThread, started daemon 9620)>, <_MainThread(MainThread, started 4852)>]
>>>def demo(v):
    print(v)
>>>t=threading.Timer(3, demo, args=(5,))      #创建线程
>>>t.start()                                  #启动线程,3s 之后调用 demo 函数
>>>t.cancel()                                 #如果仍在等待时间到达,则取消
```

10.1.1 线程创建与管理

标准库 threading 中的 Thread 类用来创建和管理线程对象，支持使用两种方法来创建线程：①直接使用 Thread 类实例化一个线程对象并传递一个可调用对象作为参数；②继承 Thread 类并在派生类中重写 __init__()和 run()方法。创建了线程对象以后，可以调用其 start()方法来启动，该方法自动调用该类对象的 run()方法，此时该线程处于 alive 状态，直至线程的 run()方法运行结束。Thread 对象主要成员如表 10-2 所示。

表 10-2 Thread 对象成员

成员	说明
start()	自动调用 run()方法，启动线程，执行线程代码
run()	线程代码，用来实现线程的功能与业务逻辑，可以在子类中重写该方法来自定义线程的行为
__init__(self, group=None, target=None, name=None, args=(), kwargs=None, verbose=None)	构造函数
name	用来读取或设置线程的名字
ident	线程标识，非 0 数字或 None(线程未被启动)
is_alive()、isAlive()	测试线程是否处于 alive 状态
daemon	布尔值，表示线程是否为守护线程
join(timeout=None)	等待线程结束或超时返回

(1) join([timeout])。

阻塞当前线程，等待被调线程结束或超时后再继续执行当前线程的后续代码，参数 timeout 用来指定最长等待时间，单位是秒。

例 10-1 线程对象的 join()方法。

```
from threading import Thread
import time

def func1(x, y):
    for i in range(x, y):
        print(i, end=' ')
    print()
    time.sleep(10)                              #等待 10s

t1=Thread(target=func1, args=(15, 20))          #创建线程对象,args 是传递给函数的参数
t1.start()                                       #启动线程
t1.join(5)                                       #等待线程 t1 运行结束或等待 5s
t2=Thread(target=func1, args=(5, 10))
t2.start()
```

保存并运行上面的程序,首先输出 15 至 19 这 5 个整数,然后程序暂停 5s 以后又继续输出 5 至 9 这 5 个整数。如果把 t1.join(5)这一行注释或删除之后再次运行,两个线程的输出将会重叠在一起,这是因为两个线程并发运行,而不是等待第一个结束以后再运行第二个。如果把 time.sleep(10)这一行注释或删除再运行,会发现两个线程的输出之间没有时间间隔,这是因为线程对象的 join()方法当线程运行结束或超时之后返回,虽然指定了超时时间为 5s,而实际上线程函数瞬间就执行结束了。

拓展知识:以创建并启动线程的方式来执行一个函数可以实现多个函数或功能代码并发或同时运行,而直接调用函数的话会阻塞当前线程,直到函数执行结束返回后才能继续执行当前线程的代码。

拓展知识:前面我们已经多次用过 Python 标准库 time 中的 sleep()函数,它的功能是暂停(或者说阻塞当前线程)指定时间(单位是秒)。从多线程编程和调度算法上来讲,该函数的真正功能是告诉操作系统在一定的时间内不要再调度自己了,主动把时间片让出来给别的线程。也就是说,虽然系统中同时会存在大量的线程,但是由于优先级问题以及主动阻塞等原因,真正处于可调度状态的线程数量并不是非常多,系统也不会给暂时没事可做的线程分配 CPU 时间。

(2)isAlive()。

这个方法用来测试线程是否处于运行状态,如果仍在运行则返回 True,如果尚未启动或运行已结束则返回 False。

例 10-2 线程状态检测。

```
from threading import Thread
import time
def func1():
    time.sleep(10)

t1=Thread(target=func1)
print('t1:',t1.isAlive())              #线程还没有运行,返回 False
t1.start()
print('t1:',t1.isAlive())              #线程还在运行,返回 True
t1.join(5)                             #join()方法因超时而结束
print('t1:',t1.isAlive())              #线程还在运行,返回 True
t1.join()                              #等待线程结束
print('t1:',t1.isAlive())              #线程已结束,返回 False
```

上面程序的输出结果为

```
t1: False
t1: True
t1: True
t1: False
```

(3) daemon 属性。

在脚本运行过程中有一个主线程,若在主线程中创建了子线程,当主线程结束时根据子线程 daemon 属性值的不同可能会发生下面的两种情况之一。

① 如果某个子线程的 daemon 属性为 False，主线程结束时会检测该子线程是否结束，如果该子线程还在运行，则主线程会等待它完成后再退出。

② 如果某个子线程的 daemon 属性为 True，主线程运行结束时不对这个子线程进行检查而直接退出，同时所有 daemon 值为 True 的子线程将随主线程一起结束，而不论是否运行完成。

daemon 属性的值默认为 False，如果需要修改，则必须在调用 start() 方法启动线程之前进行设置。

以上论述不适用于 IDLE 环境中的交互模式或脚本运行模式，因为在该环境中的主线程只有在退出 Python IDLE 时才终止。

例 10-3 线程对象的 daemon 属性。

```python
import threading
import time

class mythread(threading.Thread):                #继承 Thread 类，创建自定义线程类
    def __init__(self, num, threadname):
        threading.Thread.__init__(self, name=threadname)
        self.num=num
    def run(self):                               #重写 run() 方法
        time.sleep(self.num)
        print(self.num)

t1=mythread(1, 't1')                             #创建自定义线程类对象，daemon 默认为 False
t2=mythread(5, 't2')
t2.daemon=True                                   #设置线程对象 t2 的 daemon 属性为 True
print(t1.daemon)
print(t2.daemon)
t1.start()                                       #启动线程
t2.start()
```

把上面的代码存储为 ThreadDaemon.py 文件，在 IDLE 环境中运行结果如图 10-2 所示，在命令提示符环境中运行结果如图 10-3 所示。可以看到，在命令提示符环境中执行该程序时，线程 t2 没有执行结束就跟随主线程一同结束了，因此并没有输出数字 5。

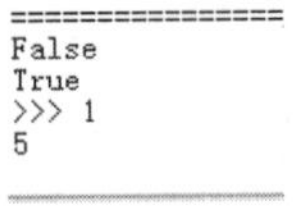

图 10-2　在 IDLE 环境中运行 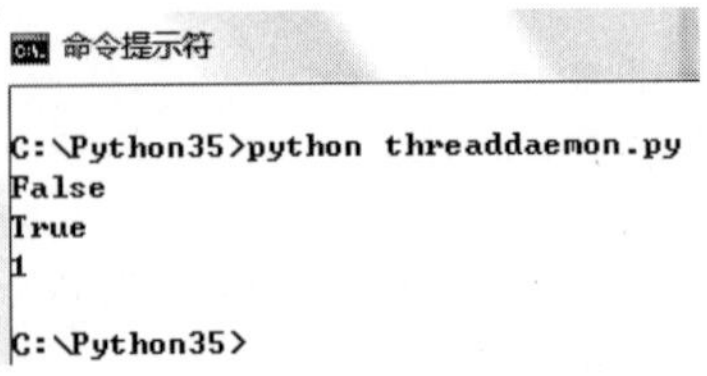　　　　图 10-3　在命令提示符环境中运行

10.1.2 线程同步技术

俗话说，人多力量大，众人拾柴火焰高。一般情况下这是成立的，如果团队中每个人都能够互不干扰地工作，整体效率会达到最佳状态。但这在实际中是不可能实现的，随着队伍的壮大，团队成员之间需要协调和同步才能有条不紊地完成整个任务，随之而来的是多人合作时的管理和沟通代价的增加，而当成员人数超过一定阈值之后，可能会带来不可估量的困难。另外，如果多人使用同一个资源，就更需要很好地控制对资源的访问。例如，服装店里的试衣间或者火车上的卫生间，里面有人时会有一定的标记，里面没人时下一个人才可以进入并使用。多线程编程也是同样的道理，不仅要合理控制线程数量，还需要在多个线程之间进行有效地同步和调度才能发挥最大功效。

多线程的主要作用之一是为了充分利用硬件资源，尤其是提高 CPU 的利用率，提高系统的任务处理速度和吞吐量，让各个部件都处于高速运转和忙碌状态。把任务拆分成互相协作的多个线程同时运行，那么属于同一个任务的多个线程之间必然会有交互和同步以便能够互相协作地完成任务，例如，使用多线程技术从网络上下载文件可以提高整体下载速度，但需要合理安排每个线程负责下载的内容才行。

拓展知识：多线程技术的提出并不仅仅是为了提高处理速度和硬件资源利用率，还有就是为了增加系统的可扩展性（采用多线程技术编写的代码移植到多处理器平台上继续不需要改写就能立刻适应新的平台）和提高用户体验。对于单核 CPU 计算机而言，使用多线程并不能提高任务完成速度，但有些场合必须要使用多线程技术，或者采用多线程技术可以让整个系统的设计更加人性化。例如，在执行另一段代码的同时还想接收用户的键盘或鼠标事件以提高用户体验，这时候就只能以线程的形式来运行另一个函数的代码。Windows 操作系统的 Windows Indexing Services 创建了一个低优先级的线程，该线程定期被唤醒并对磁盘上的特定区域的文件内容进行索引以提高用户搜索速度。打开 Photoshop、3ds Max 这样的大型软件时需要加载很多模块和动态链接库，软件启动时间会比较长。这时候可以使用一个线程来显示一个小动画来表示当前软件正在启动，而当后台线程加载完所有的模块和库之后，结束该动画的播放并打开软件主界面，这也是多线程同步的一个典型应用。字处理软件可以使用一个优先级高的线程来接收用户键盘输入，而使用一些低优先级线程来进行拼写检查、语法检查、分页以及字数统计之类的功能并实时将结果显示在状态栏上，这无疑会极大方便用户的使用，对于提高用户体验有重要帮助。图 10-4 展示了字处理软件 WPS 启动之后创建的部分线程，图 10-5 中列出了作者的计算机上某个时刻所有进程拥有的线程数量。

1. Lock/RLock 对象

Lock 是比较低级的同步原语，当被锁定以后不属于特定的线程。一个锁有两种状态：locked 和 unlocked。如果锁处于 unlocked 状态，acquire()方法将其修改为 locked 并立即返回；如果锁已处于 locked 状态，则阻塞当前线程并等待其他线程释放锁，然后将其修改为 locked 并立即返回。release()方法用来将锁的状态由 locked 修改为 unlocked 并

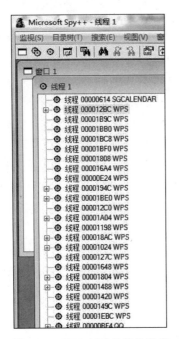

图 10-4　WPS 创建的部分线程　　图 10-5　系统中每个进程的线程数量

立即返回,如果锁状态本来已经是 unlocked,调用该方法将会抛出异常。

可重入锁 RLock 对象也是一种常用的线程同步原语,可被同一个线程 acquire()多次。当处于 locked 状态时,某线程拥有该锁;当处于 unlocked 状态时,该锁不属于任何线程。RLock 对象的 acquire()/release()调用对可以嵌套,仅当最后一个或者最外层的 release()执行结束后,锁才会被设置为 unlocked 状态。

例 10-4　使用 Lock/RLock 对象实现线程同步。

```
import threading
import time

#自定义线程类
class mythread(threading.Thread):
    def __init__(self):
        threading.Thread.__init__(self)
    #重写 run()方法
    def run(self):
        global x
        #获取锁,如果成功则进入临界区
        lock.acquire()
        x=x+3
```

```
            print(x)
            #退出临界区,释放锁
            lock.release()

lock=threading.RLock()
#也可以使用 Lock 类实现加锁和线程同步
#lock=threading.Lock()

#存放多个线程的列表
tl=[]
for i in range(10):
    #创建线程并添加到列表
    t=mythread()
    tl.append(t)

#多个线程互斥访问的变量
x=0
#启动列表中的所有线程
for i in tl:
    i.start()
```

保存并运行上面的程序,依次输出 3、6、9、12、15、18、21、24、27、30 这几个数字。把 lock.acquire()和 lock.release()这两行注释或删除之后再运行,会发现多个线程之间没有任何"默契",输出结果变得杂乱无章,并且每次运行都会得到不同的结果,完全不可再现。

注意：多线程同步时如果需要获得多个锁才能进入临界区,可能会发生死锁,在多线程编程时一定要注意并认真检查和避免这种情况。例如,下面的代码在计算机非常繁忙的情况下就有可能发生死锁,感兴趣的读者可以搜索一下"哲学家就餐问题"了解更多内容。

```
import threading
import time

class mythread1(threading.Thread):
    def __init__(self):
        threading.Thread.__init__(self)
    def run(self):
        lock1.acquire()                    #获取一个锁
        lock2.acquire()                    #获取另一个锁
        #实际功能代码(略)
        lock2.release()
        lock1.release()
```

```python
class mythread2(threading.Thread):
    def __init__(self):
        threading.Thread.__init__(self)
    def run(self):
        lock2.acquire()
        lock1.acquire()
        #实际功能代码(略)
        lock1.release()
        lock2.release()

lock1=threading.RLock()
lock2=threading.RLock()
t1=mythread1()
t2=mythread2()
t1.start()
t2.start()
```

2. Condition 对象

使用 Condition 对象可以在某些事件触发后才处理数据或执行特定的功能代码,可以用于不同线程之间的通信或通知,以实现更高级别的同步。Condition 对象除了具有 acquire()和 release()方法之外,还有 wait()、notify()、notify_all()等方法。下面通过经典的生产者/消费者问题来演示 Condition 对象的用法,程序中生产者线程和消费者线程共享一个列表,生产者负责在列表尾部追加元素,消费者则从列表首部获取并删除元素。如果列表长度到了 20 表示已满,生产者等待,如果列表已空则消费者等待。

例 10-5　使用 Condition 对象实现线程同步。

```python
import threading
from random import randint
from time import sleep

#自定义生产者线程类
class Producer(threading.Thread):
    def __init__(self, threadname):
        threading.Thread.__init__(self,name=threadname)
    def run(self):
        global x
        while True:
            #获取锁
            con.acquire()
            #假设共享列表中最多能容纳 20 个元素
            if len(x)==20:
                #如果共享列表已满,生产者等待
                con.wait()
```

```python
            print('Producer is waiting…')
        else:
            print('Producer:', end=' ')
            #产生新元素,添加至共享列表
            x.append(randint(1, 1000))
            print(x)
            sleep(1)
            #唤醒等待条件的线程
            con.notify()
        #释放锁
        con.release()

#自定义消费者线程类
class Consumer(threading.Thread):
    def __init__(self, threadname):
        threading.Thread.__init__(self, name =threadname)
    def run(self):
        global x
        while True:
            #获取锁
            con.acquire()
            if not x:
                #等待
                con.wait()
                print('Consumer is waiting…')
            else:
                print(x.pop(0))
                print(x)
                sleep(2)
                con.notify()
            con.release()

#创建 Condition 对象以及生产者线程和消费者线程
con=threading.Condition()
x=[]
p=Producer('Producer')
c=Consumer('Consumer')
p.start()
c.start()
p.join()
c.join()
```

该程序的运行结果如图 10-6 所示。

```
>>>
====================== RESTART: C:\Python 3.5\demo.py ========
>>> Producer: [842]
Producer: [842, 755]
Producer: [842, 755, 713]
Producer: [842, 755, 713, 73]
Producer: [842, 755, 713, 73, 166]
842
[755, 713, 73, 166]
Producer: [755, 713, 73, 166, 902]
Producer: [755, 713, 73, 166, 902, 921]
755
[713, 73, 166, 902, 921]
Producer: [713, 73, 166, 902, 921, 448]
713
[73, 166, 902, 921, 448]
Producer: [73, 166, 902, 921, 448, 53]
Producer: [73, 166, 902, 921, 448, 53, 865]
73
[166, 902, 921, 448, 53, 865]
```

图 10-6　使用 Condition 实现线程同步

3．Queue 对象

queue 模块（在 Python 2 中为 Queue 模块）的 Queue 对象实现了多生产者/多消费者队列，尤其适合需要在多个线程之间进行信息交换的场合，实现了多线程编程所需要的所有锁语义。

例 10-6　使用 queue 对象实现线程同步。

```python
import threading
import time
import queue

#自定义生产者线程类
class Producer(threading.Thread):
    def __init__(self, threadname):
        threading.Thread.__init__(self, name=threadname)
    def run(self):
        global myqueue
        #在队列尾部追加元素
        myqueue.put(self.getName())
        print(self.getName(), ' put ', self.getName(), ' to queue.')


class Consumer(threading.Thread):
    def __init__(self, threadname):
        threading.Thread.__init__(self, name=threadname)
    def run(self):
        global myqueue
        #在队列首部获取元素
        print(self.getName(), ' get ', myqueue.get(), ' from queue.')


myqueue=queue.Queue()

#创建生产者线程和消费者线程
```

```python
plist=[]
clist=[]
for i in range(10):
    p=Producer('Producer'+str(i))
    plist.append(p)
    c=Consumer('Consumer'+str(i))
    clist.append(c)

#依次启动生产者线程和消费者线程
for p, c in zip(plist, clist):
    p.start()
    p.join()
    c.start()
    c.join()
```

上面的程序运行结果如下:

```
Producer0  put  Producer0  to queue.
Consumer0  get  Producer0  from queue.
Producer1  put  Producer1  to queue.
Consumer1  get  Producer1  from queue.
Producer2  put  Producer2  to queue.
Consumer2  get  Producer2  from queue.
Producer3  put  Producer3  to queue.
Consumer3  get  Producer3  from queue.
Producer4  put  Producer4  to queue.
Consumer4  get  Producer4  from queue.
Producer5  put  Producer5  to queue.
Consumer5  get  Producer5  from queue.
Producer6  put  Producer6  to queue.
Consumer6  get  Producer6  from queue.
Producer7  put  Producer7  to queue.
Consumer7  get  Producer7  from queue.
Producer8  put  Producer8  to queue.
Consumer8  get  Producer8  from queue.
Producer9  put  Producer9  to queue.
Consumer9  get  Producer9  from queue.
```

4. Event 对象

Event 对象是一种简单的线程通信技术,一个线程设置 Event 对象,另一个线程等待 Event 对象。Event 对象的 set()方法可以设置 Event 对象内部的信号标志为真;clear() 方法可以清除 Event 对象内部的信号标志,将其设置为假;isSet()方法用来判断其内部信号标志的状态;wait()方法在其内部信号状态为真时会立刻执行并返回,若 Event 对象的内部信号标志为假，wait()方法就一直等待至超时或者内部信号状态为真。

例 10-7 使用 Event 对象实现线程同步。

```python
import threading

#自定义线程类
class mythread(threading.Thread):
    def __init__(self, threadname):
        threading.Thread.__init__(self, name=threadname)

    def run(self):
        global myevent
        #根据 Event 对象是否已设置做出不同的响应
        if myevent.isSet():
            #清除标志
            myevent.clear()
            #等待
            myevent.wait()
            print(self.getName()+' set')
        else:
            print(self.getName()+' not set')
            #设置标志
            myevent.set()

myevent=threading.Event()
#设置标志
myevent.set()

for i in range(10):
    t=mythread(str(i))
    t.start()
```

将上面的代码保存为 demo.py 文件并多次运行，会发现每次运行结果略有不同，图 10-7 是其中两次的运行结果。

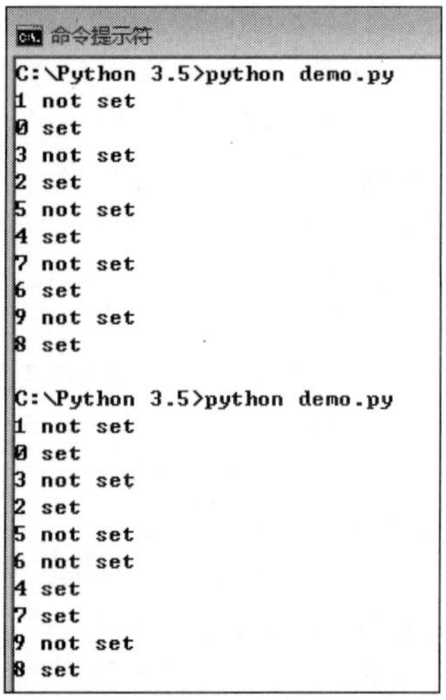

图 10-7 使用 Event 对象实现线程同步

10.2 多进程编程

进程是正在执行中的应用程序。一个进程是一个执行中的文件使用资源的总和，包括虚拟地址空间、代码、数据、对象句柄、环境变量和执行单元等。一个应用程序同时打开并执行多次，就会创建多个进程。

Python 标准库 multiprocessing 用来实现进程的创建与管理以及进程间的同步与数据交换，用法与 threading 类似，是支持并行处理的重要模块。标准库 multiprocessing 同时支持本地并发与远程并发,有效避免了全局解释器锁(Global Interpreter Lock,GIL)问题,可以更有效地利用 CPU 资源,尤其适合多核或多 CPU 环境。

10.2.1 进程创建与管理

与使用 threading 创建和启动线程类似,可以通过创建 Process 对象来创建一个进程,然后通过调用进程对象的 start()方法来启动,通过调用 join()方法等待一个进程执行结束。

例 10-8 进程创建与启动。

```
from multiprocessing import Process
import os

def f(name):
    print('module name:', __name__)
    print('parent process:', os.getppid())    #查看父进程 ID
    print('process id:', os.getpid())         #查看当前进程 ID
    print('hello', name)

if __name__ == '__main__':
    p=Process(target=f, args=('bob',))        #创建进程
    p.start()                                  #启动进程
    p.join()                                   #等待进程运行结束
```

💡**小提示**:本节的很多程序需要在命令提示符环境中运行。

除了支持与 threading 管理线程相似的接口之外,multiprocessing 还提供了 Pool 对象支持数据的并行操作。例如,下面的代码可以并发计算二维数组每行的平均值。

```
from multiprocessing import Pool
from statistics import mean

def f(x):
    return mean(x)

if __name__ == '__main__':
    x=[list(range(10)), list(range(20,30)),
       list(range(50,60)), list(range(80,90))]
    with Pool(5) as p:                        #创建包含 5 个进程的进程池
        print(p.map(f, x))                    #并发运行
```

10.2.2 进程间数据交换

例 10-9 使用 Queue 对象在进程间交换数据,一个进程把数据放入 Queue 对象,另一个进程从 Queue 对象中获取数据。

```python
import multiprocessing as mp

def foo(q):
    q.put('hello world!')                        #把数据放入队列

if __name__ == '__main__':
    mp.set_start_method('spawn')                 #Windows系统创建子进程的默认方式
    q=mp.Queue()
    p=mp.Process(target=foo, args=(q,))          #创建进程,把Queue对象作为参数传递
    p.start()
    p.join()
    print(q.get())                               #从队列中获取数据
```

也可以使用上下文对象context的Queue对象实现不同进程间的数据交换。

```python
import multiprocessing as mp

def foo(q):
    q.put('hello world')

if __name__ == '__main__':
    ctx=mp.get_context('spawn')
    q=ctx.Queue()
    p=ctx.Process(target=foo, args=(q,))
    p.start()
    p.join()
    print(q.get())
```

例10-10 使用管道实现进程间数据交换。管道有两个端,一个接收端和一个发送端,相当于在两个进程之间建立了一个用于传输数据的通道。

```python
from multiprocessing import Process, Pipe

def f(conn):
    conn.send('hello world')                     #向管道中发送数据
    conn.close()                                 #关闭管道

if __name__ == '__main__':
    parent_conn, child_conn=Pipe()               #创建管道对象
    p=Process(target=f, args=(child_conn,))      #将管道的一方作为参数传递给子进程
    p.start()
    p.join()
    print(parent_conn.recv())                    #通过管道的另一方获取数据
    parent_conn.close()
```

例10-11 使用共享内存实现进程间数据传递,比较适合大量数据的场合。

```python
from multiprocessing import Process, Value, Array

def f(n, a):
    n.value=3.1415927
    for i in range(len(a)):
        a[i]=a[i] * a[i]

if __name__ == '__main__':
    num=Value('d', 0.0)                         #实型
    arr=Array('i', range(10))                   #整型数组
    p=Process(target=f, args=(num, arr))        #创建进程对象
    p.start()
    p.join()
    print(num.value)
    print(arr[:])
```

例 10-12 使用 Manager 对象实现进程间数据交换。

Manager 对象控制一个拥有 list、dict、Lock、RLock、Semaphore、BoundedSemaphore、Condition、Event、Barrier、Queue、Value、Array、Namespace 等对象的服务端进程，并且允许其他进程访问这些对象。

```python
from multiprocessing import Process, Manager

def f(d, l, t):
    d['name']='Dong Fuguo'
    d['age']=38
    d['sex']='Male'
    d['affiliation']='SDIBT'
    l.reverse()
    t.value=3

if __name__ == '__main__':
    with Manager() as manager:
        d=manager.dict()
        l=manager.list(range(10))
        t=manager.Value('i', 0)
        p=Process(target=f, args=(d, l, t))
        p.start()
        p.join()
        for item in d.items():
            print(item)
        print(l)
        print(t.value)
```

10.2.3 进程同步技术

在需要协同工作完成大型任务时，多个进程间的同步非常重要。进程同步方法与线程同步方法类似，稍微改写一些代码即可，下面以 Lock 对象和 Event 对象为例简单演示其用法。

例 10-13 使用 Lock 对象实现进程同步。

```
from multiprocessing import Process, Lock

def f(l, i):
    l.acquire()                              #获取锁
    try:
        print('hello world', i)
    finally:
        l.release()                          #释放锁

if __name__ == '__main__':
    lock=Lock()                              #创建锁对象
    for num in range(10):
        Process(target=f, args=(lock, num)).start()
```

例 10-14 使用 Event 对象实现进程同步。

```
from multiprocessing import Process, Event

def f(e, i):
    if e.is_set():
        e.wait()
        print('hello world', i)
        e.clear()
    else:
        e.set()

if __name__ == '__main__':
    e=Event()
    for num in range(10):
        Process(target=f, args=(e,num)).start()
```

注意：与多线程编程类似，在编写多进程程序时，一定要控制好进程之间的同步，避免"死锁"。

第 11 章 大数据处理

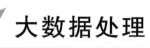

11.1 大数据简介

历史上有个著名的故事叫"草船借箭",故事的主人公诸葛亮对天象的观察实际上就是对风、云、温度、湿度、光照和所处节气等大量多元化的非结构数据进行综合分析,最终通过复杂的计算得出了正确的结论,正是他精准的预测才能有"万事俱备,只欠东风"的从容,最终为决策提供了有力支持,这可以看作是大数据的一个经典应用。

大数据的概念自从提出来以后,迅速在各行各业得到广泛应用。如饭店选址、客户口味分析、菜品销量预测、食材供应商原材料质量分析、企业运作的内在规律挖掘、调度管理、物流优化、社交网络、智能交通、城市规划、客户关系管理、智能推荐系统、智能定制广告、信息安全、个人生活,等等。大数据不仅仅是对历史数据进行分析,更重要的是通过分析历史数据对未来进行精准预测,未雨绸缪,挖掘潜在的商机,预测并尽量避免可能的危机。

目前在学术界公认的大数据有四大特征。

(1) 数据量巨大。在过去的 20 年里全球数据量增长了 100 多倍,并且以越来越快的速度持续增长,数据量的单位从 TB 级别跃升到 PB 甚至 EB、ZB 级别。根据 IDC 的"数字宇宙"的报告,预计到 2020 年,全球数据使用量将达到 35.2ZB。

(2) 数据类型繁多。非结构化数据越来越多,如邮件、网络日志、音频、微信、微博、视频、图片、地理位置信息等,这对数据处理能力和算法提出了非常高的要求。

(3) 价值密度低。例如,没有任何意外事件发生时,连续不间断的监控视频是没有任何价值的,而发生意外事件时连续若干小时的监控视频中真正有价值的数据很可能只有几秒钟。

(4) 要求处理速度快。大数据时代的数据产生速度非常快,例如,1 分钟的时间里新浪微博大概增加 2 万条信息,Twitter 产生超过 10 万条推文,Facebook 会产生 600 万次访问记录。可以说,在如此海量并且飞速增加的数据面前,处理数据的效率就是企业的生命,在某些企业秒级的延迟已经是能够容忍的极限。

一般而言,进行大数据处理时,首先要分析原始数据的质量,尤其是缺失值、异常值、重复数据、特殊符号等,通过精选数据样本,提高原始数据的相关性、可靠性、有效性,不仅能够节省系统资源,更重要的是提高探寻数据内在规律的准确性。然后再分析采样的数据的分布特征与类型、周期性、贡献度、相关性等各项指标。如何通过强大的计算能力和高效的算法更迅速地完成数据的价值"提纯",是目前大数据背景下亟待解决的难题和重要的研究热点之一。另外,数据的来源直接导致分析结果的准确性和真实性,如果数据来源是完整的并且是真实的,最终的分析结果会更加准确,并且可以大幅度提高处理速度。

11.2 MapReduce 框架

MapReduce 编程思路非常简单,首先对大数据进行分割,切分为一定大小的数据,然后把分割的数据交给多个 Mapper 函数进行处理,Mapper 函数处理后将产生一组规模较小的数据,多个规模较小的数据再提交给 Reducer 函数进行处理,得到一个更小规模的数据或最终结果。对于不同的具体应用,需要根据特定的要求来编写不同的 Mapper 和 Reducer 代码,并且可能会需要多次迭代来最终完成任务,如图 11-1 所示。

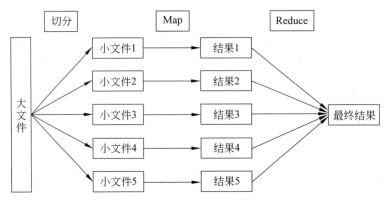

图 11-1　MapReduce 流程

了解了基本原理以后,接下来我们通过一个例子来演示一下 MapReduce 思路的应用。Windows 系统的升级日志文件一般较大,现假设要求统计一下日志文件中与不同日期有关的记录条数。首先将大文件切分成多个小的文件,然后对每个小文件进行 map 处理,然后对得到的处理结果再进行 reduce 处理,最终得到所需要的数据和结论。

1. 大文件切分:FileSplit.py

```
import os
import os.path
import time

def FileSplit(sourceFile, targetFolder):
    if not os.path.isfile(sourceFile):              #源文件必须存在
        print(sourceFile, ' does not exist.')
        return
    if not os.path.isdir(targetFolder):             #目标文件夹不存在则创建
        os.mkdir(targetFolder)
    tempData=[]                                     #用来存放临时数据
    number=1000                                     #切分后的每个小文件包含 1000 行
```

```python
            fileNum=1                                          #切分后的文件编号
            with open(sourceFile, 'r')as srcFile:
                dataLine=srcFile.readline().strip()
                while dataLine:
                    for i in range(number):                    #读取1000行文本
                        tempData.append(dataLine)
                        dataLine=srcFile.readline()
                        if not dataLine:
                            break
                    desFile=os.path.join(targetFolder, sourceFile[0:-4]+str(fileNum)
                                    +'.txt')
                    with open(desFile, 'a+')as f:              #创建一个小文件
                        f.writelines(tempData)
                    tempData=[]
                    fileNum=fileNum+1                          #小文件编号加1

if __name__=='__main__':
    sourceFile='test.txt'                                      #指定源文件
    targetFolder='test'                                        #指定存放切分后小文件的文件夹
    FileSplit(sourceFile, targetFolder)
```

2. Mapper 代码：Map.py

```python
import os
import re
import threading
import time

def Map(sourceFile):                                           #这段代码仅适用于配套文件
    if not os.path.exists(sourceFile):                         #或者类似的 Windows 升级日志
        print(sourceFile, ' does not exist.')
        return
    pattern=re.compile(r'[0-9]{1,2}/[0-9]{1,2}/[0-9]{4}')
    result={}
    with open(sourceFile, 'r')as srcFile:
        for dataLine in srcFile:
            r=pattern.findall(dataLine)                        #查找符合日期格式的字符串
            if r:
                result[r[0]]=result.get(r[0], 0)+1
    desFile=sourceFile[0:-4]+'_map.txt'
    with open(desFile, 'a+')as fp:                             #中间临时结果
        for k, v in result.items():
            fp.write(k+':'+str(v)+'\n')
```

```python
if __name__=='__main__':
    desFolder='test'
    files=os.listdir(desFolder)
    def Main(i):                                    #使用多线程
        Map(desFolder+'\\'+files[i])
    fileNumber=len(files)
    for i in range(fileNumber):
        t=threading.Thread(target=Main, args = (i,))
        t.start()
```

3. Reducer 代码：Reduce.py

```python
from os.path import isdir
from os import listdir

def Reduce(sourceFolder, targetFile):
    if not isdir(sourceFolder):
        print(sourceFolder, ' does not exist.')
        return
    result={}
    #Deal only with the mapped files
    allFiles=[sourceFolder+'\\'+f for f in listdir(sourceFolder)
              if f.endswith('_map.txt')]
    for f in allFiles:
        with open(f, 'r')as fp:
            for line in fp:
                line=line.strip()
                if not line:
                    continue
                key, value=line.split(':')          #结合 Map.py 代码理解这个地方
                result[key]=result.get(key,0)+int(value)
    with open(targetFile, 'w')as fp:                #创建结果文件
        for k,v in result.items():
            fp.write(k+':'+str(v)+'\n')

if __name__=='__main__':
    Reduce('test', 'test\\result.txt')
```

保存并运行上述程序，首先运行 FileSplit.py，将文件切分，生成若干小文件，如图 11-2 所示。

然后运行 Map.py 程序，得到中间结果，如图 11-3 所示。

最后运行 Reduce.py 程序，得到最终结果，如图 11-4 所示。

也可以使用下面的思路来解决上面这个问题，不需要对大文件进行切分，改写 Map.py 程序代码如下（文件名为 Hadoop_Map.py）：

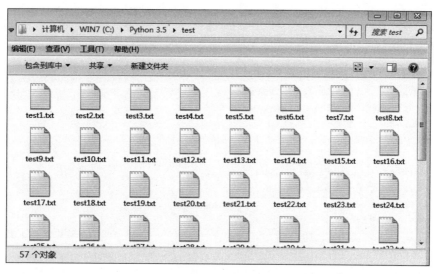

图 11-2 文件切分结果

图 11-3 Map 结果

图 11-4 Reduce 结果

```python
import os
import re
import time

def Map(sourceFile):
    if not os.path.exists(sourceFile):
        print(sourceFile, ' does not exist.')
        return
    pattern=re.compile(r'[0-9]{1,2}/[0-9]{1,2}/[0-9]{4}')
    result={}
    with open(sourceFile, 'r')as srcFile:
        for dataLine in srcFile:
            r=pattern.findall(dataLine)
            if r:
                print(r[0], ',', 1)              #将中间结果输出到标准控制台
Map('test.txt')
```

然后将 Reduce.py 程序代码改写如下（文件名为 Hadoop_Reduce.py）：

```python
import os
import sys

def Reduce(targetFile):
    result={}
    for line in sys.stdin:                       #从标准控制台中获取中间结果数据
        riqi, shuliang=line.strip().split(',')
        result[riqi]=result.get(riqi, 0)+1
    with open(targetFile, 'w')as fp:
        for k,v in result.items():
            fp.write(k+':'+str(v)+'\n')
Reduce('result.txt')
```

最后在命令提示符环境中执行下面的命令，其中竖线（|）表示管道，前一个程序的输出直接作为后面程序的输入。

```
python Hadoop_Map.py test.txt | python Hadoop_Reduce.py
```

假设测试样本文件 test.txt 在当前文件夹中，命令执行结束后，在当前文件夹生成结果文件 result.txt，内容与图 11-4 完全一致。

> **拓展知识**：在 Win 7 系统中，单击"开始"→"计算机"→"管理"命令，然后展开"事件查看器"，选择感兴趣的事件类别并右击，选择"将所有事件另存为"，就可以把系统日志保存为记事本文档。上面例子中用到的系统日志就是这么导出的。

11.3 Spark 应用开发

Hadoop 是 Apache 软件基金会旗下的一个开源分布式计算平台，以分布式文件系统 HDFS 和 MapReduce 为核心，为用户提供了系统底层细节透明的分布式基础框架，允许

用户部署在低端硬件上,并且不需要了解底层细节就可以开发并行应用程序。Hadoop 的缺点是其 Shuffle 过程对本地硬盘 I/O 操作过多,并且不适合 SQL 交互式查询、实时流处理以及机器学习等应用。

Spark 是一个基于内存的开源计算框架,其活跃度在 Apache 基金会所有开源项目中排第三位,其特点在于基于内存计算,适合迭代计算,兼容多种应用场景,同时还兼容 Hadoop 生态系统中的组件,并且具有非常强的容错性。Spark 的设计目的是全栈式解决批处理、结构化数据查询、流计算、图计算和机器学习等业务和应用。

Spark 集成了 Spark SQL(分布式 SQL 查询引擎,提供了一个 DataFrame 编程抽象)、Spark Streaming(把流式计算分解成一系列短小的批处理计算,并且提供高可靠和吞吐量服务)、MLlib(提供机器学习服务)、GraphX(提供图计算服务)、SparkR(R on Spark)等子框架,为不同应用领域的从业者提供了全新的大数据处理方式,越来越便捷、轻松。

为了适应迭代计算,Spark 把经常被重用的数据缓存到内存中以提高数据读取和操作速度,Spark 比 Hadoop 快近百倍,支持 Java、Scala、Python、R 等多种语言,除 map 和 reduce 之外,还支持 filter、foreach、reduceByKey、aggregate 以及 SQL 查询、流式查询等。

旧时王谢堂前燕,飞入寻常百姓家。随着普通家用计算机(手机也早已进入多核时代,但如何在手机上搭建 Spark 环境不在本书讨论范围之内)进入多处理器和多核时代,完全可以在自己家的计算机上搭建 Spark 环境。当然,如果数据量大到一定程度,还是要在集群或云平台上部署的 Spark 环境中进行处理和计算。进行 Spark 应用开发时一般是先在本地进行开发和测试,通过测试后再提交到集群执行。下面我们以 Win 7 平台为例介绍 Spark 环境的搭建和简单使用。首先安装 JDK 并配置环境变量 path,下载安装 Scala 语言包并配置系统环境变量 path,下载安装 Spark 并配置系统环境变量 HADOOP_HOME 和 SPARK_HOME 的值为 Spark 安装目录,使用 pip 工具安装扩展库 py4j,到 http://public-repo-1.hortonworks.com/hdp-win-alpha/winutils.exe 网址下载 winutils.exe 放到 Spark 安装目录的 bin 文件夹中,最后切换至命令提示符环境并切换到 F:\spark-1.6.1-bin-hadoop2.6\bin 文件夹,执行命令 pyspark.cmd,进入 Python 开发环境,如图 11-5 所示。可以看到,不仅可以使用 pyspark 库,还可以使用 Python 标准库和已安装的扩展库。

另外,在 Spark 的 bin 文件夹中还提供了 spark-submit.cmd 文件,这个文件是用来执行 Python 程序的,使用任意 Python 开发环境编写程序文件 hello.py,其中只有一行代码:

```
print('Hello world')
```

然后在命令提示符环境中提交该程序即可执行,如图 11-6 所示。

下面的 Python 程序文件 pi.py 用来估算圆周率的值,保存至 Spark 安装目录中的 bin 目录,可以使用命令 spark-submit.cmd pi.py 运行程序并输出圆周率的值。

```
from pyspark import SparkConf, SparkContext
from pyspark.sql import SQLContext
```

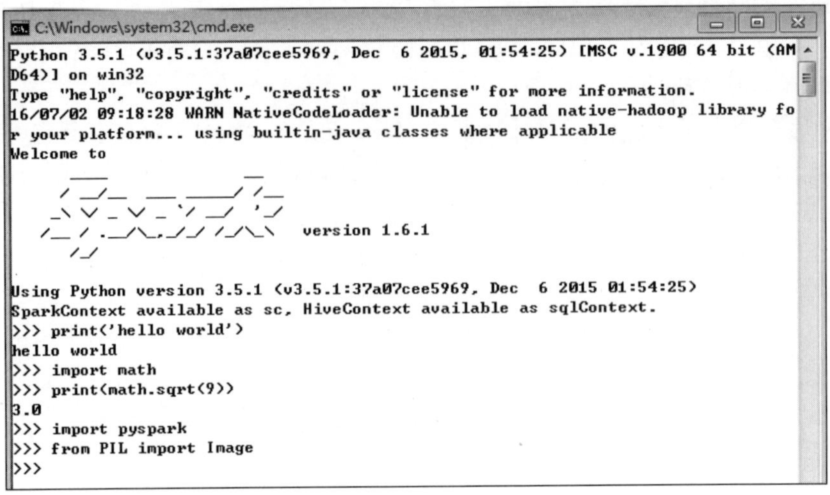

图 11-5　pyspark 开发界面

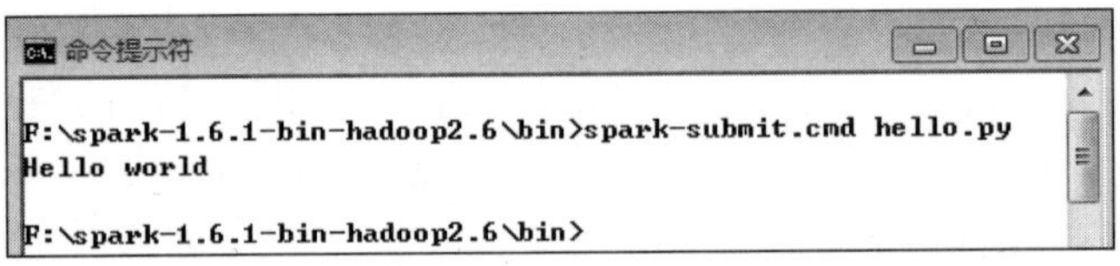

图 11-6　执行 Python 程序

```
from random import random

conf=SparkConf().setAppName("pi")
sc=SparkContext(conf=conf)
sqlCtx=SQLContext(sc)

def sample(p):
    x, y=random(), random()
    return 1 if x*x+y*y<1 else 0

NUM_SAMPLES=100000                              #数值越大结果越准确
count=sc.parallelize(range(NUM_SAMPLES))
count=count.map(sample).reduce(lambda a, b: a+b)
print('='*30)
print("Pi is roughly %f" %(4.0 * count / NUM_SAMPLES))
print('='*30)
```

下面的代码使用 Spark 来统计 100 000 000 以内的素数数量，在 6GB RAM、双核 CPU 的 64 位 Win 7＋Spark 单机平台上运行时间为 765.015 428s。

```
from pyspark import SparkConf, SparkContext
from pyspark.sql import SQLContext
```

```python
from random import random

conf=SparkConf().setAppName("isPrime")
sc=SparkContext(conf=conf)
sqlCtx=SQLContext(sc)
def isPrime(n):
    if n<2:
        return False
    if n==2:
        return True
    if not n&1:
        return False
    for i in range(3, int(n**0.5)+2, 2):
        if n%i ==0:
            return False
    return True
rdd=sc.parallelize(range(100000000))
result=rdd.filter(isPrime).count()
print('=' * 30)
print(result)
```

下面的代码在相同的平台上使用传统的方式来求解同一个问题，运行时间为 1666.616 000 18 秒，由此可见，即使是在单机多核环境下，spark 也会获得速度上的很大提高，提高的比例和处理器的数量有一定的关系。

```python
import time

def isPrime(n):
    if n<2:
        return False
    if n==2:
        return True
    if not n&1:
        return False
    for i in range(3, int(n**0.5)+2, 2):
        if n%i ==0:
            return False
    return True

num=0
start=time.time()
for n in range(100000000):
    if isPrime(n):
        num +=1
print(num)
```

```
print(time.time()-start)
```

下面的代码演示了 pyspark 的很少一部分功能和用法,更加详细的函数介绍请参考网址 http://spark.apache.org/docs/latest/api/python/pyspark.html。

```
>>>from pyspark import SparkFiles
>>>path='test.txt'
>>>with open(path, 'w') as fp:                         #创建文件
       fp.write('100')
>>>sc.addFile(path)                                    #提交文件
>>>def func(iterator):
       with open(SparkFiles.get('test.txt')) as fp:    #打开文件
           Val=int(fp.readline())                      #读取文件内容
           return [x * Val for x in iterator]
>>>sc.parallelize([1, 2, 3, 4, 5]).mapPartitions(func).collect()
                                                       #并行处理
[100, 200, 300, 400, 500]
>>>sc.parallelize([2, 3, 4]).count()                   #count()用来返回RDD中元素的个数
3
>>>rdd=sc.parallelize([1, 2])
>>>sorted(rdd.cartesian(rdd).collect())                #collect()返回包含RDD中元素的列表
[(1, 1), (1, 2), (2, 1), (2, 2)]                       #笛卡儿积
>>>rdd=sc.parallelize([1, 2, 3, 4, 5])
>>>rdd.filter(lambda x: x%2==0).collect()              #只保留符合条件的元素
[2, 4]
>>>sorted(sc.parallelize([1, 1, 2, 3]).distinct().collect())
                                                       #返回唯一元素
[1, 2, 3]
>>>rdd=sc.parallelize(range(10))
>>>rdd.map(lambda x: str(x)).collect()                 #映射
>>>rdd=sc.parallelize([1.0, 5.0, 43.0, 10.0])
>>>rdd.max()                                           #最大值
43.0
>>>rdd.max(key=str)
5.0
```

小技巧:读者运行上面的代码会发现,屏幕上会出现非常详细的执行过程,而实际上很多时候我们并不需要那些信息,只想关心代码的执行结果。如果想关闭这些详细信息的显示,可以把 Spark 安装文件夹的 conf 文件夹中 log4j.properties.template 文件复制一份保存到 conf 文件夹中并改名为 log4j.properties,然后使用记事本打开新文件,把里面的 INFO 都改为 WARN,关闭后重启 pyspark.cmd 就可以了,这个操作和设置对使用 spark-submit.cmd 提交并执行的程序也同样有效。

第 12 章 图形编程与图像处理

12.1 图形编程

计算机图形学主要研究如何使用计算机来生成具有真实感的图形，涉及的内容主要包括三维建模、图形几何变换、光照模型、纹理映射、阴影模型等内容，在机械制造、虚拟现实、游戏开发、漫游系统设计、产品展示等多个领域具有重要的应用。随着3D打印机的诞生，只要有模型就能够快速生成实物，无疑这将会大大扩展计算机图形学的应用范围，例如，可以使用计算机图形学制作出各种可爱的模型，然后参照这些模型使用 3D 打印机批量生产各种食品、玩偶、饰品等。目前大部分计算机图形学的书籍都是基于 OpenGL 的，Python 也提供了相应的扩展库 PyOpenGL，这极大方便了编写图形学程序的 Python 程序员。OpenGL 的功能非常强大，提供了图形学编程所需要的所有 API 函数，本书仅选取很小一部分进行简单介绍，有了这些入门知识以后，读者参考 OpenGL 书籍和计算机图形学知识，很容易写出更多更好的程序。

12.1.1 绘制三维图形

在 OpenGL 中绘制图形的代码需要放在 glBegin(mode) 和 glEnd() 这一对函数的调用之间，其中，mode 表示绘图类型，取值范围如表 12-1 所示。

表 12-1 mode 取值

取值	说明
GL_POINTS	绘制点
GL_LINES	绘制直线
GL_LINE_STRIP	绘制连续直线,不封闭
GL_LINE_LOOP	绘制封闭的连续直线
GL_TRIANGLES	绘制三角形
GL_TRIANGLE_STRIP	绘制三角形串
GL_TRIANGLE_FAN	绘制三角扇形
GL_QUADS	绘制四边形
GL_QUAD_STRIP	绘制四边形串
GL_POLYGON	绘制多边形

下面的代码可以绘制一个彩色三角形和一条彩色直线。在这段代码中,首先设置绘制模式为多边形,然后依次绘制该多边形的顶点,绘制每个顶点之前设置顶点颜色,然后修改绘制模式为直线并指定直线段的端点颜色和位置。需要注意的是,使用 glColor3f( )函数设置颜色之后,直到下一次使用该函数改变颜色之前,绘制的所有顶点都使用这个颜色。或者说,OpenGL 采用的是"状态机"工作方式,一旦设置了某种状态之后,除非显式修改该状态,否则该状态将一直保持。

例 12-1 绘制三维直线、二维三角形和圆。

```python
import sys
from math import pi as PI
from math import sin, cos
from OpenGL.GL import *
from OpenGL.GLU import *
from OpenGL.GLUT import *

class MyPyOpenGLTest:
    #重写构造函数,初始化 OpenGL 环境,指定显示模式以及用于绘图的函数
    def __init__(self, width=640, height=480,
                    title='MyPyOpenGLTest'.encode('gbk')):
        glutInit(sys.argv)
        glutInitDisplayMode(GLUT_RGBA | GLUT_DOUBLE | GLUT_DEPTH)
        glutInitWindowSize(width, height)
        self.window=glutCreateWindow(title)
        #指定绘制函数
        glutDisplayFunc(self.Draw)
        glutIdleFunc(self.Draw)
        self.InitGL(width, height)
    
    #根据特定的需要,进一步完成 OpenGL 的初始化
    def InitGL(self, width, height):
        #初始化窗口背景为白色
        glClearColor(1.0, 1.0, 1.0, 0.0)
        glClearDepth(1.0)
        glDepthFunc(GL_LESS)
        #光滑渲染
        glEnable(GL_BLEND)
        glShadeModel(GL_SMOOTH)
        glEnable(GL_POINT_SMOOTH)
        glEnable(GL_LINE_SMOOTH)
        glEnable(GL_POLYGON_SMOOTH)
        glMatrixMode(GL_PROJECTION)
        #反走样,也称为抗锯齿
        glHint(GL_POINT_SMOOTH_HINT,GL_NICEST)
```

```python
        glHint(GL_LINE_SMOOTH_HINT,GL_NICEST)
        glHint(GL_POLYGON_SMOOTH_HINT,GL_FASTEST)
        glLoadIdentity()
        #透视投影变换
        gluPerspective(45.0, float(width)/float(height), 0.1, 100.0)
        glMatrixMode(GL_MODELVIEW)

    #定义自己的绘图函数
    def Draw(self):
        glClear(GL_COLOR_BUFFER_BIT | GL_DEPTH_BUFFER_BIT)
        glLoadIdentity()
        #平移
        glTranslatef(-3.0, 2.0, -8.0)
        #绘制二维图形,z坐标为0
        #指定模式,绘制多边形
        glBegin(GL_POLYGON)
        #设置顶点颜色
        glColor3f(1.0, 0.0, 0.0)
        #绘制多边形顶点
        glVertex3f(0.0, 1.0, 0.0)
        glColor3f(0.0, 1.0, 0.0)
        glVertex3f(1.0, -1.0, 0.0)
        glColor3f(0.0, 0.0, 1.0)
        glVertex3f(-1.0, -1.0, 0.0)
        #结束本次绘制
        glEnd()

        glTranslatef(3, -1, 0.0)

        #绘制三维线段
        glBegin(GL_LINES)
        glColor3f(1.0, 0.0, 0.0)
        glVertex3f(1.0, 1.0, -1.0)
        glColor3f(0.0, 1.0, 0.0)
        glVertex3f(-1.0, -1.0, 3.0)
        glEnd()

        glTranslatef(-0.3, 1, 0)

        #使用折线段绘制圆
        glBegin(GL_LINE_LOOP)
        n=100
        theta=2*PI/n
        r=0.8
```

```
        for i in range(100):
            x=r * cos(i * theta)
            y=r * sin(i * theta)
            glVertex3f(x, y, 0)
        glEnd()

        glutSwapBuffers()

    #消息主循环
    def MainLoop(self):
        glutMainLoop()

if __name__ == '__main__':
    #实例化窗口对象,运行程序,启动消息主循环
    w=MyPyOpenGLTest()
    w.MainLoop()
```

程序运行结果如图 12-1 所示。

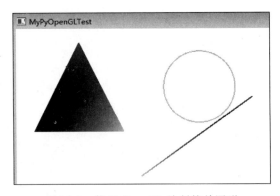

图 12-1　使用 OpenGL 绘制简单图形

12.1.2　绘制三次贝塞尔曲线

贝塞尔曲线/曲面和 B 样条曲线/曲面被广泛应用于汽车、轮船、飞机、高铁以及其他大量物体的外形设计。由于高次贝塞尔曲线/曲面计算量过大,所以一般常使用三次贝塞尔曲线/曲面,而很少使用更高次的,对于复杂外形则常使用 B 样条曲线/曲面。下面给出一段三次贝塞尔曲线的绘制过程,对于稍微复杂的情况可以使用多段三次贝塞尔曲线进行拼接。

在 12.1.1 节中绘制简单图形的程序框架基础上,首先增加一个成员方法计算三次贝塞尔曲线上给定参数对应的点的坐标:

```
def getBezier(self, P0, P1, P2, P3, t):
    a0 = (1-t)**3
    a1=3 * (1-t)**2 * t
```

```
        a2=3 * t**2 * (1-t)
        a3=t**3

        x=a0 * P0[0]+a1 * P1[0]+a2 * P2[0]+a3 * P3[0]
        y=a0 * P0[1]+a1 * P1[1]+a2 * P2[1]+a3 * P3[1]
        z=a0 * P0[2]+a1 * P1[2]+a2 * P2[2]+a3 * P3[2]

        return(x, y, z)
```

然后修改绘图方法 Draw()中的代码如下,即可实现三次贝塞尔曲线的绘制。

```
    def Draw(self):
        glClear(GL_COLOR_BUFFER_BIT | GL_DEPTH_BUFFER_BIT)
        glLoadIdentity()
        #平移
        glTranslatef(-3.0, 0.0, -8.0)
        #指定三次贝塞尔曲线的 4 个控制点坐标
        P0 = (-4, -2, -9)
        P1 = (-0.5, 3, 0)
        P2 = (2, -3, 0)
        P3 = (4.5, 2, 0)
        #指定模式,绘制连续的折线段
        glBegin(GL_LINE_STRIP)
        #设置顶点颜色
        glColor3f(0.0, 0.0, 0.0)
        #使用 100 段直线段的拼接来逼近三次贝塞尔曲线
        for i in range(101):
            #参数 t 必须是介于[0,1]区间的实数
            t=i/100.0
            p=self.getBezier(P0, P1, P2, P3, t)
            glVertex3f(*p)

        #结束本次绘制
        glEnd()

        glutSwapBuffers()
```

拓展知识:对于贝塞尔曲线而言,其特点在于第一个控制点恰好是曲线的起点,最后一个控制点是曲线的终点,其他控制点并不在曲线上,而是起到控制曲线形状的作用。另外,曲线的起点处与前两个控制点构成的线段相切,而曲线的终点处与最后两个控制点构成的线段相切。

12.1.3 纹理映射

在现实中,人们主要通过物体表面丰富的纹理细节来区分具有相同形状的不同物体。

在三维建模时也往往通过纹理映射来简化建模的工作量,可以在保证图形具有较强真实感的前提下大幅度提高渲染效率。

简单地说,纹理映射就是为物体表面进行贴图以使其呈现出特定的视觉效果。这需要首先准备好纹理,然后构建物体空间坐标和纹理坐标之间的对应关系来完成贴图。可以使用函数来生成一些规则或不规则的纹理,例如,粗布纹理、棋盘纹理、随机纹理等,也可以将拍摄或通过网络搜索下载的图片作为纹理映射到物体表面上。进行纹理映射之前,首先要读取并设置纹理数据。

例 12-2 纹理映射与计算机动画。

```
import sys
from OpenGL.GL import *
from OpenGL.GLUT import *
from OpenGL.GLU import *
from PIL import Image

class MyPyOpenGLTest:
    def __init__(self, width=640, height=480,
                 title='MyPyOpenGLTest'.encode('gbk')):
        glutInit(sys.argv)
        glutInitDisplayMode(GLUT_RGBA | GLUT_DOUBLE | GLUT_DEPTH)
        glutInitWindowSize(width, height)
        self.window=glutCreateWindow(title)
        glutDisplayFunc(self.Draw)
        glutIdleFunc(self.Draw)
        self.InitGL(width, height)
        #绕各坐标轴旋转的角度
        self.x=0.0
        self.y=0.0
        self.z=0.0

    #绘制图形
    def Draw(self):
        glClear(GL_COLOR_BUFFER_BIT | GL_DEPTH_BUFFER_BIT)
        glLoadIdentity()
        #沿 z 轴平移
        glTranslate(0.0, 0.0, -5.0)
        #分别绕 x、y、z 轴旋转
        glRotatef(self.x, 1.0, 0.0, 0.0)
        glRotatef(self.y, 0.0, 1.0, 0.0)
        glRotatef(self.z, 0.0, 0.0, 1.0)

        #开始绘制立方体的每个面,同时设置纹理映射
        #绘制四边形
```

```python
    glBegin(GL_QUADS)
    #设置纹理坐标
    glTexCoord2f(0.0, 0.0)
    #绘制顶点
    glVertex3f(-1.0, -1.0, 1.0)
    glTexCoord2f(1.0, 0.0)
    glVertex3f(1.0, -1.0, 1.0)
    glTexCoord2f(1.0, 1.0)
    glVertex3f(1.0, 1.0, 1.0)
    glTexCoord2f(0.0, 1.0)
    glVertex3f(-1.0, 1.0, 1.0)
    #其他几个面的代码类似,详见配套资源
    ⋮
    #结束绘制
    glEnd()
    #刷新屏幕,产生动画效果
    glutSwapBuffers()
    #修改各坐标轴的旋转角度
    self.x +=0.2
    self.y +=0.3
    self.z +=0.1

#加载纹理
def LoadTexture(self):
    img=Image.open('sample_texture.bmp')
    width, height=img.size
    img=img.tobytes('raw', 'RGBX', 0, -1)
    glBindTexture(GL_TEXTURE_2D, glGenTextures(1))
    glPixelStorei(GL_UNPACK_ALIGNMENT,1)
    glTexImage2D(GL_TEXTURE_2D, 0, 4, width, height, 0, GL_RGBA,
              GL_UNSIGNED_BYTE,img)
    glTexParameterf(GL_TEXTURE_2D, GL_TEXTURE_WRAP_S, GL_CLAMP)
    glTexParameterf(GL_TEXTURE_2D, GL_TEXTURE_WRAP_T, GL_CLAMP)
    glTexParameterf(GL_TEXTURE_2D, GL_TEXTURE_WRAP_S, GL_REPEAT)
    glTexParameterf(GL_TEXTURE_2D, GL_TEXTURE_WRAP_T, GL_REPEAT)
    glTexParameterf(GL_TEXTURE_2D, GL_TEXTURE_MAG_FILTER, GL_NEAREST)
    glTexParameterf(GL_TEXTURE_2D, GL_TEXTURE_MIN_FILTER, GL_NEAREST)
    glTexEnvf(GL_TEXTURE_ENV, GL_TEXTURE_ENV_MODE, GL_DECAL)

def InitGL(self, width, height):
    self.LoadTexture()
    glEnable(GL_TEXTURE_2D)
    glClearColor(1.0, 1.0, 1.0, 0.0)
    glClearDepth(1.0)
```

```
            glDepthFunc(GL_LESS)
            glShadeModel(GL_SMOOTH)
            #背面剔除,消隐
            glEnable(GL_CULL_FACE)
            glCullFace(GL_BACK)
            glEnable(GL_POINT_SMOOTH)
            glEnable(GL_LINE_SMOOTH)
            glEnable(GL_POLYGON_SMOOTH)
            glMatrixMode(GL_PROJECTION)
            glHint(GL_POINT_SMOOTH_HINT,GL_NICEST)
            glHint(GL_LINE_SMOOTH_HINT,GL_NICEST)
            glHint(GL_POLYGON_SMOOTH_HINT,GL_FASTEST)
            glLoadIdentity()
            gluPerspective(45.0, float(width)/float(height), 0.1, 100.0)
            glMatrixMode(GL_MODELVIEW)
       def MainLoop(self):
            glutMainLoop()

if __name__=='__main__':
    w=MyPyOpenGLTest()
    w.MainLoop()
```

程序运行后,一个带有贴图的立方体在不停地旋转,某个时刻的运行效果如图 12-2 所示。如果读者在运行时遇到错误,很可能是图片尺寸的问题,程序中使用的图片文件大小是 256×256,大家可以尝试着修改一下自己的图片尺寸再运行。

图 12-2 纹理映射

12.1.4 响应键盘事件

在计算机动画或者游戏中,经常需要和用户或玩家交互,这就需要接收并响应用户的鼠标和键盘操作。在 12.1.3 节的代码中稍做修改就可以实现用户通过键盘控制的旋转立方体，首先在类中的 Draw()方法中把最后三行修改各坐标轴旋转角度的代码删除，然后在构造函数__init__()中增加一行代码：

```
glutKeyboardFunc(self.KeyPress)
```

然后在类中增加一个方法 KeyPress()，用来接收并响应键盘输入，用户按下键盘上的几个字母键时，立方体会沿不同的轴进行一定角度的旋转。

```
def KeyPress(self, key, x, y):
    if key=='a':
        self.x +=0.3
    elif key=='s':
        self.x -=0.3
    elif key=='j':
        self.y +=0.3
    elif key=='k':
        self.y -=0.3
    elif key=='g':
        self.z +=0.3
    elif key=='h':
        self.z -=0.3
```

12.1.5　光照模型

光照是增强图形真实感的重要技术之一。光线投射到物体表面上会同时发生反射、透射和吸收等几种情况，其中进入人眼的反射或透射光线使得物体可见。另外，物体表面的材质属性决定了对光线每个分量的反射系数，不同的材质属性导致物体表面呈现出不同的颜色。可以说，光源和材质的属性共同决定了物体表面的视觉效果。

例 12-3　法向量与光照模型。

```
import sys
from OpenGL.GL import *
from OpenGL.GLU import *
from OpenGL.GLUT import *

class MyPyOpenGLTest:
    #重写构造函数，初始化 OpenGL 环境，指定显示模式以及用于绘图的函数
    def __init__(self, width=640, height=480, title=b'Normal_Light'):
        glutInit(sys.argv)
        glutInitDisplayMode(GLUT_RGBA | GLUT_DOUBLE | GLUT_DEPTH)
        glutInitWindowSize(width, height)
        self.window=glutCreateWindow(title)
        #指定绘制函数
        glutDisplayFunc(self.Draw)
```

```python
        glutIdleFunc(self.Draw)
        self.InitGL(width, height)

    #根据特定的需要,进一步完成OpenGL的初始化
    def InitGL(self, width, height):
        #初始化窗口背景为白色
        glClearColor(1.0, 1.0, 1.0, 0.0)
        glClearDepth(1.0)
        glDepthFunc(GL_LESS)
        #设置灯光与材质属性
        mat_sp = (1.0, 1.0, 1.0, 1.0)
        mat_sh= [50.0]
        light_position = (-0.5, 1.5, 1, 0)
        yellow_l = (1, 1, 0, 1)
        ambient = (0.1, 0.8, 0.2, 1.0)
        glMaterialfv(GL_FRONT, GL_SPECULAR, mat_sp)
        glMaterialfv(GL_FRONT, GL_SHININESS, mat_sh)
        glLightfv(GL_LIGHT0, GL_POSITION, light_position)
        glLightfv(GL_LIGHT0, GL_DIFFUSE, yellow_l)
        glLightfv(GL_LIGHT0, GL_SPECULAR, yellow_l)
        glLightModelfv(GL_LIGHT_MODEL_AMBIENT, ambient)
        #启用光照模型
        glEnable(GL_LIGHTING)
        glEnable(GL_LIGHT0)
        glEnable(GL_DEPTH_TEST)
        #光滑渲染
        glEnable(GL_BLEND)
        glShadeModel(GL_SMOOTH)
        glEnable(GL_POINT_SMOOTH)
        glEnable(GL_LINE_SMOOTH)
        glEnable(GL_POLYGON_SMOOTH)
        glMatrixMode(GL_PROJECTION)
        #反走样,也称为抗锯齿
        glHint(GL_POINT_SMOOTH_HINT,GL_NICEST)
        glHint(GL_LINE_SMOOTH_HINT,GL_NICEST)
        glHint(GL_POLYGON_SMOOTH_HINT,GL_FASTEST)
        glLoadIdentity()
        #透视投影变换
        gluPerspective(45.0, float(width)/float(height), 0.1, 100.0)
        glMatrixMode(GL_MODELVIEW)

    #定义自己的绘图函数
    def Draw(self):
        glClear(GL_COLOR_BUFFER_BIT | GL_DEPTH_BUFFER_BIT)
        glLoadIdentity()
        #平移
```

```
        glTranslatef(-1.5, 2.0, -8.0)
        #绘制三维线段
        glBegin(GL_LINES)
        #设置顶点颜色
        glColor3f(1.0, 0.0, 0.0)
        #设置顶点法向量
        glNormal3f(1.0, 1.0, 1.0)
        glVertex3f(1.0, 1.0, -1.0)
        glColor3f(0.0, 1.0, 0.0)
        glNormal3f(-1.0, -1.0, -1.0)
        glVertex3f(-1.0, -1.0, 3.0)
        glEnd()

        #球
        glColor3f(0.8, 0.3, 1.0)
        glTranslatef(0, -1.5, 0)
        #第一个参数是球的半径,后面两个参数是分段数
        glutSolidSphere(1.0,40,40)

        glutSwapBuffers()

    #消息主循环
    def MainLoop(self):
        glutMainLoop()

if __name__=='__main__':
    #实例化窗口对象,运行程序,启动消息主循环
    w=MyPyOpenGLTest()
    w.MainLoop()
```

运行结果如图 12-3 所示。

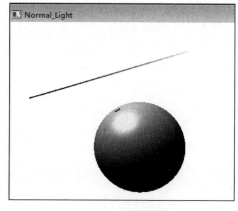

图 12-3　光照模型

12.2 图像处理

Python Imaging Library(PIL)是 Python 的图像处理扩展库,提供了非常强大的图像处理功能,支持多种图像格式。PIL 模块需要单独进行安装后才能使用,在 PIL 中主要提供了 Image、ImageChops、ImageColor、ImageDraw、ImagePath、ImageFile、ImageEnhance、PSDraw 以及其他一些模块来支持图像的处理,而 ImageGrab 模块还支持对指定区域进行截图。PIL 对 Python 3.x 支持不是特别好,可以使用扩展库 pillow 替代,本书重点介绍 pillow 的用法。

12.2.1 pillow 模块基本用法

Image 是 pillow 库中一个非常重要的模块,提供了大量用于图像处理的方法。使用该模块时,首先需要导入。

```
>>>from PIL import Image
```

接下来,我们通过几个示例来简单演示一下这个模块的用法。

(1) 打开图像文件。

```
>>>im=Image.open('sample.jpg')
```

(2) 显示图像。

```
>>>im.show()
```

(3) 查看图像信息。

```
>>>im.format                    #查看图像格式
'JPEG'
>>>im.size                      #查看图像大小,格式为(宽度,高度)
(200, 100)
>>>im.height                    #查看图像高度
100
>>>im.width                     #查看图像宽度
200
```

(4) 查看图像直方图。

```
>>>im.histogram()               #如果图像包含多个通道,则返回所有通道的直方图
>>>im.histogram()[:256]         #查看第一个通道的直方图
```

(5) 读取像素值。

```
>>>im.getpixel((150, 80))       #参数必须是元组,两个元素分别表示 x 和 y 坐标
(255, 248, 220)                 #返回值分别表示红、绿、蓝三原色分量的值
```

(6) 设置像素值，通过读取和修改图像像素值可以实现图像点运算。

```
>>>im.putpixel((100,50),(128,30,120))    #第二个参数用来指定目标像素的颜色值
```

💡**小提示**：在使用时应注意图像文件的格式，这里演示的是 24 位颜色深度的图像，如果是 256 色的图像文件，那么 getpixel()的返回值只是一个数字，而 putpixel()的第二个参数也只需要一个数字就可以了。

(7) 保存图像文件。

```
>>>im.save('sample1.jpg')                #可以把图像保存为另一个文件
>>>im.save('sample.bmp')                 #通过该方法也可以进行格式转换
>>>def img2jpg(imgFile):                 #转换图像文件格式
    if type(imgFile)==str and imgFile.endswith(('.bmp', '.gif', '.png')):
        with Image.open(imgFile) as im:
            im.convert('RGB').save(imgFile[:-3]+'jpg')
>>>img2jpg('1.gif')
>>>img2jpg('1.bmp')
>>>img2jpg('1.png')
```

(8) 图像缩放。

```
>>>im=im.resize((100,100))               #参数表示图像的新尺寸,分别表示宽度和高度
```

(9) 旋转图像，rotate()方法支持任意角度的旋转，而 transpose()方法支持部分特殊角度的旋转，如 90°、180°、270°旋转以及水平、垂直翻转等。

```
>>>im=im.rotate(90)                              #逆时针旋转 90°
>>>im=im.transpose(Image.ROTATE_180)             #逆时针旋转 180°
>>>im=im.transpose(Image.FLIP_LEFT_RIGHT)        #水平翻转
>>>im=im.transpose(Image.FLIP_TOP_BOTTOM)        #垂直翻转
```

(10) 图像裁剪与粘贴。

```
>>>box = (120, 194, 220, 294)                    #定义裁剪区域
>>>region=im.crop(box)                           #裁剪
>>>region=region.transpose(Image.ROTATE_180)
>>>im.paste(region,box)                          #粘贴
```

例如，图 12-4 是 lena 图形，而图 12-5 是将其中一部分逆时针旋转 180°以后的结果，请注意左下角区域图像的变化。

(11) 图像通道分离与合并。

```
>>>r, g, b=im.split()                    #将彩色图像分离为同样大小的红、绿、蓝三分量子图
>>>imNew=Image.merge(im.mode,(r,g,b))
```

(12) 创建缩略图。

```
>>>im.thumbnail((50, 20))                #参数为缩略图尺寸
```

```
>>>im.save('2.jpg')                              #保存缩略图
```

图 12-4　原始 lena 图像　　　　　图 12-5　部分区域被旋转 180°以后的 lena 图像

(13) 屏幕截图。

```
>>>from PIL import ImageGrab
>>>im=ImageGrab.grab((0,0,800,200))              #截取屏幕指定区域的图像
>>>im=ImageGrab.grab()                           #不带参数表示全屏幕截图
```

(14) 图像增强。

```
>>>from PIL import ImageFilter
>>>im=im.filter(ImageFilter.DETAIL)              #创建滤波器,使用不同的卷积核
>>>im=im.filter(ImageFilter.EDGE_ENHANCE)        #边缘增强
>>>im=im.filter(ImageFilter.EDGE_ENHANCE_MORE)   #边缘增强
```

(15) 图像模糊。

```
>>>im=im.filter(ImageFilter.BLUR)
>>>im=im.filter(ImageFilter.GaussianBlur)        #高斯模糊
>>>im.filter(ImageFilter.MedianFilter)           #中值滤波
```

(16) 图像边缘提取。

```
>>>im=im.filter(ImageFilter.FIND_EDGES)
```

(17) 图像点运算。

```
>>>im=im.point(lambda i:i * 1.3)                 #整体变亮
>>>im=im.point(lambda i:i * 0.7)                 #整体变暗
>>>im=im.point(lambda i: i * 1.8 if i<100 else i * 0.7)   #自定义调整图像明暗度
```

也使用图像增强模块来实现上面类似的功能,例如:

```
>>>from PIL import ImageEnhance
>>>enh=ImageEnhance.Brightness(im)
>>>enh.enhance(1.3).show()
```

(18) 图像冷暖色调整。

```
>>>r, g, b=im.split()                    #分离图像
>>>r=r.point(lambda i:i*1.3)             #红色分量变为原来的1.3倍
>>>g=g.point(lambda i:i*0.9)             #绿色分量变为原来的0.9
>>>b=b.point(lambda i:0)                 #把蓝色分量变为0
>>>im=Image.merge(im.mode,(r,g,b))       #合并图像
>>>im.show()
```

(19) 图像对比度增强。

```
>>>from PIL import ImageEnhance
>>>im=ImageEnhance.Contrast(im)
>>>im=im.enhance(1.3)                    #对比度增强为原来的1.3倍
```

12.2.2 计算椭圆中心

本节案例用来计算和确定任意形状椭圆的中心，使用 pillow 扩展库实现。

例 12-4 计算椭圆中心。

```
from PIL import Image
import os

def searchLeft(width, height, im):
    for w in range(width):                    #从左向右扫描
        for h in range(height):               #从下向上扫描
            color=im.getpixel((w, h))         #获取图像指定位置的像素颜色
            if color != (255, 255, 255):
                return w                      #遇到并返回椭圆边界最左端的x坐标

def searchRight(width, height, im):
    for w in range(width-1, -1, -1):          #从右向左扫描
        for h in range(height):
            color=im.getpixel((w, h))
            if color != (255, 255, 255):
                return w                      #遇到并返回椭圆边界最右端的x坐标

def searchTop(width, height, im):
    for h in range(height-1, -1, -1):
        for w in range(width):
            color=im.getpixel((w,h))
            if color != (255, 255, 255):
                return h                      #遇到并返回椭圆边界最上端的y坐标

def searchBottom(width, height, im):
```

```
        for h in range(height):
            for w in range(width):
                color=im.getpixel((w,h))
                if color != (255, 255, 255):
                    return h                    #遇到并返回椭圆边界最下端的 y 坐标

#遍历指定文件夹中所有 bmp 图像文件,假设图像为白色背景,椭圆为其他任意颜色
images=[f for f in os.listdir('testimages') if f.endswith('.bmp')]
for f in images:
    f='testimages\\'+f
    im=Image.open(f)
    width, height=im.size                       #获取图像大小
    x0, x1=searchLeft(width, height, im), searchRight(width, height, im)
    y0, y1=searchBottom(width, height, im), searchTop(width, height, im)
    center = ((x0+x1)//2, (y0+y1)//2)
    im.putpixel(center,(255,0,0))               #把椭圆中心像素画成红色
    im.save(f[0:-4]+'_center.bmp')              #保存为新图像文件
    im.close()
```

12.2.3 动态生成比例分配图

本节使用 pillow 实现另一个案例,具体功能为:使用 3 种颜色填充横条矩形区域,并在每段中分别居中输出字母 A、B、C,要求 A、B、C 各自所占比例可动态调整。

例 12-5 动态生成比例分配图。

```
from PIL import Image, ImageDraw, ImageFont

def redraw(f, v1, v2):
    start=int(600 * v1)
    end=int(600 * v2)
    im=Image.open(f)
    for w in range(start):                      #绘制红色区域
        for h in range(36, 61):                 #具体数值需要根据图片大小进行调整
            im.putpixel((w,h),(255, 0, 0))
    for w in range(start, end):                 #绘制绿色区域
        for h in range(36, 61):
            im.putpixel((w,h),(0, 255, 0))
    for w in range(end, 600):                   #绘制品红色区域
        for h in range(36, 61):
            im.putpixel((w,h),(255, 0, 255))
    draw=ImageDraw.Draw(im)
    font=ImageFont.truetype('simsun.ttc', 18)
    draw.text((start//2,38), 'A',(0,0,0), font=font)
                                                #在各自区域内居中显示字母
```

```
draw.text(((end-start)//2+start,38), 'B',(0,0,0), font=font)
draw.text(((600-end)//2+end,38), 'C',(0,0,0), font=font)
im.save(f)                               #保存图片

redraw(r'd:\biaotou1.png', 0.1, 0.9)
```

程序运行结果如图 12-6 所示。图中上面浅蓝色部分的百分比和文字描述是提前做好的,下面 3 种颜色的矩形区域和 A、B、C 是由本程序动态生成,可以根据图片大小修改代码中的数值。

图 12-6　比例分配图

12.2.4　生成验证码图片

验证码在网络应用开发中占有重要地位,广泛应用于用户注册、登录、留言、购物、网络支付等场合,可以有效阻止恶意用户频繁的非法数据提交。图片验证码是比较传统的验证码形式,图片中除了经过平移、旋转、错切、缩放等基本变换的字母和数字之外,还有一些线条或其他干扰因素。另外,还有问答型验证码,验证码是一个简单的问题,用户需要输入正确的答案才能进行后续的操作。某些系统的验证码系统更加复杂,实现了基于内容的图像识别功能甚至拼图功能,题目难度较大,在一定程度上也阻碍了用户的正常使用。

例 12-6　生成验证码图片。

```
from PIL import Image, ImageDraw, ImageFont
import random
import string

#所有可能的字符,主要是英文字母和数字
characters=string.ascii_letters+string.digits

#获取指定长度的字符串
def selectedCharacters(length):
    '''length:the number of characters to show'''
    result=""
    for i in range(length):
        result+=random.choice(characters)
    return result

def getColor():
    '''get a random color'''
    r=random.randint(0,255)
```

```python
        g=random.randint(0,255)
        b=random.randint(0,255)
        return(r,g,b)

def main(size=(200,100), characterNumber=6, bgcolor=(255,255,255)):
    imageTemp=Image.new('RGB', size, bgcolor)
    #设置字体和字号
    font=ImageFont.truetype('c:\\windows\\fonts\\TIMESBD.TTF', 48)
    draw=ImageDraw.Draw(imageTemp)
    text=selectedCharacters(characterNumber)
    width, height=draw.textsize(text, font)
    #绘制验证码字符串
    offset=2
    for i in range(characterNumber):
        offset +=width//characterNumber
        position = (offset,(size[1]-height)//2+random.randint(-10,10))
        draw.text(xy=position, text=text[i], font=font, fill=getColor())
    #对验证码图片进行简单变换,这里采用简单的点运算
    imageFinal=Image.new('RGB', size, bgcolor)
    pixelsFinal=imageFinal.load()
    pixelsTemp=imageTemp.load()
    for y in range(0, size[1]):
        offset=random.randint(-1,1)
        for x in range(0, size[0]):
            newx=x+offset
            if newx>=size[0]:
                newx=size[0]-1
            elif newx<0:
                newx=0
            pixelsFinal[newx,y]=pixelsTemp[x,y]
    draw=ImageDraw.Draw(imageFinal)
    #绘制干扰噪点像素
    for i in range(int(size[0] * size[1] * 0.07)):
        draw.point((random.randint(0,size[0]), random.randint(0,size[1])), fill=getColor())
    #绘制干扰线条
    for i in range(8):
        start = (0, random.randint(0, size[1]-1))
        end = (size[0], random.randint(0, size[1]-1))
        draw.line([start, end], fill=getColor(), width=1)
    #绘制干扰弧线
    for i in range(8):
        start = (-50, -50)
        end = (size[0]+10, random.randint(0, size[1]+10))
```

```
        draw.arc(start+end, 0, 360, fill=getColor())
    #保存验证码图片
    imageFinal.save("result.jpg")
    imageFinal.show()

if __name__=="__main__":
    main((200,100), 8,(255,255,255))
```

将上面的程序保存并运行,即可生成验证码图片,如图 12-7 所示。

(a) 验证码图片（一）

(b) 验证码图片（二）

(c) 验证码图片（三）

图 12-7　验证码图片

12.2.5　gif 动态图像分离与生成

GIF(Graphics Interchange Format)是比较常见的一种动态图像,在一个文件中可以存储多幅图像,把这些图像依次读出并显示,可以得到简单的动画效果,因其体积较小并且清晰度还不错而得到广泛应用,大家在网上见到的很多搞笑动态图片都是 GIF 格式的。

例 12-7　GIF 动态图像分离与生成。

下面的代码可以把动态图像 test.gif 分离得到每帧图片：

```
from PIL import Image
import os

gifFileName='test.gif'
#使用 Image 模块的 open()方法打开 GIF 动态图像时,默认是第一帧
im=Image.open(gifFileName)
pngDir=gifFileName[:-4]
#创建存放每帧图片的文件夹
os.mkdir(pngDir)

try:
    while True:
        #保存当前帧图片
        current=im.tell()
        im.save(pngDir+'\\'+str(current)+'.png')
        #获取下一帧图片
        im.seek(current+1)
except EOFError:
```

```
        pass
```

下面的代码可以把上面代码分离 GIF 文件得到的所有 png 图像文件再重新合并为一个 GIF 文件，如果无法正常执行的话，除了需要使用 pip 安装 images2gif 扩展库，很可能还需要找到扩展库 images2gif 的主文件 images2gif.py，然后把第 426 行代码

```
    palettes.append(getheader(im)[1])
```

改为

```
    palettes.append(im.palette.getdata()[1])
```

然后再执行下面的程序：

```python
import os
import os.path
from PIL import Image
import images2gif

def pngs2gif(gifName, path, duration=0.1, np=0.1):
    pngFiles=[f for f in os.listdir(path)]
    pngFiles.sort(key=lambda f: int(f[:-4]))
    pngFiles=[os.path.join(path, f) for f in pngFiles]
    images=[]
    for f in pngFiles:
        images.append(Image.open(f))
    images2gif.writeGif(gifName, images, duration, np)

pngs2gif('abc.gif', 'test')
```

12.2.6 材质贴图

在 12.1.3 节中介绍了计算机图形学中的纹理映射，通过纹理映射或者材质贴图可以让同一个表面呈现出不同的视觉效果。在贴图时，很难保证原始图片恰好和目标物体表面的尺寸完全相同，这时候就必须要对原始图片进行必要的缩放和像素采样。

例 12-8 把固定大小的图片进行缩放并映射到任意大小物体表面。

```python
from PIL import Image
from math import floor

def textureMap(srcTextureFile, dstSurfaceFile, dstWidth, dstHeight):
    '''srcTextureFile:原始图片
       dstSurfaceFile:模拟目标物体表面
       dstWidth:目标物体表面宽度
       dstHeight:目标物体表面高度
    '''
```

```
#打开原始图片
srcTexture=Image.open(srcTextureFile)
#创建指定尺寸的目标物体表面
dstSurface=Image.new('RGBA',(dstWidth, dstHeight))
srcWidth, srcHeight=srcTexture.size
#根据目标物体表面尺寸,计算并获取原始图片中对应位置的像素值
for w in range(dstWidth):
    for h in range(dstHeight):
        x, y=floor(w/dstWidth * srcWidth), floor(h/dstHeight * srcHeight)
        dstSurface.putpixel((w,h), srcTexture.getpixel((x,y)))
dstSurface.save(dstSurfaceFile)
dstSurface.close()
srcTexture.close()
#也可以尝试下面的写法,更简单一些
'''
srcTexture=Image.open(srcTextureFile)
srcTexture=srcTexture.resize((dstWidth,dstHeight))
srcTexture.save(dstSurfaceFile)
srcTexture.close()
'''

#测试
textureMap('sample.jpg', r'new.jpg', 200, 250)
```

12.2.7 图像融合

图像融合(Image Fusion)是指将多源信道所采集到的关于同一目标的图像数据经过图像处理和计算机技术等,最大限度地提取各自信道中的有利信息,最后综合成高质量的图像,提高图像信息的利用率、改善计算机解译精度和可靠性、提升原始图像的空间分辨率和光谱分辨率,利于监测。待融合源图像必须已配准并且像素位宽一致的,否则会影响融合效果。

图像融合技术可以简单地分为像素级融合、特征级融合和决策级融合。像素级融合又有空域融合算法和变换域融合算法,下面的代码演示了空域融合算法的一种,也就是灰度加权平均法。

例 12-9 使用 pillow 进行图像空域融合。

```
from random import randint
from PIL import Image

#根据原始24位色BMP图像文件,生成指定数量含有随机噪点的临时图像
def addNoise(fileName, num):
    if not fileName.endswith('.bmp'):
        print('Must be bmp image')
```

```python
        return
    for i in range(num):
        im=Image.open(fileName)
        width, height=im.size
        n=randint(1, 20)
        for j in range(n):
            w=randint(0, width-1)
            h=randint(0, height-1)
            im.putpixel((w,h),(0,0,0))
        im.save(fileName[:-4]+'_'+str(i+1)+'.bmp')

#根据多个含有随机噪点的图像,对应位置像素计算平均值,生成结果图像
def mergeOne(fileName, num):
    if not fileName.endswith('.bmp'):
        print('Must be bmp image')
        return
    ims=[Image.open(fileName[:-4]+'_'+str(i+1)+'.bmp') for i in range(num)]
    im=Image.new('RGB', ims[0].size,(255,255,255))
    for w in range(im.size[0]):
        for h in range(im.size[1]):
            r, g, b=[0] * 3
            for tempIm in ims:
                value=tempIm.getpixel((w,h))
                r +=value[0]
                g +=value[1]
                b +=value[2]
            r=r//num
            g=g//num
            b=b//num
            im.putpixel((w,h),(r,g,b))
    im.save(fileName[:-4]+'_result.bmp')

#对比合并后的图像和原始图像之间的相似度
def compare(fileName):
    im1=Image.open(fileName)
    im2=Image.open(fileName[:-4]+'_result.bmp')
    width, height=im1.size
    total=width * height
    right=0
    expectedRatio=0.05
    for w in range(width):
        for h in range(height):
            r1, g1, b1=im1.getpixel((w,h))
            r2, g2, b2=im2.getpixel((w,h))
```

```
            if(abs(r1-r2),abs(g1-g2),abs(b1-b2))<(255*expectedRatio,)*3:
                right +=1
    return(total, right)

if __name__=='__main__':
    #生成32个临时图像,然后进行融合,并对比融合后的图像与原始图像的相似度
    addNoise('test.bmp', 32)
    mergeOne('test.bmp', 32)
    result=compare('test.bmp')
    print('Total number of pixels:{0[0]},right number:{0[1]}'.format(result))
```

12.2.8 棋盘纹理生成

国际象棋棋盘由 8 行 8 列黑白相间的颜色区域组成,共 64 个小格子。在制作棋盘纹理时,首先生成一个指定大小的白色图像,然后再进行区域颜色的填充。代码运行成功后,会在当前文件夹中生成图像文件 qipan.jpg。

例 12-10 使用 pillow 生成国际象棋棋盘纹理。

```
from PIL import Image

def qipan(fileName, width, height, color1, color2):
    #生成空白图像
    im=Image.new('RGB',(width,height))
    for h in range(height):
        for w in range(width):
            #填充颜色交叉的图案
            if(int(h/height * 8)+int(w/width * 8))%2==0:
                im.putpixel((w,h), color1)
            else:
                im.putpixel((w,h), color2)
    #保存图像文件
    im.save(fileName)

if __name__=='__main__':
    fileName='qipan.jpg'
    qipan(fileName, 500, 500,(128,128,128),(10,10,10))
```

第 13 章 数据分析与科学计算可视化

用于数据分析与科学计算可视化的 Python 模块非常多，如 numpy、scipy、pandas、statistics、matplotlib、sympy、traits、traitsUI、Chaco、TVTK、Mayavi、VPython、OpenCV。其中，numpy 模块是科学计算包，提供了 Python 中没有的数组对象，支持 N 维数组运算、处理大型矩阵、成熟的广播函数库、矢量运算、线性代数、傅里叶变换以及随机数生成等功能，可与 C++、FORTRAN 等语言无缝结合，树莓派 Python v3 默认安装就已包含了 numpy。scipy 模块依赖于 numpy，提供了更多的数学工具，包括矩阵运算、线性方程组求解、积分、优化等。matplotlib 是比较常用的绘图模块，可以快速地将各种计算结果以各种图形形式展示出来。大部分扩展库都可以使用 pip 命令直接安装，如果有不能安装或者安装之后无法正常工作的扩展库，可以登录下面的网页选择合适的版本下载和安装：

http://www.lfd.uci.edu/~gohlke/pythonlibs/

13.1 扩展库 numpy 简介

根据 Python 社区的习惯，首先使用下面的方式来导入 numpy 模块：

```
>>>import numpy as np
```

1. 生成数组

```
>>>np.array((1, 2, 3, 4, 5))             #把 Python 列表转换成数组
array([1, 2, 3, 4, 5])
>>>np.array(range(5))                     #把 Python 的 range 对象转换成数组
array([0, 1, 2, 3, 4])
>>>np.array([[1, 2, 3], [4, 5, 6]])
array([[1, 2, 3],
       [4, 5, 6]])
>>>np.linspace(0, 10, 11)                 #生成等差数组
array([ 0.,  1.,  2.,  3.,  4.,  5.,  6.,  7.,  8.,  9., 10.])
>>>np.linspace(0, 1, 11)
array([ 0. ,  0.1,  0.2,  0.3,  0.4,  0.5,  0.6,  0.7,  0.8,  0.9,  1. ])
>>>np.logspace(0, 100, 10)                #对数数组
```

```
array([  1.00000000e+000,   1.29154967e+011,   1.66810054e+022,
         2.15443469e+033,   2.78255940e+044,   3.59381366e+055,
         4.64158883e+066,   5.99484250e+077,   7.74263683e+088,
         1.00000000e+100])
>>>np.zeros((3,3))                    #全 0 二维数组
[[ 0.  0.  0.]
 [ 0.  0.  0.]
 [ 0.  0.  0.]]
>>>np.zeros((3,1))                    #全 0 一维数组
array([[ 0.],
       [ 0.],
       [ 0.]])
>>>np.zeros((1,3))
array([[ 0.,  0.,  0.]])
>>>np.ones((3,3))                     #全 1 二维数组
array([[ 1.,  1.,  1.],
       [ 1.,  1.,  1.],
       [ 1.,  1.,  1.]])
>>>np.ones((1,3))                     #全 1 一维数组
array([[ 1.,  1.,  1.]])
>>>np.identity(3)                     #单位矩阵
array([[ 1.,  0.,  0.],
       [ 0.,  1.,  0.],
       [ 0.,  0.,  1.]])
>>>np.identity(2)
array([[ 1.,  0.],
       [ 0.,  1.]])
>>>np.empty((3,3))                    #空数组,只申请空间而不初始化,元素值是不确定的
array([[ 0.,  0.,  0.],
       [ 0.,  0.,  0.],
       [ 0.,  0.,  0.]])
```

2. 数组与数值的算术运算

```
>>>x=np.array((1, 2, 3, 4, 5))        #创建数组对象
>>>x
array([1, 2, 3, 4, 5])
>>>x * 2                              #数组与数值相乘,所有元素与数值相乘
array([ 2, 4, 6, 8, 10])
>>>x / 2                              #数组与数值相除
array([ 0.5, 1. , 1.5, 2. , 2.5])
>>>x // 2                             #数组与数值整除
array([0, 1, 1, 2, 2], dtype=int32)
>>>x ** 3                             #幂运算
```

```
array([1, 8, 27, 64, 125], dtype=int32)
>>>x+2                                      #数组与数值相加
array([3, 4, 5, 6, 7])
>>>x %3                                     #余数
array([1, 2, 0, 1, 2], dtype=int32)
```

3. 数组与数组的算术运算

```
>>>a=np.array((1, 2, 3))
>>>b=np.array(([1, 2, 3], [4, 5, 6], [7, 8, 9]))
>>>c=a * b                                  #数组与数组相乘
>>>c                                        #a 中的每个元素乘以 b 中的每一列元素
array([[ 1,  4,  9],
       [ 4, 10, 18],
       [ 7, 16, 27]])
>>>c / b                                    #数组之间的除法运算
array([[ 1.,  2.,  3.],
       [ 1.,  2.,  3.],
       [ 1.,  2.,  3.]])
>>>c / a
array([[ 1.,  2.,  3.],
       [ 4.,  5.,  6.],
       [ 7.,  8.,  9.]])
>>>a+a                                      #数组之间的加法运算
array([2, 4, 6])
>>>a * a                                    #数组之间的乘法运算
array([1, 4, 9])
>>>a-a                                      #数组之间的减法运算
array([0, 0, 0])
>>>a / a                                    #数组之间的除法运算
array([ 1.,  1.,  1.])
```

4. 二维数组转置

```
>>>b=np.array(([1, 2, 3], [4, 5, 6], [7, 8, 9]))
>>>b
array([[1, 2, 3],
       [4, 5, 6],
       [7, 8, 9]])
>>>b.T                                      #转置
array([[1, 4, 7],
       [2, 5, 8],
       [3, 6, 9]])
>>>a=np.array((1, 2, 3, 4))
```

```
>>>a
array([1, 2, 3, 4])
>>>a.T                                    #一维数组转置以后和原来是一样的
array([1, 2, 3, 4])
```

5．向量内积

```
>>>a=np.array((5, 6, 7))
>>>b=np.array((6, 6, 6))
>>>a.dot(b)                               #向量内积
108
>>>np.dot(a,b)
108
>>>c=np.array(([1,2,3],[4,5,6],[7,8,9]))  #二维数组
>>>cT=c.T                                 #转置
>>>c.dot(a)                               #二维数组的每行与一维向量计算内积
array([ 38, 92, 146])
>>>c[0].dot(a)                            #两个一维向量计算内积
38
>>>c[1].dot(a)
92
>>>c[2].dot(a)
146
>>>a.dot(c)                               #一维向量与二维向量的每列计算内积
array([ 78, 96, 114])
>>>a.dot(cT[0])
78
>>>a.dot(cT[1])
96
>>>a.dot(cT[2])
114
```

6．数组元素访问

```
>>>b=np.array(([1,2,3],[4,5,6],[7,8,9]))
>>>b
array([[1, 2, 3],
       [4, 5, 6],
       [7, 8, 9]])
>>>b[0]                                   #第 0 行
array([1, 2, 3])
>>>b[0][0]                                #第 0 行第 0 列的元素值
1
```

数组元素还支持多元素同时访问，例如：

```
>>>x=np.arange(0, 100, 10, dtype=np.floating)    #创建等差数组
>>>x
array([0., 10., 20., 30., 40., 50., 60., 70., 80., 90.])
>>>index=np.random.randint(0, len(x), 5)         #生成5个随机整数作为下标
>>>index
array([5, 4, 1, 2, 9])
>>>x[index]                                       #同时访问多个元素的值
array([50., 40., 10., 20., 90.])
>>>x[index]=[1, 2, 3, 4, 5]                      #同时修改多个下标指定的元素值
>>>x
array([0., 3., 4., 30., 2., 1., 60., 70., 80., 5.])
>>>x[[1,2,3]]                                    #同时访问多个元素的值
array([3., 4., 30.])
```

7. 对数组进行函数运算

```
>>>x=np.arange(0, 100, 10, dtype=np.floating)
>>>np.sin(x)                                     #一维数组中所有元素求正弦值
array([ 0.        , -0.54402111,  0.91294525, -0.98803162,  0.74511316,
       -0.26237485, -0.30481062,  0.77389068, -0.99388865,  0.89399666])
>>>b=np.array(([1, 2, 3], [4, 5, 6], [7, 8, 9]))
>>>np.cos(b)                                     #二维数组中所有元素求余弦值
array([[ 0.54030231, -0.41614684, -0.9899925 ],
       [-0.65364362,  0.28366219,  0.96017029],
       [ 0.75390225, -0.14550003, -0.91113026]])
>>>np.round(_)                                   #四舍五入
array([[ 1., -0., -1.],
       [-1.,  0.,  1.],
       [ 1., -0., -1.]])
>>>x=np.random.rand(10)                          #包含10个随机数的数组
>>>x=x * 10
>>>x
array([6.03635335, 3.90542305, 0.05402166, 0.97778005, 8.86122047,
       8.68849771, 8.43456386, 6.10805351, 1.01185534, 5.52150462])
>>>np.floor(x)                                   #所有元素向下取整
array([6., 3., 0., 0., 8., 8., 8., 6., 1., 5.])
>>>np.ceil(x)                                    #所有元素向上取整
array([7., 4., 1., 1., 9., 9., 9., 7., 2., 6.])
```

8. 对矩阵不同维度上的元素进行计算

```
>>>x=np.arange(0,10).reshape(2,5)                #创建二维数组
>>>x
array([[0, 1, 2, 3, 4],
```

```
       [5, 6, 7, 8, 9]])
>>>np.sum(x)                                    #二维数组所有元素求和
45
>>>np.sum(x, axis=0)                            #二维数组纵向求和
array([ 5,  7,  9, 11, 13])
>>>np.sum(x, axis=1)                            #二维数组横向求和
array([10, 35])
>>>np.mean(x, axis=0)                           #二维数组纵向计算算术平均值
array([ 2.5, 3.5, 4.5, 5.5, 6.5])
>>>weight=[0.3, 0.7]                            #权重
>>>np.average(x, axis=0, weights=weight)        #二维数组纵向计算加权平均值
array([ 3.5, 4.5, 5.5, 6.5, 7.5])
>>>np.max(x)                                    #所有元素最大值
9
>>>np.max(x, axis=0)                            #每列元素的最大值
array([5, 6, 7, 8, 9])
>>>x=np.random.randint(0, 10, size=(3,3))       #创建二维数组
>>>x
array([[4, 9, 1],
       [7, 4, 9],
       [8, 9, 1]])
>>>np.std(x)                                    #所有元素的标准差
3.1544599036840864
>>>np.std(x, axis=1)                            #每行元素的标准差
array([3.29983165, 2.05480467, 3.55902608])
>>>np.var(x, axis=0)                            #每列元素的标准差
array([2.88888889, 5.55555556, 14.22222222])
>>>np.sort(x, axis=0)                           #纵向排序
array([[4, 4, 1],
       [7, 9, 1],
       [8, 9, 9]])
>>>np.sort(x, axis=1)                           #横向排序
array([[1, 4, 9],
       [4, 7, 9],
       [1, 8, 9]])
```

9. 改变数组大小

```
>>>a=np.arange(1, 11, 1)
>>>a
array([1, 2, 3, 4, 5, 6, 7, 8, 9, 10])
>>>a.shape=2, 5                                 #改为 2 行 5 列
>>>a
array([[ 1,  2,  3,  4,  5],
```

```
       [ 6,  7,  8,  9, 10]])
>>>a.shape=5,-1                              #-1表示自动计算
>>>a
array([[ 1,  2],
       [ 3,  4],
       [ 5,  6],
       [ 7,  8],
       [ 9, 10]])
>>>b=a.reshape(2,5)                          #reshape()方法返回新数组
>>>b
array([[ 1,  2,  3,  4,  5],
       [ 6,  7,  8,  9, 10]])
```

10. 切片操作

```
>>>a=np.arange(10)
>>>a
array([0, 1, 2, 3, 4, 5, 6, 7, 8, 9])
>>>a[::-1]                                   #反向切片
array([9, 8, 7, 6, 5, 4, 3, 2, 1, 0])
>>>a[::2]                                    #隔一个取一个元素
array([0, 2, 4, 6, 8])
>>>a[:5]                                     #前5个元素
array([0, 1, 2, 3, 4])
>>>c=np.arange(25)                           #创建数组
>>>c.shape=5,5                               #修改数组大小
>>>c
array([[ 0,  1,  2,  3,  4],
       [ 5,  6,  7,  8,  9],
       [10, 11, 12, 13, 14],
       [15, 16, 17, 18, 19],
       [20, 21, 22, 23, 24]])
>>>c[0, 2:5]                                 #第0行中下标[2,5)之间的元素值
array([2, 3, 4])
>>>c[1]                                      #第0行所有元素
array([5, 6, 7, 8, 9])
>>>c[2:5, 2:5]                               #行下标和列下标都介于[2,5)之间的元素值
array([[12, 13, 14],
       [17, 18, 19],
       [22, 23, 24]])
```

11. 布尔运算

```
>>>x=np.random.rand(10)                      #包含10个随机数的数组
```

```
>>>x
array([ 0.56707504,  0.07527513,  0.0149213 ,  0.49157657,  0.75404095,
        0.40330683,  0.90158037,  0.36465894,  0.37620859,  0.62250594])
>>>x>0.5                                          #比较数组中每个元素值是否大于0.5
array([ True, False, False, False,  True, False,  True, False, False,  True],
dtype=bool)
>>>x[x>0.5]                                       #获取数组中大于0.5的元素
array([ 0.56707504,  0.75404095,  0.90158037,  0.62250594])
>>>a=np.array([1, 2, 3])
>>>b=np.array([3, 2, 1])
>>>a>b                                            #两个数组中对应位置上的元素比较
array([False, False,  True], dtype=bool)
>>>a[a>b]
array([3])
>>>a==b
array([False,  True, False], dtype=bool)
>>>a[a==b]
array([2])
```

12. 广播

```
>>>a=np.arange(0,60,10).reshape(-1,1)             #列向量
>>>b=np.arange(0,6)                               #行向量
>>>a
array([[ 0],
       [10],
       [20],
       [30],
       [40],
       [50]])
>>>b
array([0, 1, 2, 3, 4, 5])
>>>a+b                                            #广播
array([[ 0,  1,  2,  3,  4,  5],
       [10, 11, 12, 13, 14, 15],
       [20, 21, 22, 23, 24, 25],
       [30, 31, 32, 33, 34, 35],
       [40, 41, 42, 43, 44, 45],
       [50, 51, 52, 53, 54, 55]])
>>>a * b
array([[  0,   0,   0,   0,   0,   0],
       [  0,  10,  20,  30,  40,  50],
       [  0,  20,  40,  60,  80, 100],
       [  0,  30,  60,  90, 120, 150],
```

```
       [  0,  40,  80, 120, 160, 200],
       [  0,  50, 100, 150, 200, 250]])
```

13. 分段函数

```
>>>x=np.random.randint(0, 10, size=(1,10))
>>>x
array([[0, 4, 3, 3, 8, 4, 7, 3, 1, 7]])
>>>np.where(x<5, 0, 1)                           #小于5的元素值对应0,其他对应1
array([[0, 0, 0, 0, 1, 0, 1, 0, 0, 1]])
#小于4的元素乘以2,大于7的元素乘以3,其他元素变为0
>>>np.piecewise(x,[x<4, x>7],[lambda x:x*2,lambda x:x*3])
array([[ 0,  0,  6,  6, 24,  0,  0,  6,  2,  0]])
```

14. 计算唯一值以及出现次数

```
>>>x=np.random.randint(0,10,7)
>>>x
array([8, 7, 7, 5, 3, 8, 0])
>>>np.bincount(x)                                #元素出现的次数,0表示出现1次
array([1, 0, 0, 1, 0, 1, 0, 2, 2], dtype=int64)  #1、2表示没出现,3表示出现1次,以此类推
>>>np.sum(_)                                     #所有元素出现次数之和等于数组长度
7
>>>len(x)
7
>>>np.unique(x)                                  #返回唯一元素值
array([0, 3, 5, 7, 8])
>>>x=np.random.randint(0,10,2)
>>>x
array([2, 1])
>>>np.bincount(x)                                #结果数组的长度取决于原始数组中最大元素值
array([0, 1, 1], dtype=int64)
>>>x=np.random.randint(0, 10, 10)
>>>x
array([3, 6, 4, 5, 2, 9, 7, 0, 9, 0])
>>>y=np.random.rand(10)                          #随机小数,模拟权重
>>>y=np.round_(y, 1)                             #保留一位小数
>>>y
array([ 0.6,  0.8,  0.8,  0. ,  0.6,  0.1,  0. ,  0.2,  0.8,  0.7])
>>>np.sum(x*y)/np.sum(np.bincount(x))            #加权总和/出现总次数或元素个数
2.9199999999999999
```

15. 矩阵运算

```
>>>a_list=[3, 5, 7]
```

```
>>>a_mat=np.matrix(a_list)                    #创建矩阵
>>>a_mat
matrix([[3, 5, 7]])
>>>a_mat.T                                    #矩阵转置
matrix([[3],
        [5],
        [7]])
>>>a_mat.shape                                #矩阵形状
(1, 3)
>>>a_mat.size
3
>>>b_mat=np.matrix((1, 2, 3))
>>>b_mat
matrix([[1, 2, 3]])
>>>a_mat * b_mat.T                            #矩阵相乘
matrix([[34]])
>>>a_mat.mean()                               #元素平均值
5.0
>>>a_mat.sum()                                #所有元素之和
15
>>>a_mat.max()
7
>>>c_mat=np.matrix([[1, 5, 3], [2, 9, 6]])    #创建二维矩阵
>>>c_mat
matrix([[1, 5, 3],
        [2, 9, 6]])
>>>c_mat.argsort(axis=0)                      #纵向排序后的元素序号
matrix([[0, 0, 0],
        [1, 1, 1]], dtype=int64)
>>>c_mat.argsort(axis=1)                      #横向排序后的元素序号
matrix([[0, 2, 1],
        [0, 2, 1]], dtype=int64)
>>>d_mat=np.matrix([[1, 2, 3], [4, 5, 6], [7, 8, 9]])
>>>d_mat.diagonal()                           #矩阵对角线元素
matrix([[1, 5, 9]])
>>>d_mat.flatten()                            #矩阵平铺
matrix([[1, 2, 3, 4, 5, 6, 7, 8, 9]])
```

13.2 科学计算扩展库 scipy

scipy 是专门为科学计算和工程应用设计的 Python 工具包，在 numpy 的基础上增加了大量用于科学计算以及工程计算的模块，包括统计、优化、整合、线性代数、常微分方程数值求解、信号处理、图像处理、稀疏矩阵等。scipy 工具包的主要模块如表 13-1 所示。

表 13-1　scipy 工具包的主要模块

模　　块	说　　明
constants	常数
special	特殊函数
optimize	数值优化算法，如最小二乘拟合（leastsq）、函数最小值（fmin 系列）、非线性方程组求解（fsolve）等
interpolate	插值（interp1d、interp2d 等）
integrate	数值积分
signal	信号处理
ndimage	图像处理，包括滤波器模块 filters、傅里叶变换模块 fourier、图像插值模块 interpolation、图像测量模块 measurements、形态学图像处理模块 morphology 等
stats	统计
misc	提供了读取图像文件的方法和一些测试图像
io	提供了读取 Matlab 和 Fortran 文件的方法

13.2.1　数学、物理常用常数与单位模块 constants

scipy 工具包的常数模块 constants 包含大量用于科学计算的常数，下面给出其中几个，更多的可以查看 http://docs.scipy.org/doc/scipy/reference/constants.html。

例如，可以使用下面的方法来访问该模块中预定义的常数。

```
>>>from scipy import constants as C
>>>C.pi                               #圆周率
3.141592653589793
>>>C.golden                           #黄金比例
1.618033988749895
>>>C.c                                #真空中的光速
299792458.0
>>>C.h                                #普朗克常数
6.62606896e-34
>>>C.mile                             #一英里等于多少米
1609.3439999999998
>>>C.inch                             #一英寸等于多少米
0.0254
>>>C.degree                           #一度等于多少弧度
0.017453292519943295
>>>C.minute                           #一分钟等于多少秒
60.0
>>>C.g                                #标准重力加速度
9.80665
```

13.2.2 特殊函数模块 special

scipy 工具包的 special 模块包含了大量函数库，包括基本数学函数和很多特殊函数。

```
>>>from scipy import special as S
>>>S.cbrt(8)                              #立方根
2.0
>>>S.exp10(3)                             #10**3
1000.0
>>>S.sindg(90)                            #正弦函数，参数为角度
1.0
>>>S.round(3.1)                           #四舍五入函数
3.0
>>>S.round(3.5)
4.0
>>>S.round(3.499)
3.0
>>>S.comb(5,3)                            #从 5 个中任选 3 个的组合数
10.0
>>>S.perm(5,3)                            #排列数
60.0
>>>S.gamma(4)                             #gamma 函数
6.0
>>>S.beta(10, 200)                        #beta 函数
2.839607777781333e-18
>>>S.sinc(0)                              #sinc 函数
1.0
```

13.2.3 信号处理模块 signal

signal 模块包含大量滤波函数、B 样条插值算法等。下面的代码演示了一维信号的卷积运算：

```
>>>import numpy as np
>>>x=np.array([1, 2, 3])
>>>h=np.array([4, 5, 6])
>>>import scipy.signal
>>>scipy.signal.convolve(x, h)            #一维卷积运算
array([ 4, 13, 28, 27, 18])
```

下面的代码演示了二维图像卷积运算，运行结果如图 13-1 所示。

```
import numpy as np
from scipy import signal, misc
import matplotlib.pyplot as plt
```

```
image=misc.lena()                                #二维图像数组,lena 图像
w=np.zeros((50, 50))                             #全 0 二维数组,卷积核
w[0][0]=1.0                                      #修改参数,调整滤波器
w[49][25]=1.0                                    #可以根据需要调整
image_new=signal.fftconvolve(image, w)           #使用 FFT 算法进行卷积

plt.figure()
plt.imshow(image_new)                            #显示滤波后的图像
plt.gray()
plt.title('Filtered image')
plt.show()
```

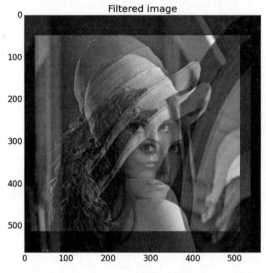

图 13-1　lena 图像处理结果

下面的代码对 lena 图像进行模糊,运行结果如图 13-2 所示。

```
image=misc.lena()
w=signal.gaussian(50, 10.0)
image_new=signal.sepfir2d(image, w, w)
```

中值滤波是数字信号处理、数字图像处理中常用的预处理技术,特点是将信号中每个值都替换为其邻域内的中值,即邻域内所有值排序后中间位置上的值。下面的代码演示了 scipy 模块的中值滤波算法的用法。

```
>>>import random
>>>import numpy as np
>>>import scipy.signal as signal
>>>x=np.arange(0,100,10)
>>>random.shuffle(x)                             #打乱顺序
```

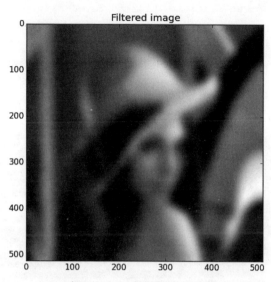

图 13-2 lena 图像模糊处理结果

```
>>>x
array([40, 0, 60, 20, 50, 70, 80, 90, 30, 10])
>>>signal.medfilt(x,3)                              #中值滤波
array([ 0., 40., 20., 50., 50., 70., 80., 80., 30., 10.])
```

13.2.4 图像处理模块 ndimage

模块 ndimage 提供了大量用于 N 维图像处理的方法,下面仅选取一部分进行演示,更多的用法可以参考官方文档。

1. 图像滤波

```
>>>from scipy import misc
>>>from scipy import ndimage
>>>import matplotlib.pyplot as plt
>>>face=misc.face()                                 #face 是测试图像之一
>>>plt.figure()                                     #创建图形
>>>plt.imshow(face)                                 #绘制测试图像
>>>plt.show()                                       #原始图像,如图 13-3 所示
>>>blurred_face=ndimage.gaussian_filter(face, sigma=7)
                                                    #高斯滤波
>>>plt.imshow(blurred_face)
>>>plt.show()                                       #高斯滤波图像,如图 13-4 所示
>>>blurred_face1=ndimage.gaussian_filter(face, sigma=1)
                                                    #以下 3 行为边缘锐化
```

```
>>>blurred_face3=ndimage.gaussian_filter(face,sigma=3)
>>>sharp_face=blurred_face3+6*(blurred_face3-blurred_face1)
>>>plt.imshow(sharp_face)                    #见图13-5
>>>plt.show()
>>>median_face=ndimage.median_filter(face,7)  #中值滤波
>>>plt.imshow(median_face)                    #见图13-6
>>>plt.show()
```

图 13-3 原始图像

图 13-4 高斯滤波结果

图 13-5　边缘锐化结果

图 13-6　中值滤波结果

2. 数学形态学

```
>>>square=np.zeros((32,32))                          #全 0 数组
>>>square[10:20, 10:20]=1                            #把其中一部分设置为 1
>>>x, y = (32 * np.random.random((2, 15))).astype(np.int)
                                                     #随机位置
>>>square[x,y]=1                                     #把随机位置设置为 1
>>>plt.imshow(square)                                #原始随机图像,见图 13-7
>>>plt.show()
>>>open_square=ndimage.binary_opening(square)        #开运算
>>>plt.imshow(open_square)                           #开运算结果,见图 13-8
>>>plt.show()
```

```
>>>eroded_square=ndimage.binary_erosion(square)      #膨胀运算
>>>plt.imshow(eroded_square)                         #膨胀运算结果,见图 13-9
>>>plt.show()
>>>closed_square=ndimage.binary_closing(square)      #闭运算
>>>plt.imshow(closed_square)                         #闭运算结果,见图 13-10
>>>plt.show()
```

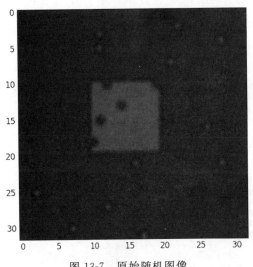

图 13-7　原始随机图像

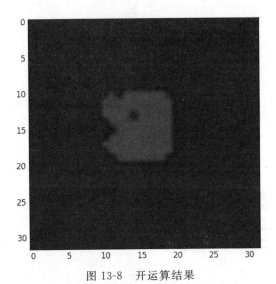

图 13-8　开运算结果

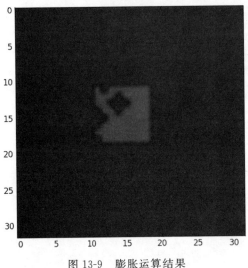

图 13-9　膨胀运算结果

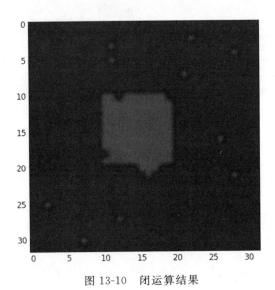

图 13-10　闭运算结果

3. 图像测量

```
>>>ndimage.measurements.maximum(face)                #最大值
255
>>>ndimage.measurements.maximum_position(face)       #最大值位置
```

```
(242, 560, 2)
>>>ndimage.measurements.mean(face)              #平均值
110.16274388631184
>>>ndimage.measurements.median(face)            #中值
109.0
>>>ndimage.measurements.sum(face)
259906521
>>>ndimage.measurements.variance(face)
3307.17544034096
>>>ndimage.measurements.standard_deviation(face)
57.508046744268405
>>>ndimage.measurements.histogram(face, 0, 255, 256)
```

13.3 扩展库 pandas 简介

pandas(Python Data Analysis Library)是基于 numpy 的数据分析模块，提供了大量标准数据模型和高效操作大型数据集所需要的工具，可以说 pandas 是使得 Python 能够成为高效且强大的数据分析环境的重要因素之一。

pandas 主要提供了 3 种数据结构：①Series，带标签的一维数组；②DataFrame，带标签且大小可变的二维表格结构；③Panel，带标签且大小可变的三维数组。

可以在命令提示符环境使用 pip 工具下载和安装 pandas，然后按照 Python 社区的习惯，使用下面的语句导入：

```
>>>import pandas as pd
```

1. 生成一维数组

```
>>>import numpy as np
>>>x=pd.Series([1, 3, 5, np.nan])
```

2. 生成二维数组

```
>>>dates=pd.date_range(start='20130101', end='20131231', freq='D')
                                                #间隔为天
>>>dates=pd.date_range(start='20130101', end='20131231', freq='M')
                                                #间隔为月
>>>df=pd.DataFrame(np.random.randn(12,4), index=dates, columns=list('ABCD'))
>>>df=pd.DataFrame([[np.random.randint(1,100) for j in range(4)] for i in range
(12)], index=dates, columns=list('ABCD'))       #4 列随机数
>>>df=pd.DataFrame({'A':[np.random.randint(1,100) for i in range(4)],
                    'B':pd.date_range(start='20130101', periods=4, freq='D'),
                    'C':pd.Series([1, 2, 3, 4], index=list(range(4)), dtype=
                    'float32'),
```

```
                    'D':np.array([3] * 4,dtype='int32'),
                    'E':pd.Categorical(["test","train","test","train"]),
                    'F':'foo'})
>>>df=pd.DataFrame({'A':[np.random.randint(1,100)for i in range(4)],
        'B':pd.date_range(start='20130101', periods=4, freq='D'),
        'C':pd.Series([1, 2, 3, 4],index=['zhang', 'li', 'zhou', 'wang'],dtype
        ='float32'),
        'D':np.array([3] * 4,dtype='int32'),
        'E':pd.Categorical(["test","train","test","train"]),
        'F':'foo'})
```

3. 二维数据查看

```
>>>df.head()                                        #默认显示前5行
>>>df.head(3)                                       #查看前3行
>>>df.tail(2)                                       #查看最后2行
```

4. 查看二维数据的索引、列名和数据

```
>>>df.index
>>>df.columns
>>>df.values
```

5. 查看数据的统计信息

```
>>>df.describe()                  #返回平均值、标准差、最小值、最大值等信息
```

6. 二维数据转置

```
>>>df.T
```

7. 排序

```
>>>df.sort_index(axis=0, ascending=False)           #对轴进行排序
>>>df.sort_index(axis=1, ascending=False)
>>>df.sort_values(by='A')                           #对数据进行排序
>>>df.sort_values(by='A', ascending=False)          #降序排列
```

8. 数据选择

```
>>>df['A']                                          #选择列
>>>df[0:2]                                          #使用切片选择多行
>>>df.loc[:, ['A', 'C']]                            #选择多列
>>>df.loc[['zhang', 'zhou'], ['A', 'D', 'E']]       #同时指定多行与多列进行选择
>>>df.loc['zhang', ['A', 'D', 'E']]
```

```
>>>df.at['zhang', 'A']                              #查询指定行、列位置的数据值
>>>df.at['zhang', 'D']
>>>df.iloc[3]                                       #查询第 3 行数据
>>>df.iloc[0:3, 0:4]                                #查询前 3 行、前 4 列数据
>>>df.iloc[[0, 2, 3], [0, 4]]                       #查询指定的多行、多列数据
>>>df.iloc[0,1]                                     #查询指定行、列位置的数据值
>>>df.iloc[2,2]
>>>df[df.A>50]                                      #按给定条件进行查询
```

9. 数据修改与设置

```
>>>df.iat[0, 2]=3                                   #修改指定行、列位置的数据值
>>>df.loc[:, 'D']=[np.random.randint(50, 60)for i in range(4)]
                                                    #修改某列的值
>>>df['C']=-df['C']                                 #对指定列数据取反
```

10. 缺失值处理（缺失值和异常值处理是大数据预处理环节中很重要的一个步骤）

```
>>>df1=df.reindex(index=['zhang', 'li', 'zhou', 'wang'], columns=list(df.columns)+['G'])
>>>df1.iat[0, 6]=3              #修改指定位置的元素值,该列其他元素为缺失值 NaN
>>>pd.isnull(df1)               #测试缺失值,返回值为 True/False 阵列
>>>df1.dropna()                                     #返回不包含缺失值的行
>>>df1['G'].fillna(5, inplace=True)                 #使用指定值填充缺失值
```

11. 数据操作

```
>>>df1.mean()                                       #平均值,自动忽略缺失值
>>>df.mean(1)                                       #横向计算平均值
>>>df1.shift(1)                                     #数据移位
>>>df1['D'].value_counts()                          #直方图统计
>>>df2=pd.DataFrame(np.random.randn(10, 4))
>>>p1=df2[:3]                                       #数据行拆分
>>>p2=df2[3:7]
>>>p3=df2[7:]
>>>df3=pd.concat([p1, p2, p3])                      #数据行合并
>>>df2 ==df3                    #测试两个二维数据是否相等,返回 True/False 阵列
>>>df4=pd.DataFrame({'A':[np.random.randint(1,5)for i in range(8)],
                     'B':[np.random.randint(10,15)for i in range(8)],
                     'C':[np.random.randint(20,30)for i in range(8)],
                     'D':[np.random.randint(80,100)for i in range(8)]})
>>>df4.groupby('A').sum()                           #数据分组计算
>>>df4.groupby(['A','B']).mean()
```

12. 结合 matplotlib 绘图

```
>>>import pandas as pd
>>>import numpy as np
>>>import matplotlib.pyplot as plt
>>>df=pd.DataFrame(np.random.randn(1000, 2), columns=['B', 'C']).cumsum()
>>>df['A']=pd.Series(list(range(len(df))))
>>>plt.figure()
>>>df.plot(x='A')
>>>plt.show()
```

代码运行结果如图 13-11 所示。

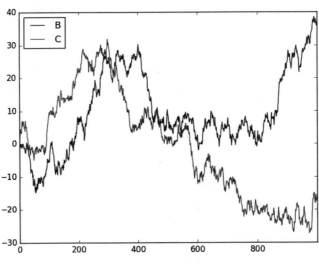

图 13-11　绘制曲线图结果

下面的代码用来绘制柱状图，结果如图 13-12 所示。

```
>>>df=pd.DataFrame(np.random.rand(10, 4), columns=['a', 'b', 'c', 'd'])
>>>df.plot(kind='bar')
>>>plt.show()
```

将上面代码中的绘图语句改为

```
>>>df.plot(kind='barh', stacked=True)
```

运行结果如图 13-13 所示。

13. 文件读写

```
>>>df5.to_excel('d:\\test.xlsx', sheet_name='dfg')      #将数据保存为 Excel 文件
>>>df6=pd.read_excel('d:\\test.xlsx', 'dfg', index_col=None, na_values=['NA'])
>>>df6.to_csv('d:\\test.csv')                           #将数据保存为 csv 文件
>>>df7=pd.read_csv('d:\\test.csv')                      #读取 csv 文件中的数据
```

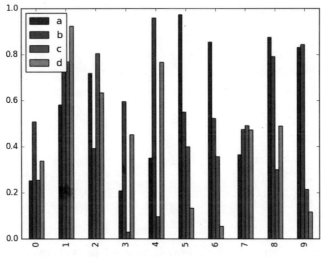

图 13-12　绘制柱状图结果

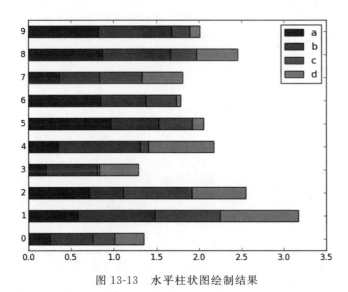

图 13-13　水平柱状图绘制结果

13.4　统计分析标准库 statistics 用法简介

Python 标准库 statistics 提供了大量方法用于计算数值数据的数理统计信息。

1. 计算平均数函数 mean()

```
>>>import statistics
>>>statistics.mean([1, 2, 3, 4, 5, 6, 7, 8, 9])         #使用包含整数的列表做参数
5.0
>>>statistics.mean(range(1,10))                          #使用 range 对象做参数
```

```
5.0
>>>import fractions
>>>x=[(3, 7),(1, 21),(5, 3),(1, 3)]
>>>y=[fractions.Fraction(*item)for item in x]     #创建包含分数的列表
>>>y
[Fraction(3, 7), Fraction(1, 21), Fraction(5, 3), Fraction(1, 3)]
>>>statistics.mean(y)                              #使用包含分数的列表做参数
Fraction(13, 21)
>>>import decimal
>>>x = ('0.5', '0.75', '0.625', '0.375')
>>>y=map(decimal.Decimal, x)
>>>statistics.mean(y)
Decimal('0.5625')
```

2. 中位数函数 median()、median_low()、median_high()、median_grouped()

```
>>>statistics.median([1, 3, 5, 7])        #偶数个样本时取中间两个数的平均数
4.0
>>>statistics.median_low([1, 3, 5, 7])    #偶数个样本时取中间两个数的较小者
3
>>>statistics.median_high([1, 3, 5, 7])   #偶数个样本时取中间两个数的较大者
5
>>>statistics.median(range(1,10))
5
>>>statistics.median_low([5, 3, 7]), statistics.median_high([5, 3, 7])
(5, 5)
>>>statistics.median_grouped([5, 3, 7])
5.0
>>>statistics.median_grouped([52, 52, 53, 54])
52.5
>>>statistics.median_grouped([1, 3, 3, 5, 7])
3.25
>>>statistics.median_grouped([1, 2, 2, 3, 4, 4, 4, 4, 5])
3.7
>>>statistics.median_grouped([1, 2, 2, 3, 4, 4, 4, 4, 5], interval=2)
3.4
```

3. 返回最常见数据或出现次数最多的数据的函数 mode()

```
>>>statistics.mode([1, 3, 5, 7])          #无法确定出现次数最多的唯一元素
statistics.StatisticsError: no unique mode; found 4 equally common values
>>>statistics.mode([1, 3, 5, 7, 3])
3
>>>statistics.mode(["red", "blue", "blue", "red", "green", "red", "red"])
'red'
```

4. pstdev()

返回总体标准差。

```
>>>statistics.pstdev([1.5, 2.5, 2.5, 2.75, 3.25, 4.75])
0.986893273527251
>>>statistics.pstdev(range(20))
5.766281297335398
```

5. pvariance()

返回总体方差或二次矩。

```
>>>statistics.pvariance([1.5, 2.5, 2.5, 2.75, 3.25, 4.75])
0.9739583333333334
>>>x=[1, 2, 3, 4, 5, 10, 9, 8, 7, 6]
>>>mu=statistics.mean(x)
>>>mu
5.5
>>>statistics.pvariance([1, 2, 3, 4, 5, 10, 9, 8, 7, 6], mu)
8.25
>>>statistics.pvariance(range(20))
33.25
>>>statistics.pvariance((random.randint(1,10000) for i in range(30)))
10903549.933333334
```

6. variance()、stdev()

计算样本方差和样本标准差也称为均方差。

```
>>>statistics.variance(range(20))
35.0
>>>statistics.stdev(range(20))
5.916079783099616
>>>_ * _
35.0
>>>statistics.variance([3, 3, 3, 3, 3, 3]), statistics.stdev([3, 3, 3, 3, 3, 3])
(0.0, 0.0)
```

13.5 matplotlib

matplotlib 模块依赖于 numpy 模块和 tkinter 模块,可以绘制多种形式的图形,包括线图、直方图、饼状图、散点图、误差线图等,图形质量可满足出版要求,是计算结果可视化的重要工具。

13.5.1 绘制正弦曲线

```
import numpy as np
import pylab as pl

t=np.arange(0.0, 2.0*np.pi, 0.01)    #生成数组,0~2π之间,以 0.01 为步长
s=np.sin(t)                          #对数组中的所有元素求正弦值,得到新数组
pl.plot(t,s)                         #画图,以 t 为横坐标,s 为纵坐标
pl.xlabel('x')                       #设置坐标轴标签
pl.ylabel('y')
pl.title('sin')                      #设置图形标题
pl.show()                            #显示图形
```

运行结果如图 13-14 所示。

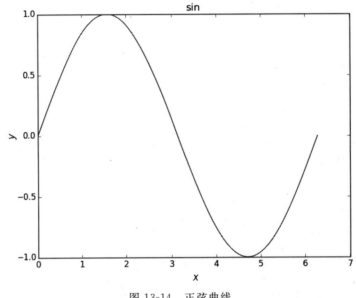

图 13-14　正弦曲线

13.5.2 绘制散点图

下面的代码绘制了余弦曲线的散点图,运行结果如图 13-15 所示。

```
import numpy as np
import pylab as pl

a=np.arange(0, 2.0*np.pi, 0.1)
b=np.cos(a)
pl.scatter(a,b)                      #绘制散点图
```

```
pl.show()
```

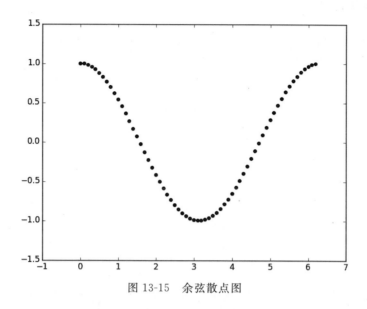

图 13-15　余弦散点图

散点图是分析数据相关性常用的方法，下面的代码使用随机数生成数值然后生成散点图，并根据数值大小来计算散点的大小，运行结果如图 13-16 所示。

```
import matplotlib.pylab as pl
import numpy as np

x=np.random.random(100)
y=np.random.random(100)
pl.scatter(x,y,s=x*500,c=u'r',marker=u'*')     #s指大小,c指颜色,marker指符号形状
pl.show()
```

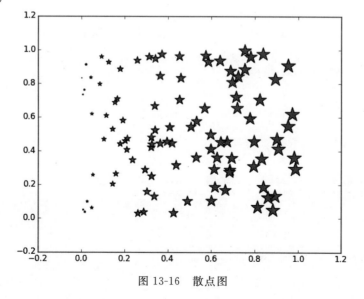

图 13-16　散点图

13.5.3 绘制饼状图

```python
import numpy as np
import matplotlib.pyplot as plt

#The slices will be ordered and plotted counter-clockwise.
labels='Frogs', 'Hogs', 'Dogs', 'Logs'
sizes=[15, 30, 45, 10]
colors=['yellowgreen', 'gold', '#FF0000', 'lightcoral']
explode = (0, 0.1, 0, 0.1)                    #使饼状图中第2片和第4片裂开

fig=plt.figure()
ax=fig.gca()
ax.pie(np.random.random(4), explode=explode, labels=labels, colors=colors,
       autopct='%1.1f%%', shadow=True, startangle=90,
       radius=0.25, center=(0, 0), frame=True)
ax.pie(np.random.random(4), explode=explode, labels=labels, colors=colors,
       autopct='%1.1f%%', shadow=True, startangle=90,
       radius=0.25, center=(1, 1), frame=True)
ax.pie(np.random.random(4), explode=explode, labels=labels, colors=colors,
       autopct='%1.1f%%', shadow=True, startangle=90,
       radius=0.25, center=(0, 1), frame=True)
ax.pie(np.random.random(4), explode=explode, labels=labels, colors=colors,
       autopct='%1.1f%%', shadow=True, startangle=90,
       radius=0.25, center=(1, 0), frame=True)
ax.set_xticks([0, 1])                         #设置坐标轴刻度
ax.set_yticks([0, 1])
ax.set_xticklabels(["Sunny", "Cloudy"])       #设置坐标轴刻度上显示的标签
ax.set_yticklabels(["Dry", "Rainy"])
ax.set_xlim((-0.5, 1.5))                      #设置坐标轴跨度
ax.set_ylim((-0.5, 1.5))
#Set aspect ratio to be equal so that pie is drawn as a circle.
ax.set_aspect('equal')

plt.show()
```

程序运行结果如图13-17所示。

13.5.4 绘制带有中文标签和图例的图

```python
import numpy as np
import pylab as pl
import matplotlib.font_manager as fm
```

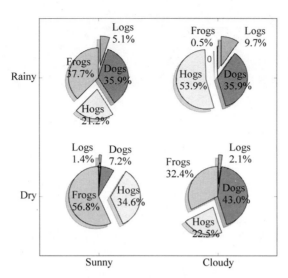

图 13-17　绘制饼状图

```
myfont=fm.FontProperties(fname=r'C:\Windows\Fonts\STKAITI.ttf')
                                        #设置字体
t=np.arange(0.0, 2.0*np.pi, 0.01)       #自变量取值范围
s=np.sin(t)                             #计算正弦函数值
z=np.cos(t)                             #计算余弦函数值
pl.plot(t, s, label='正弦')
pl.plot(t, z, label='余弦')
pl.xlabel('x-变量', fontproperties='STKAITI', fontsize=24)
                                        #设置 x 标签
pl.ylabel('y-正弦余弦函数值', fontproperties='STKAITI', fontsize=24)
pl.title('sin-cos 函数图像', fontproperties='STKAITI', fontsize=32)
                                        #图形标题
pl.legend(prop=myfont)                  #设置图例
pl.show()
```

运行结果如图 13-18 所示。

13.5.5　绘制图例标签中带有公式的图

```
import numpy as np
import matplotlib.pyplot as plt

x=np.linspace(0, 2*np.pi, 500)
y=np.sin(x)
z=np.cos(x*x)
plt.figure(figsize=(8,5))
#标签前后加$将使用内嵌的 LaTex 引擎将其显示为公式
plt.plot(x,y,label='$sin(x)$',color='red',linewidth=2)        #红色,2 个像素宽
```

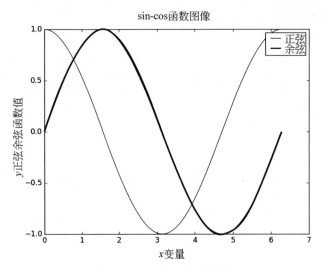

图 13-18　中文标签和图例

```
plt.plot(x,z,'b--',label='$cos(x^2)$')          #蓝色,虚线
plt.xlabel('Time(s)')
plt.ylabel('Volt')
plt.title('Sin and Cos figure using pyplot')
plt.ylim(-1.2,1.2)
plt.legend()                                     #显示图例
plt.show()
```

运行结果如图 13-19 所示。

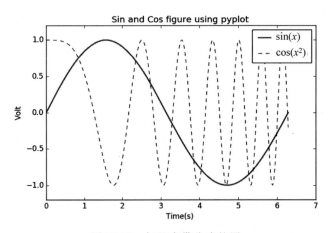

图 13-19　标签中带公式的图

13.5.6　使用 pyplot 绘制,多个图形单独显示

```
import numpy as np
import matplotlib.pyplot as plt
```

```
x=np.linspace(0, 2 * np.pi, 500)        #创建自变量数组
y1=np.sin(x)                            #创建函数值数组
y2=np.cos(x)
y3=np.sin(x * x)
plt.figure(1)                           #创建图形
#create three axes
ax1=plt.subplot(2,2,1)                  #第一行第一列图形
ax2=plt.subplot(2,2,2)                  #第一行第二列图形
ax3=plt.subplot(2,1,2)                  #第二行
plt.sca(ax1)                            #选择 ax1
plt.plot(x,y1,color='red')              #绘制红色曲线
plt.ylim(-1.2,1.2)                      #限制 y 坐标轴范围
plt.sca(ax2)                            #选择 ax2
plt.plot(x,y2,'b--')                    #绘制蓝色曲线
plt.ylim(-1.2,1.2)
plt.sca(ax3)                            #选择 ax3
plt.plot(x,y3,'g--')
plt.ylim(-1.2,1.2)
plt.show()
```

运行结果如图 13-20 所示。

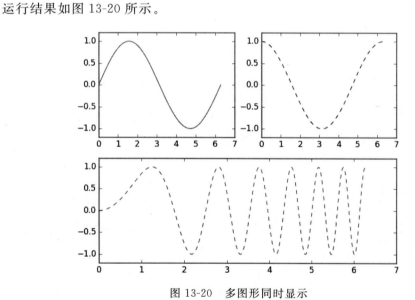

图 13-20 多图形同时显示

13.5.7 绘制三维参数曲线

```
import matplotlib as mpl
from mpl_toolkits.mplot3d import Axes3D
import numpy as np
import matplotlib.pyplot as plt
```

```
mpl.rcParams['legend.fontsize']=10          #图例字号
fig=plt.figure()
ax=fig.gca(projection='3d')                 #三维图形
theta=np.linspace(-4 * np.pi, 4 * np.pi, 100)
z=np.linspace(-4, 4, 100) * 0.3             #测试数据
r=z**3+1
x=r * np.sin(theta)
y=r * np.cos(theta)
ax.plot(x, y, z, label='parametric curve')
ax.legend()
plt.show()
```

程序运行结果如图 13-21 所示。

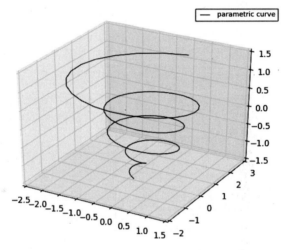

图 13-21　绘制三维参数曲线

13.5.8　绘制三维图形

```
import numpy as np
import matplotlib.pyplot as plt
import mpl_toolkits.mplot3d

x,y=np.mgrid[-2:2:20j, -2:2:20j]
z=50 * np.sin(x+y)                          #测试数据
ax=plt.subplot(111, projection='3d')        #三维图形
ax.plot_surface(x,y,z,rstride=2, cstride=1, cmap=plt.cm.Blues_r)
ax.set_xlabel('X')                          #设置坐标轴标签
ax.set_ylabel('Y')
ax.set_zlabel('Z')
plt.show()
```

运行结果如图 13-22 所示,在绘图窗口中可用鼠标来旋转绘制图形。

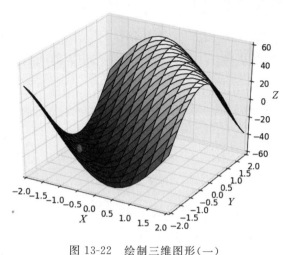

图 13-22　绘制三维图形(一)

下面的代码绘制了另一个略加复杂的三维图形,运行结果如图 13-23 所示。

```
import pylab as pl
import numpy as np
import mpl_toolkits.mplot3d

rho, theta=np.mgrid[0:1:40j, 0:2*np.pi:40j]
z=rho**2
x=rho*np.cos(theta)
y=rho*np.sin(theta)
ax=pl.subplot(111, projection='3d')
ax.plot_surface(x,y,z)
pl.show()
```

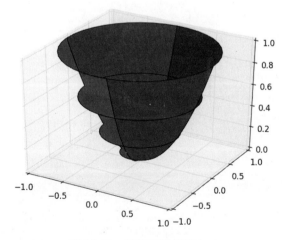

图 13-23　绘制三维图形(二)

13.5.9 使用指令绘制自定义图形

下面的代码使用 matplotlib.path 的 Path 对象根据设定的指令和坐标绘制任意图形，绘图结果如图 13-24 所示。

```python
from matplotlib.path import Path
from matplotlib.patches import PathPatch
import matplotlib.pyplot as plt


fig, ax=plt.subplots()

#定义绘图指令与控制点坐标
#其中,MOVETO 表示将绘制起点移到指定坐标
#CURVE4 表示使用 4 个控制点绘制 3 次贝塞尔曲线
#CURVE3 表示使用 3 个控制点绘制 2 次贝塞尔曲线
#LINETO 表示从当前位置绘制直线到指定位置
#CLOSEPOLY 表示从当前位置绘制直线到指定位置,并闭合多边形
path_data=[
            (Path.MOVETO,(1.58, -2.57)),
            (Path.CURVE4,(0.35, -1.1)),
            (Path.CURVE4,(-1.75, 2.0)),
            (Path.CURVE4,(0.375, 2.0)),
            (Path.LINETO,(0.85, 1.15)),
            (Path.CURVE4,(2.2, 3.2)),
            (Path.CURVE4,(3, 0.05)),
            (Path.CURVE4,(2.0, -0.5)),
            (Path.CURVE3,(3.5, -1.8)),
            (Path.CURVE3,(2, -2)),
            (Path.CLOSEPOLY,(1.58, -2.57)),
            ]
codes, verts=zip(*path_data)
path=Path(verts, codes)
#按指令和坐标进行绘图
patch=PathPatch(path, facecolor='r', alpha=0.9)
ax.add_patch(patch)

#绘制控制多边形和连接点
x, y=zip(*path.vertices)
line,=ax.plot(x, y, 'go-')

#显示网格
ax.grid()
#设置坐标轴刻度大小一致,可以更真实地显示图形
```

```
ax.axis('equal')
plt.show()
```

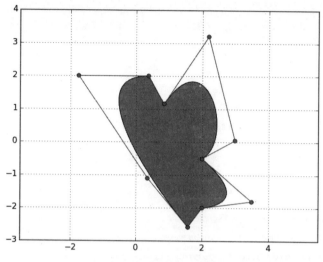

图 13-24　使用指令绘制任意图形

13.5.10　在 tkinter 中使用 matplotlib

强大的 matplotlib 不仅提供了显示图形的窗口，还支持把绘制的结果嵌入 Python 的 tkinter 程序界面中。下面的代码把 matplotlib 绘制的图形嵌入 tkinter 程序中，运行结果如图 13-25 所示，关于 tkinter 的内容请参考第 15 章。

```
import sys
import tkinter as Tk
import matplotlib
from numpy import arange, sin, pi
from matplotlib.backends.backend_tkagg import FigureCanvasTkAgg
from matplotlib.backends.backend_tkagg import NavigationToolbar2TkAgg
from matplotlib.backend_bases import key_press_handler
from matplotlib.figure import Figure

matplotlib.use('TkAgg')

root=Tk.Tk()
root.title("matplotlib in TK")
#设置图形尺寸与质量
f=Figure(figsize=(5, 4), dpi=100)
a=f.add_subplot(111)
t=arange(0.0, 3, 0.01)
s=sin(2*pi*t)
```

```
#绘制图形
a.plot(t, s)


#把绘制的图形显示到 tkinter 窗口上
canvas=FigureCanvasTkAgg(f, master=root)
canvas.show()
canvas.get_tk_widget().pack(side=Tk.TOP, fill=Tk.BOTH, expand=1)
#把 matplotlib 绘制图形的导航工具栏显示到 tkinter 窗口上
toolbar=NavigationToolbar2TkAgg(canvas, root)
toolbar.update()
canvas._tkcanvas.pack(side=Tk.TOP, fill=Tk.BOTH, expand=1)

#定义并绑定键盘事件处理函数
def on_key_event(event):
    print('you pressed %s' %event.key)
    key_press_handler(event, canvas, toolbar)
canvas.mpl_connect('key_press_event', on_key_event)

#按钮单击事件处理函数
def _quit():
    #结束事件主循环,并销毁应用程序窗口
    root.quit()
    root.destroy()
button=Tk.Button(master=root, text='Quit', command=_quit)
button.pack(side=Tk.BOTTOM)

Tk.mainloop()
```

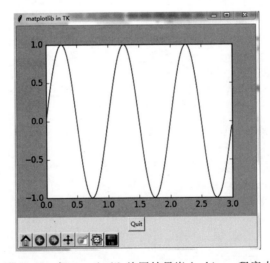

图 13-25　把 matplotlib 绘图结果嵌入 tkinter 程序中

13.5.11 使用 matplotlib 提供的组件实现交互式图形显示

工具箱 matplotlib 除了丰富的绘图功能,还提供了单选按钮、复选框、按钮等一系列组件,可以实现更加强大的图形交互式显示功能。下面的代码运行后,可以通过左侧的三组单选按钮来设置正弦曲线的频率、颜色和线型,并根据新的设置来绘制正弦曲线,每次单击图形下方的按钮,由系统随机设置频率、颜色和线型并绘制新图形,同时根据随机选择的值来设置三组单选按钮的选中项。运行结果如图 13-26 所示。

```
from random import choice
import numpy as np
import matplotlib.pyplot as plt
from matplotlib.widgets import RadioButtons, Button

t=np.arange(0.0, 2.0, 0.01)
s0=np.sin(2 * np.pi * t)
s1=np.sin(4 * np.pi * t)
s2=np.sin(8 * np.pi * t)

fig, ax=plt.subplots()
l,=ax.plot(t, s0, lw=2, color='red')
plt.subplots_adjust(left=0.3)

#定义允许的几种频率,并创建单选按钮组件
#其中,[0.05, 0.7, 0.15, 0.15]表示组件在窗口上的归一化位置和大小
axcolor='lightgoldenrodyellow'
rax=plt.axes([0.05, 0.7, 0.15, 0.15], axisbg=axcolor)
radio=RadioButtons(rax,('2 Hz', '4 Hz', '8 Hz'))
hzdict={'2 Hz': s0, '4 Hz': s1, '8 Hz': s2}
def hzfunc(label):
    ydata=hzdict[label]
    l.set_ydata(ydata)
    plt.draw()
radio.on_clicked(hzfunc)

#定义允许的几种颜色,并创建单选按钮组件
rax=plt.axes([0.05, 0.4, 0.15, 0.15], axisbg=axcolor)
colors = ('red', 'blue', 'green')
radio2=RadioButtons(rax, colors)
def colorfunc(label):
    l.set_color(label)
    plt.draw()
radio2.on_clicked(colorfunc)

#定义允许的几种线型,并创建单选按钮组件
rax=plt.axes([0.05, 0.1, 0.15, 0.15], axisbg=axcolor)
```

```
styles = ('-', '--', '-.', 'steps', ':')
radio3=RadioButtons(rax, styles)
def stylefunc(label):
    l.set_linestyle(label)
    plt.draw()
radio3.on_clicked(stylefunc)

#定义按钮单击事件处理函数,并在窗口上创建按钮
def randomFig(event):
    #随机选择一个频率,同时设置单选按钮的选中项
    hz=choice(tuple(hzdict.keys()))
    hzLabels=[label.get_text() for label in radio.labels]
    radio.set_active(hzLabels.index(hz))
    l.set_ydata(hzdict[hz])
    #随机选择一个颜色,同时设置单选按钮的选中项
    c=choice(colors)
    radio2.set_active(colors.index(c))
    l.set_color(c)
    #随机选择一个线型,同时设置单选按钮的选中项
    style=choice(styles)
    radio3.set_active(styles.index(style))
    l.set_linestyle(style)
    #根据设置的属性绘制图形
    plt.draw()
axRnd=plt.axes([0.5, 0.015, 0.2, 0.045])
buttonRnd=Button(axRnd, 'Random Figure')
buttonRnd.on_clicked(randomFig)
#显示图形
plt.show()
```

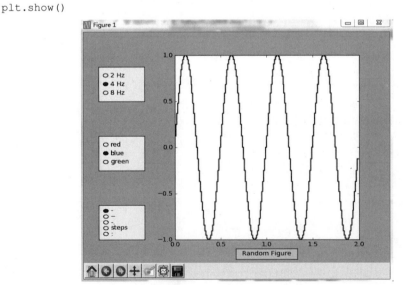

图 13-26　matplotlib 组件的应用

13.5.12 根据实时数据动态更新图形

在开发与数据监测和数据可视化有关的系统时，我们需要根据最新的数据对图形进行更新。下面的代码模拟了这种情况，单击 Start 按钮时会更新数据并重新绘制图形使得曲线看上去在移动一样，单击 Stop 按钮则停止更新数据，运行结果如图 13-27 所示。代码中用到了多线程编程技术，请参考第 10 章内容。

```python
from time import sleep
from threading import Thread
import numpy as np
import matplotlib.pyplot as plt
from matplotlib.widgets import Button

fig, ax=plt.subplots()
#设置图形的显示位置
plt.subplots_adjust(bottom=0.2)
#实验数据
range_start, range_end, range_step=0, 1, 0.005
t=np.arange(range_start, range_end, range_step)
s=np.sin(4 * np.pi * t)
l,=plt.plot(t, s, lw=2)
#自定义类，用来封装两个按钮的单击事件处理函数
class ButtonHandler:
    def __init__(self):
        self.flag=True
        self.range_s, self.range_e, self.range_step=0, 1, 0.005
    #线程函数，用来更新数据并重新绘制图形
    def threadStart(self):
        while self.flag:
            sleep(0.02)
            self.range_s +=self.range_step
            self.range_e +=self.range_step
            t=np.arange(self.range_s, self.range_e, self.range_step)
            ydata=np.sin(4 * np.pi * t)
            #更新数据
            l.set_xdata(t-t[0])
            l.set_ydata(ydata)
            #重新绘制图形
            plt.draw()
    def Start(self, event):
        self.flag=True
        #创建并启动新线程
        t=Thread(target=self.threadStart)
```

```
            t.start()
    def Stop(self, event):
        self.flag=False

callback=ButtonHandler()
#创建按钮并设置单击事件处理函数
axprev=plt.axes([0.81, 0.05, 0.1, 0.075])
bprev=Button(axprev, 'Stop')
bprev.on_clicked(callback.Stop)
axnext=plt.axes([0.7, 0.05, 0.1, 0.075])
bnext=Button(axnext, 'Start')
bnext.on_clicked(callback.Start)

plt.show()
```

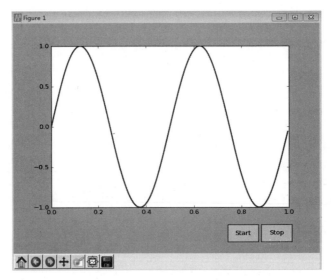

图 13-27　根据最新数据动态更新图形

13.5.13　使用 Slider 组件调整曲线参数

在进行数据可视化时，经常需要调整一些参数，前面 13.5.11 节中通过三组单选按钮实现了曲线的频率、颜色和线型的实时调整，下面的代码使用 matplotlib 库提供的 Slider 组件实现正弦曲线振幅和频率的调整，比使用单选按钮更加灵活。运行结果如图 13-28 所示，当然，也可以在此基础上增加用来调整曲线颜色三分量的 Slider 组件，大家不妨尝试一下。

```
import numpy as np
import matplotlib.pyplot as plt
from matplotlib.widgets import Slider, Button, RadioButtons
```

```python
fig, ax=plt.subplots()
#设置图形的区域位置
plt.subplots_adjust(left=0.1, bottom=0.25)
t=np.arange(0.0, 1.0, 0.001)
#初始振幅与频率,并绘制初始图形
a0, f0=5, 3
s=a0 * np.sin(2 * np.pi * f0 * t)
l,=plt.plot(t, s, lw=2, color='red')
#设置坐标轴刻度范围
plt.axis([0, 1, -10, 10])

axColor='lightgoldenrodyellow'
#创建两个 Slider 组件,分别设置位置/尺寸、背景色和初始值
axfreq=plt.axes([0.1, 0.1, 0.75, 0.03], axisbg=axColor)
sfreq=Slider(axfreq, 'Freq', 0.1, 30.0, valinit=f0)
axamp=plt.axes([0.1, 0.15, 0.75, 0.03], axisbg=axColor)
samp=Slider(axamp, 'Amp', 0.1, 10.0, valinit=a0)
#为 Slider 组件设置事件处理函数
def update(event):
    #获取 Slider 组件的当前值,并以此来更新图形
    amp=samp.val
    freq=sfreq.val
    l.set_ydata(amp * np.sin(2 * np.pi * freq * t))
    plt.draw()
    #fig.canvas.draw_idle()
sfreq.on_changed(update)
samp.on_changed(update)

#创建 Adjust 按钮,设置大小、位置和事件处理函数
def adjustSliderValue(event):
    ampValue=samp.val+0.05
    if ampValue>10:
        ampValue=0.1
    samp.set_val(ampValue)
    
    freqValue=sfreq.val+0.05
    if freqValue>30:
        freqValue=0.1
    sfreq.set_val(freqValue)
    update(event)
axAdjust=plt.axes([0.6, 0.025, 0.1, 0.04])
buttonAdjust=Button(axAdjust, 'Adjust', color=axColor, hovercolor='red')
buttonAdjust.on_clicked(adjustSliderValue)
```

```
#创建按钮组件,用来恢复初始值
resetax=plt.axes([0.8, 0.025, 0.1, 0.04])
button=Button(resetax, 'Reset', color=axColor, hovercolor='yellow')
def reset(event):
    sfreq.reset()
    samp.reset()
button.on_clicked(reset)

plt.show()
```

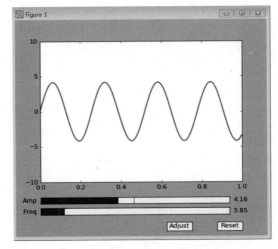

图 13-28　使用 Slider 组件调整曲线参数

第 14 章

密码学编程

信息加密和信息隐藏是实现信息安全与保密的主要手段。从古到今人类发明了大量的加密算法和隐藏信息的方法。例如,把纸条螺旋缠绕在一根木棍上然后往上写字,展开后通过一定的渠道把纸条传递给对方,对方把纸条螺旋缠绕到同样粗细的木棍上就可以正常阅读信息,其他人即使知道这样的方法,如果不知道木棍的直径也无法解密信息,可以说木棍的直径是这种加密方法中的密钥。再如,古代武林高手把一些秘籍通过特定的手段写到羊皮上,只有使用正确的方法才能看到上面的内容,战争年代聪明的老百姓也发明了在煮熟的鸡蛋清上写字的方法来传递情报,电视剧《连城诀》中的终极秘密则是隐藏在一本《唐诗 300 首》书中。

除了可以设计自己的加密算法或者自己编写程序实现经典的加密解密算法之外,还可以充分利用 Python 标准库和扩展库提供的丰富功能。Python 标准库 hashlib 实现了 SHA1、SHA224、SHA256、SHA384、SHA512 以及 MD5 等多个安全哈希算法,标准库 zlib 提供了 adler32 和 crc32 算法的实现,标准库 hmac 实现了 HMAC 算法。在众多的 Python 扩展库中,pycrypto 可以说是密码学编程模块中最成功也是最成熟的一个,具有很高的市场占有率。另外,cryptography 也有一定数量的用户在使用。扩展库 pycrypto 和 cryptography 提供了 SHA 系列算法和 RIPEMD160 等多个安全哈希算法,以及 DES、AES、RSA、DSA、ElGamal 等多个加密算法和数字签名算法的实现。

14.1 经典密码算法

很多经典密码算法的安全性很低,甚至很容易被词频分析这样的简易攻击手段破解,因此已经很少再使用。尽管如此,经典密码算法的一些思想还是很有借鉴意义和参考价值的。

14.1.1 恺撒密码算法

恺撒密码作为古老的加密算法之一,在古罗马的时候就已经广泛使用了。恺撒密码算法的基本思想是:通过把字母移动一定的位数来实现加密和解密。明文中的所有字母都在字母表上向后(或向前)按照一个固定数目进行偏移后被替换成密文。例如,当偏移量是 3 的时候,所有的字母 A 将被替换成 D,B 变成 E,以此类推,X 将变成 A,Y 变成 B,

Z变成C。移动的位数就是恺撒密码加密和解密的密钥。

例 14-1 恺撒密码算法。

```python
def KaiSaEncrypt(ch, k):
    if(not isinstance(ch, str))or len(ch)!=1:
        print('The first parameter must be a character')
        return
    if(not isinstance(k, int))or(not 1<=k<=25):
        print('The second parameter must be an integer between 1 and 25')
        return
    #把英文字母变换为后面第k个字母
    if 'a'<=ch<=chr(ord('z')-k):
        return chr(ord(ch)+k)
    #把英文字母首尾相连
    elif chr(ord('z')-k)<ch<='z':
        return chr((ord(ch)-ord('a')+k)%26+ord('a'))
    elif 'A'<=ch<=chr(ord('Z')-k):
        return chr(ord(ch)+k)
    elif chr(ord('Z')-k)<ch<='Z':
        return chr((ord(ch)-ord('A')+k)%26+ord('A'))
    else:
        return ch

def encrypt(plain, k):
    return ''.join([KaiSaEncrypt(ch, k)for ch in plain])

def KaiSaDecrypt(ch, k):
    if(not isinstance(ch, str))or len(ch)!=1:
        print('The first parameter must be a character')
        return
    if(not isinstance(k, int))or(not 1<=k<=25):
        print('The second parameter must be an integer between 1 and 25')
        return
    #把英文字母首尾相连,然后把每个字母变换为前面第k个字母
    if chr(ord('a')+k)<=ch<='z':
        return chr(ord(ch)-k)
    elif 'a'<=ch<chr(ord('a')+k):
        return chr((ord(ch)-k+26))
    elif chr(ord('A')+k)<=ch<='Z':
        return chr(ord(ch)-k)
    elif 'A'<=ch<chr(ord('A')+k):
        return chr((ord(ch)-k+26))
    else:
        return ch
```

```
def decrypt(plain, k):
    return ''.join([KaiSaDecrypt(ch, k)for ch in plain])

plainText='Explicit is better than implicit.'
cipherText=encrypt(plainText, 5)
print(plainText)
print(cipherText)
print(decrypt(cipherText,5))
```

程序运行结果为

```
Explicit is better than implicit.
Jcuqnhny nx gjyyjw ymfs nruqnhny.
Explicit is better than implicit.
```

小提示：这里的代码只是恺撒密码的原理演示，实现恺撒加密其实不用这么麻烦，可以参考第 5 章中关于 maketrans()和 translate()方法的介绍。

14.1.2　维吉尼亚密码

维吉尼亚密码算法使用一个密钥和一个表来实现加密，根据明文和密钥的对应关系进行查表来决定加密结果。假设替换表如图 14-1 所示，最上面一行表示明文，最左边一列表示密钥，那么二维表格中与明文字母和密钥字母对应的字母就是加密结果。例如，单词 PYTHON 使用 ABCDEF 做密钥的加密结果为 PZVKSS。

```
  ABCDEFGHIJKLMNOPQRSTUVWXYZ
A:ABCDEFGHIJKLMNOPQRSTUVWXYZ
B:BCDEFGHIJKLMNOPQRSTUVWXYZA
C:CDEFGHIJKLMNOPQRSTUVWXYZAB
D:DEFGHIJKLMNOPQRSTUVWXYZABC
E:EFGHIJKLMNOPQRSTUVWXYZABCD
F:FGHIJKLMNOPQRSTUVWXYZABCDE
G:GHIJKLMNOPQRSTUVWXYZABCDEF
H:HIJKLMNOPQRSTUVWXYZABCDEFG
I:IJKLMNOPQRSTUVWXYZABCDEFGH
J:JKLMNOPQRSTUVWXYZABCDEFGHI
K:KLMNOPQRSTUVWXYZABCDEFGHIJ
L:LMNOPQRSTUVWXYZABCDEFGHIJK
M:MNOPQRSTUVWXYZABCDEFGHIJKL
N:NOPQRSTUVWXYZABCDEFGHIJKLM
O:OPQRSTUVWXYZABCDEFGHIJKLMN
P:PQRSTUVWXYZABCDEFGHIJKLMNO
Q:QRSTUVWXYZABCDEFGHIJKLMNOP
R:RSTUVWXYZABCDEFGHIJKLMNOPQ
S:STUVWXYZABCDEFGHIJKLMNOPQR
T:TUVWXYZABCDEFGHIJKLMNOPQRS
U:UVWXYZABCDEFGHIJKLMNOPQRST
V:VWXYZABCDEFGHIJKLMNOPQRSTU
W:WXYZABCDEFGHIJKLMNOPQRSTUV
X:XYZABCDEFGHIJKLMNOPQRSTUVW
Y:YZABCDEFGHIJKLMNOPQRSTUVWX
Z:ZABCDEFGHIJKLMNOPQRSTUVWXY
```

图 14-1　维吉尼亚密码替换表

例 14-2　维吉尼亚密码。

```
from string import ascii_uppercase as uppercase
from itertools import cycle

#创建密码表
table=dict()
for ch in uppercase:
    index=uppercase.index(ch)
    table[ch]=uppercase[index:]+uppercase[:index]

#创建解密密码表
deTable={'A':'A'}
start='Z'
for ch in uppercase[1:]:
    index=uppercase.index(ch)
    deTable[ch]=chr(ord(start)+1-index)
```

```python
#解密密钥
def deKey(key):
    return ''.join([deTable[i] for i in key])


#加密/解密
def encrypt(plainText, key):
    result=[]
    #创建cycle对象,支持密钥字母的循环使用
    currentKey=cycle(key)
    for ch in plainText:
        if 'A'<=ch<='Z':
            index=uppercase.index(ch)
            #获取密钥字母
            ck=next(currentKey)
            result.append(table[ck][index])
        else:
            result.append(ch)
    return ''.join(result)


key='DONGFUGUO'
p='PYTHON 3.5.1 PYTHON 2.7.11'
c=encrypt(p, key)
print(p)
print(c)
print(encrypt(c,deKey(key)))
```

14.1.3　换位密码算法

换位密码也是一种比较常见的经典密码算法,基本原理是先把明文按固定长度进行分组,然后对每一组的字符进行换位操作,从而实现加密。例如,字符串"Errors should never pass silently."使用密钥1432进行加密时,首先将字符串分成若干长度为4的分组,然后对每个组的字符进行换位,第1个字符和第3个字符位置不变,把第2个字符和第4个字符交换位置,得到"Eorrrs shluoden v repssa liseltny."。

例14-3　换位密码算法。

```python
def encrypt(plainText, t):
    result=[]
    length=len(t)
    #把明文分组
    temp=[plainText[i:i+length] for i in range(0,len(plainText),length)]
    #对除最后一组之外的其他进行换位处理
    for item in temp[:-1]:
        newItem=''
```

```
    for i in t:
        newItem=newItem+item[i-1]
    result.append(newItem)
return ''.join(result)+temp[-1]

p='Errors should never pass silently.'
#加密
c=encrypt(p,(1, 4, 3, 2))
print(c)
#解密
print(encrypt(c,(1, 4, 3, 2)))
```

14.2 安全哈希算法

安全哈希算法也称为报文摘要算法，对任意长度的消息可以计算得到固定长度的唯一指纹。理论上，即使是内容非常相似的消息也不会得到完全相同的指纹。安全哈希算法是不可逆的，无法从指纹还原得到原始消息，属于单向变换算法。安全哈希算法常用于数字签名领域，可以验证信息是否被篡改，很多管理信息系统把用户密码的哈希值存储到数据库中而不是直接存储密码明文，大幅度提高了系统的安全性。另外，文件完整性检查也经常用到 MD5 或其他安全哈希算法，用来验证文件发布之后是否被非法修改。

下面的代码使用 Python 标准库 hashlib 计算字符串的安全哈希值。

```
>>>import hashlib
>>>hashlib.md5('abcdefg'.encode()).hexdigest()            #使用 MD5 算法
>>>hashlib.sha512('abcdefg'.encode()).hexdigest()         #使用 SHA512 算法
>>>hashlib.sha256('abcdefg'.encode()).hexdigest()         #使用 SHA256 算法
```

Python 扩展库 pycrypto 也提供了 MD2、MD4、MD5、HMAC、RIPEMD、SHA、SHA224、SHA256、SHA384、SHA512 等多个安全哈希算法的实现。

```
>>>from Crypto.Hash import SHA256
>>>h=SHA256.SHA256Hash('abcdefg'.encode())
>>>h.hexdigest()
```

例 14-4 计算文件的 MD5 值。

```
import sys
import hashlib
import os.path

filename=sys.argv[1]
if os.path.isfile(filename):
    fp=open(filename, 'rb')
    contents=fp.read()
```

```
        fp.close()
        print(hashlib.md5(contents).hexdigest())
else:
        print('file not exists')
```

把上面的代码保存为文件 CheckMD5OfFile.py,然后计算指定文件的 MD5 值,对该文件进行微小修改后再次计算其 MD5 值,可以发现,哪怕只是修改了一点点内容,MD5 值的变化也是非常大的,如图 14-2 所示。

图 14-2 计算文件的 MD5 值

拓展知识:MD5 算法属于单向变换算法,不存在反函数,暴力测试几乎成为唯一可能的 MD5 破解方法。下面的代码演示了通过暴力测试来破解 MD5 值的思路,当然这个思路也适用于 SHA1 以及 SHA224/256/384/512 系列哈希值的破解。下面的方法使用排列算法生成指定长度和字符集的所有可能明文然后逐个尝试,也可以参考第 6 章破解 RAR 和 ZIP 文件密码的代码改写,使用字典文件进行破解。

```python
from hashlib import md5
from string import ascii_letters, digits
from itertools import permutations
from time import time

all_letters=ascii_letters+digits+'.,;'          #候选字符集

def decrypt_md5(md5_value):                      #破解 32 位 MD5 值
    if len(md5_value)!=32:
        print('error')
        return

    md5_value=md5_value.lower()                  #转换为小写 MD5 值

    for k in range(5,10):                        #预期密码长度
        for item in permutations(all_letters, k):#暴力测试
            item=''.join(item)
```

```
            print('.', end='')                                      #显示进度
            if md5(item.encode()).hexdigest()==md5_value:           #破解成功
                return item

md5_value='e7d057704ea5206d8cb61280741238f5'
start=time()
result=decrypt_md5(md5_value)
if result:
    print('\nSuccess:   '+md5_value+'==>'+result)
print('Time used:', time()-start)
```

提到 MD5 的破解,就不得不说一下我见过的一个聪明人。几年前他搭建了一个很简单的网页,页面上只提供了 MD5 加密和 MD5 解密两个按钮。当用户使用加密功能时,后台用代码调用 MD5 算法计算出用户输入信息的 MD5 值,同时把原始信息和对应的 MD5 添加到数据库中。而当用户使用解密功能时,就去数据库中查找是否存在用户输入的 MD5 值。短短几个月的时间就吸引了大量用户前来测试(实际上是免费帮忙扩充数据库),当数据库庞大到一定程度以后,作者宣布开始收费,每使用一次解密功能需要支付一块钱。

拓展知识:也可以使用 ssdeep 工具来计算文件的模糊哈希值或分段哈希值,或者编写 Python 程序调用 ssdeep 提供的 API 函数来计算文件的模糊哈希值,模糊哈希值可以用来比较两个文件的相似百分比。这个工具在 Windows 平台上安装比较麻烦,建议使用 Linux 操作系统,以 Ubuntu 为例,执行命令 sudo BUILD_LIB=1 pip3 install ssdeep 即可安装 ssdeep 和所有依赖包,当然在这之前或许还需要使用 sudo apt-get install python3-pip 命令先安装 pip3 工具。

```
>>>import ssdeep
>>>hash1=ssdeep.hash('Also called fuzzy hashes, Ctph can match inputs that have homologies.')
>>>hash1
'3:AXGBicFlgVNhBGcL6wCrFQEv:AXGHsNhxLsr2C'
>>>hash2=ssdeep.hash('Also called fuzzy hashes, CTPH can match inputs that have homologies.')
>>>hash2
'3:AXGBicFlIHBGcL6wCrFQEv:AXGH6xLsr2C'
>>>ssdeep.compare(hash1, hash2)                                    #比较两个哈希值的相似度
22
>>>import ssdeep
>>>s=ssdeep()
>>>s.hash_from_file(filename)
```

对于某些恶意软件来说,可能会对自身进行加壳或加密,真正运行时再进行脱壳或解密,这样一来,会使得磁盘文件的哈希值和内存中脱壳或解密后进程的哈希值相差很大。因此,根据磁盘文件和其相应的进程之间模糊哈希值的相似度可以判断该文件是否包含

自修改代码，并以此来判断其为恶意软件的可能性。

14.3 对称密钥密码算法 DES 和 AES

作为经典的对称密钥密码算法，DES 早在 1976 年就被美国政府采用，随后得到美国国家标准局和美国国家标准协会的认可，并成为全球范围内事实上的工业标准。DES 算法使用 56 位密钥对 64 位的数据块进行加密，并对 64 位的数据块进行 16 轮编码，最终完成变换。下面的代码演示了 Python 扩展库 pycrypto 中 DES 算法的用法。

```
>>>from Crypto.Cipher import DES
>>>des_encrypt_decrypt=DES.new('ShanDong', DES.MODE_ECB)
>>>p='Beautiful is better than ugly.'
>>>pp=p.encode()
>>>c=des_encrypt_decrypt.encrypt(pp.ljust((len(pp)//8+1) * 8, b'0'))
                                                              #按 8 字节对齐
>>>c                                                          #加密结果
b'\xc3{p\x1d\x9f\x9a\x85O\xf6:\xba=\xfc\xe1.\x8ea\xe1\x9fZ9l\xd7\xdfy\x94\xa1_
\x0e\xd8\xaf\x89'
>>>cp=des_encrypt_decrypt.decrypt(c)                          #解密
>>>cp
b'Beautiful is better than ugly.00'
>>>cp[0:len(pp)].decode()
'Beautiful is better than ugly.'
```

高级加密标准 AES(Advanced Encryption Standard)，又称为 Rijndael 算法，是美国联邦政府采用的一种区块加密标准，用来替代 DES 算法，AES 算法使用代换-置换网络，在软件及硬件上都能快速地加解密。AES 加密数据块分组长度必须为 128 位，密钥长度可以是 128/192/256 位中的任意一个(数据块及密钥长度不足时需要补齐)。

例 14-5 使用 Python 扩展库 pycrypto 提供的 AES 算法实现消息加密和解密。

```python
import string
import random
from Crypto.Cipher import AES

#生成指定长度的密钥
def keyGenerater(length):
    if length not in (16, 24, 32):
        return None
    x=string.ascii_letters+string.digits
    return ''.join([random.choice(x) for i in range(length)])

def encryptor_decryptor(key, mode):
    return AES.new(key, mode, b'0000000000000000')
```

```python
#使用指定密钥和模式对给定信息进行加密
def AESencrypt(key, mode, text):
    encryptor=encryptor_decryptor(key, mode)
    return encryptor.encrypt(text)

#使用指定密钥和模式对给定信息进行解密
def AESdecrypt(key, mode, text):
    decryptor=encryptor_decryptor(key, mode)
    return decryptor.decrypt(text)

if __name__=='__main__':
    text='山东省烟台市 Python3.5 is excellent.'
    key=keyGenerater(16)
    #随机选择AES的模式
    mode=random.choice((AES.MODE_CBC, AES.MODE_CFB, AES.MODE_ECB,
                        AES.MODE_OFB))
    if not key:
        print('Something is wrong.')
    else:
        print('key:', key)
        print('mode:', mode)
        print('Before encrypted:', text)
        #明文必须以字节串形式,且长度为16的倍数
        text_encoded=text.encode()
        text_length=len(text_encoded)
        padding_length=16-text_length%16
        text_encoded=text_encoded+b'0' * padding_length
        text_encrypted=AESencrypt(key, mode, text_encoded)
        print('After encrypted:', text_encrypted)
        text_decrypted =AESdecrypt(key, mode, text_encrypted)
        print('After decrypted:', text_decrypted.decode()[:-padding_length])
```

14.4 非对称密钥密码算法 RSA 与数字签名算法 DSA

14.4.1 RSA

RSA 是一种典型的非对称密钥密码体制,从加密密钥和解密密钥中的任何一个推导出另一个在计算上是不可行的。RSA 的安全性建立在"大数分解和素性检测"这一著名数论难题的基础上。公钥对可以完全公开,不需要进行保密,但必须提供完整性检测机制以保证不受篡改;私钥由用户自己保存。通信双方无须实现交换密钥就可以进行保密通信。

RSA 密码体制算法如下。

(1) 由用户选择两个互异并且距离较远的大素数 p 和 q。

(2) 计算 $n=p\times q$ 和 $f(n)=(p-1)\times(q-1)$。

(3) 选择正整数 e,使其与 $f(n)$ 的最大公约数为1;然后计算正整数 d,使得 $e\times d$ 对 $f(n)$ 的余数为1,即 $e\times d\equiv 1\ \text{mod}\ f(n)$,最后销毁 p 和 q。

经过以上步骤,得出公钥对 (n,e) 和私钥对 (n,d)。设 M 为明文,C 为对应的密文,则加密变换为:$C=M^e\ \text{mod}\ n$;解密变换为:$M=C^d\ \text{mod}\ n$。

Python 扩展模块 rsa 封装了 RSA 算法,可以方便地使用该算法生成密钥以及加解密。

```
>>>import rsa
>>>key=rsa.newkeys(3000)                #随机生成密钥
>>>private=key[1]                       #查看私钥分量,输出结果略
>>>print(private.d, private.e, private.n, private.p, private.q)
```

例14-6 使用 rsa 模块来实现消息加密和解密。

```
import rsa

key=rsa.newkeys(3000)                   #生成随机密钥
privateKey=key[1]                       #私钥
publicKey=key[0]                        #公钥

message='中国山东烟台.Now is better than never.'
print('Before encrypted:',message)
message=message.encode()

cryptedMessage=rsa.encrypt(message, publicKey)
print('After encrypted:\n',cryptedMessage)

message=rsa.decrypt(cryptedMessage, privateKey)
message=message.decode()
print('After decrypted:',message)
```

Python 扩展库 pycrypto 也封装了 RSA 算法以及 DSA 和 ElGamal 算法,可用于数字签名和其他相关领域。在使用之前,需要将 pycrypto 安装目录中的 Crypto/Random/OSRNG/nt.py 文件中的

```
import winrandom
```

一行改为

```
from Crypto.Random.OSRNG import winrandom
```

下面的代码演示了如何使用 pycrypto 提供的 RSA 模块进行加密和解密。

```
>>>from Crypto.PublicKey import RSA
```

```
>>>key=RSA.generate(2048)              #生成密钥,查看密钥各分量的值,输出结果略
>>>print(key.key.n, key.key.p, key.key.e, key.key.d)
>>>p='Flat is better than nested.中文测试'
>>>c=key.encrypt(p.encode(), key)
>>>cp=key.decrypt(c)
>>>cp.decode()
'Flat is better than nested.中文测试'
>>>k=key.exportKey('PEM')              #密钥导出
>>>key1=RSA.importKey(k)               #密钥导入
>>>key1 ==key
True
>>>fp=open('D:\\Key.pem', 'wb')
>>>fp.write(key.exportKey('PEM'))      #将密钥导出到文件
1674
>>>fp.close()
>>>fp=open('D:\\Key.pem','rb')
>>>key2=RSA.importKey(fp.read())       #从文件中导入密钥
>>>fp.close()
>>>key2 ==key
True
```

14.4.2 DSA

DSA 是基于公钥机制的数字签名算法,其安全性基于离散对数问题 DLP,即给定一个循环群中的元素 g 和 h,很难找到一个整数 x 使得 $g^x=h$。下面的代码简单演示了 pycrypto 扩展库中 DSA 算法的用法。

```
>>>from Crypto.Random import random
>>>from Crypto.PublicKey import DSA
>>>from Crypto.Hash import MD5
>>>message='Simple is better than complex.'
>>>key=DSA.generate(1024)              #生成密钥
>>>h=MD5.new(message.encode()).digest()  #计算消息的哈希值
>>>k=random.StrongRandom().randint(1, key.key.q-1)
>>>sig=key.sign(h, k)
>>>key.verify(h, sig)
True
>>>h1=MD5.new(message.encode()+b'3').digest()
>>>key.verify(h1, sig)
False
```

第 15 章 tkinter 编程精彩案例

Python 标准库 tkinter 是对 Tcl/Tk 的进一步封装,与 tkinter.ttk 和 tkinter.tix 共同提供了强大的跨平台图形用户界面(Graphical User Interface,GUI)编程的功能,IDLE 就是使用 tkinter 进行开发的。tkinter 提供了大量用于 GUI 编程的组件,表 15-1 列出了其中一部分。另外,tkinter.ttk 还提供了 Combobox、Progressbar 和 Treeview 等组件,tkinter.scrolledtext 提供了带滚动条的文本框,messagebox、commondialog、dialog、colorchooser、simpledialog、filedialog 等模块提供了各种对话框。本章并没有一一介绍每个组件的属性和方法,而是通过大量实用性很强的案例来演示这些组件的用法,每个案例都附加了大量注释来方便大家理解。

表 15-1 tkinter 的常用组件

组件名称	说明
Button	按钮
Canvas	画布,用于绘制直线、椭圆、多边形等各种图形
Checkbutton	复选框形式的按钮
Entry	单行文本框
Frame	框架,可作为其他组件的容器,常用来对组件进行分组
Label	标签,常用来显示单行文本
Listbox	列表框
Menu	菜单
Message	多行文本框
Radiobutton	单选按钮,同一组中的单选按钮任何时刻只能有一个处于选中状态
Scrollbar	滚动条
Toplevel	常用来创建新的窗口

15.1 用户登录界面

用户登录界面几乎无处不在,如电子邮箱、QQ、微信、论坛、微博、办公系统、各类管理信息系统等都需要登录账号才能进行后续的操作。一般而言,用户密码都是经过安全哈希算法加密之后存储到数据库中的,并不直接保存明文。下面的案例主要演示如何使用 tkinter 创建应用程序窗口,以及文本框、按钮和简单消息框等组件的用法。

例 15-1　tkinter 实现用户登录界面。

```
import tkinter
import tkinter.messagebox

#创建应用程序窗口
root=tkinter.Tk()
varName=tkinter.StringVar()
varName.set('')
varPwd=tkinter.StringVar()
varPwd.set('')
#创建标签
labelName=tkinter.Label(root, text='User Name:', justify=tkinter.RIGHT,
                       width=80)
#将标签放到窗口上
labelName.place(x=10, y=5, width=80, height=20)
#创建文本框,同时设置关联的变量
entryName=tkinter.Entry(root, width=80,textvariable=varName)
entryName.place(x=100, y=5, width=80, height=20)

labelPwd=tkinter.Label(root, text='User Pwd:', justify=tkinter.RIGHT, width=80)
labelPwd.place(x=10, y=30, width=80, height=20)
#创建密码文本框
entryPwd=tkinter.Entry(root, show='*',width=80, textvariable=varPwd)
entryPwd.place(x=100, y=30, width=80, height=20)
#登录按钮事件处理函数
def login():
    #获取用户名和密码
    name=entryName.get()
    pwd=entryPwd.get()
    if name=='admin' and pwd=='123456':
        tkinter.messagebox.showinfo(title='Python tkinter',message='OK')
    else:
        tkinter.messagebox.showerror('Python tkinter', message='Error')
#创建按钮组件,同时设置按钮事件处理函数
buttonOk=tkinter.Button(root, text='Login', command=login)
buttonOk.place(x=30, y=70, width=50, height=20)
#取消按钮的事件处理函数
def cancel():
    #清空用户输入的用户名和密码
    varName.set('')
    varPwd.set('')
buttonCancel=tkinter.Button(root, text='Cancel', command=cancel)
```

```
buttonCancel.place(x=90, y=70, width=50, height=20)

#启动消息循环
root.mainloop()
```

将上面的代码保存为 tkinter_login.pyw，运行结果如图 15-1 所示，如果用户输入用户名 admin 和密码 123456 并单击 Login 按钮，弹出密码正确对话框（见图 15-2），否则弹出密码错误对话框（见图 15-3）。单击 Cancel 按钮可以清空已输入的用户名和密码。

图 15-1 用户登录界面　　图 15-2 密码正确　　图 15-3 密码错误

15.2 选择类组件应用

下面的案例创建了一个包含文本框、单选按钮、复选框、组合框、按钮和列表框等组件的 GUI 应用程序，运行后输入学生姓名并选择年级、班级、性别以及是否是班长等信息后，单击 Add 按钮可将该学生信息添加到列表框中。在列表框中选择一项后单击 DeleteSelection 按钮可将其从列表框中删除，没有选择任何项而直接单击该按钮则提示 No Selection。

例 15-2　tkinter 单选按钮、复选框、组合框、列表框综合运用案例。

```python
import tkinter
import tkinter.messagebox
import tkinter.ttk

#创建 tkinter 应用程序
root=tkinter.Tk()
#设置窗口标题
root.title('Selection widgets')
#定义窗口大小
root['height']=400
root['width']=320
#与姓名关联的变量
varName=tkinter.StringVar()
varName.set('')
```

```python
#创建标签,然后放到窗口上
labelName=tkinter.Label(root, text='Name:',justify=tkinter.RIGHT,width=50)
labelName.place(x=10, y=5, width=50, height=20)
#创建文本框,同时设置关联的变量
entryName=tkinter.Entry(root, width=120,textvariable=varName)
entryName.place(x=70, y=5, width=120, height=20)


labelGrade=tkinter.Label(root, text='Grade:', justify=tkinter.RIGHT, width=50)
labelGrade.place(x=10, y=40, width=50, height=20)
#模拟学生所在年级,字典键为年级,字典值为班级
studentClasses={'1':['1', '2', '3', '4'],
                '2':['1', '2'],
                '3':['1', '2', '3']}
#学生班级组合框
comboGrade=tkinter.ttk.Combobox(root,width=50,
                                values=tuple(studentClasses.keys()))
comboGrade.place(x=70, y=40, width=50, height=20)
#事件处理函数
def comboChange(event):
    grade=comboGrade.get()
    if grade:
        #动态改变组合框可选项
        comboClass["values"]=studentClasses.get(grade)
    else:
        comboClass.set([])
#绑定组合框事件处理函数
comboGrade.bind('<<ComboboxSelected>>', comboChange)


labelClass=tkinter.Label(root, text='Class:', justify=tkinter.RIGHT, width=50)
labelClass.place(x=130, y=40, width=50, height=20)
#学生年级组合框
comboClass=tkinter.ttk.Combobox(root, width=50)
comboClass.place(x=190, y=40, width=50, height=20)


labelSex=tkinter.Label(root, text='Sex:', justify=tkinter.RIGHT, width=50)
labelSex.place(x=10, y=70, width=50, height=20)
#与性别关联的变量,1:男;0:女,默认为男
sex=tkinter.IntVar()
sex.set(1)
#单选按钮,男
radioMan=tkinter.Radiobutton(root,variable=sex,value=1,text='Man')
```

```python
radioMan.place(x=70, y=70, width=50, height=20)
#单选按钮,女
radioWoman=tkinter.Radiobutton(root,variable=sex,value=0,text='Woman')
radioWoman.place(x=130, y=70, width=70, height=20)
#与是否是班长关联的变量,默认当前学生不是班长
monitor=tkinter.IntVar()
monitor.set(0)
#复选框,选中时变量值为1,未选中时变量值为0
checkMonitor=tkinter.Checkbutton(root,text='Is Monitor?', variable=monitor,
                                 onvalue=1, offvalue=0)
checkMonitor.place(x=20, y=100, width=100, height=20)
#添加按钮单击事件处理函数
def addInformation():
    result='Name:'+entryName.get()
    result=result+';Grade:'+comboGrade.get()
    result=result+';Class:'+comboClass.get()
    result=result+';Sex:' + ('Man' if sex.get()else 'Woman')
    result=result+';Monitor:' + ('Yes' if monitor.get()else 'No')
    listboxStudents.insert(0, result)
buttonAdd=tkinter.Button(root, text='Add',width=40, command=addInformation)
buttonAdd.place(x=130, y=100, width=40, height=20)
#删除按钮的事件处理函数
def deleteSelection():
    selection=listboxStudents.curselection()
    if  not selection:
        tkinter.messagebox.showinfo(title='Information',
                                    message='No Selection')
    else:
        listboxStudents.delete(selection)
buttonDelete=tkinter.Button(root, text='DeleteSelection',
                            width=100, command=deleteSelection)
buttonDelete.place(x=180, y=100, width=100, height=20)
#创建列表框组件
listboxStudents=tkinter.Listbox(root, width=300)
listboxStudents.place(x=10, y=130, width=300, height=200)
#启动消息循环
root.mainloop()
```

将上面的代码保存为tkinter_selction.pyw文件,运行后效果如图15-4所示。

图 15-4　程序运行效果

15.3　简单文本编辑器

下面的案例通过设计一个文本编辑器演示了菜单、文本框、文件对话框等组件的用法，实现了打开文件、保存文件、另存文件以及文本的复制、剪切、粘贴和查找等功能。

例 15-3　使用 tkinter 实现文本编辑器。

```
import tkinter
import tkinter.filedialog
import tkinter.colorchooser
import tkinter.messagebox
import tkinter.scrolledtext

#创建应用程序窗口
app=tkinter.Tk()
app.title('My Notepad----by Dong Fuguo')
app['width']=800
app['height']=600

textChanged=tkinter.IntVar(value=0)
#当前文件名
filename=''

#创建菜单
menu=tkinter.Menu(app)
#File菜单
submenu=tkinter.Menu(menu, tearoff=0)
def Open():
    global filename
    #如果内容已改变,先保存
```

```python
        if textChanged.get():
            yesno=tkinter.messagebox.askyesno(title='Save or not?',
                                        message='Do you want to save?')
            if yesno ==tkinter.YES:
                Save()
        filename=tkinter.filedialog.askopenfilename(title='Open file',
                                        filetypes=[('Text files', '*.txt')])
        if filename:
            #清空内容,0.0是 lineNumber.Column 的表示方法
            txtContent.delete(0.0, tkinter.END)
            fp=open(filename, 'r')
            txtContent.insert(tkinter.INSERT, ''.join(fp.readlines()))
            fp.close()
            #标记为尚未修改
            textChanged.set(0)
    #创建 Open 菜单并绑定菜单事件处理函数
    submenu.add_command(label='Open', command=Open)

    def Save():
        global filename
        #如果是第一次保存新建文件,则打开"另存为"窗口
        if not filename:
            SaveAs()
        #如果内容发生改变,保存
        elif textChanged.get():
            fp=open(filename, 'w')
            fp.write(txtContent.get(0.0, tkinter.END))
            fp.close()
            textChanged.set(0)
    submenu.add_command(label='Save', command=Save)

    def SaveAs():
        global filename
        #打开"另存为"窗口
        newfilename= tkinter. filedialog. asksaveasfilename (title = ' Save As ',
                    initialdir=r'c:\\',initialfile='new.txt')
        #如果指定了文件名,则保存文件
        if newfilename:
            fp=open(newfilename, 'w')
            fp.write(txtContent.get(0.0, tkinter.END))
            fp.close()
            filename=newfilename
            textChanged.set(0)
    submenu.add_command(label='Save As', command=SaveAs)
```

```python
#添加分割线
submenu.add_separator()
def Close():
    global filename
    Save()
    txtContent.delete(0.0, tkinter.END)
    #置空文件名
    filename=''
submenu.add_command(label='Close', command=Close)
#将子菜单关联到主菜单上
menu.add_cascade(label='File', menu=submenu)

#Edit 菜单
submenu=tkinter.Menu(menu, tearoff=0)
#撤销最后一次操作
def Undo():
    #启用 undo 标志
    txtContent['undo']=True
    try:
        txtContent.edit_undo()
    except Exception as e:
        pass
submenu.add_command(label='Undo', command=Undo)

def Redo():
    txtContent['undo']=True
    try:
        txtContent.edit_redo()
    except Exception as e:
        pass
submenu.add_command(label='Redo', command=Redo)
submenu.add_separator()

def Copy():
    txtContent.clipboard_clear()
    txtContent.clipboard_append(txtContent.selection_get())
submenu.add_command(label='Copy', command=Copy)

def Cut():
    Copy()
    #删除所选内容
    txtContent.delete(tkinter.SEL_FIRST, tkinter.SEL_LAST)
submenu.add_command(label='Cut', command=Cut)
```

```python
    def Paste():
        #如果没有选中内容,则直接粘贴到鼠标位置
        #如果有所选内容,则先删除再粘贴
        try:
            txtContent.insert(tkinter.SEL_FIRST, txtContent.clipboard_get())
            txtContent.delete(tkinter.SEL_FIRST, tkinter.SEL_LAST)
            #如果粘贴成功就结束本函数,以免异常处理结构执行完成之后再次粘贴
            return
        except Exception as e:
            pass
        txtContent.insert(tkinter.INSERT, txtContent.clipboard_get())
    submenu.add_command(label='Paste', command=Paste)
    submenu.add_separator()

    def Search():
        #获取要查找的内容
        textToSearch=tkinter.simpledialog.askstring(title='Search',
                                                    prompt='What to search?')
        start=txtContent.search(textToSearch, 0.0, tkinter.END)
        if start:
            tkinter.messagebox.showinfo(title='Found', message='Ok')
    submenu.add_command(label='Search', command=Search)
    menu.add_cascade(label='Edit', menu=submenu)

    #Help 菜单
    submenu=tkinter.Menu(menu, tearoff=0)
    def About():
        tkinter.messagebox.showinfo(title='About', message='Author:Dong Fuguo')
    submenu.add_command(label='About', command=About)
    menu.add_cascade(label='Help', menu=submenu)
    #将创建的菜单关联到应用程序窗口
    app.config(menu=menu)

    #创建文本编辑组件,并自动适应窗口大小
    txtContent=tkinter.scrolledtext.ScrolledText(app, wrap=tkinter.WORD)
    txtContent.pack(fill=tkinter.BOTH, expand=tkinter.YES)
    def KeyPress(event):
        textChanged.set(1)
    txtContent.bind('<KeyPress>', KeyPress)

    app.mainloop()
```

运行结果如图 15-5 所示。

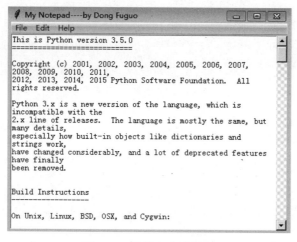

图 15-5 简单文本编辑器

15.4 简单画图程序

下面的程序实现了简单的画图功能,包括曲线、直线、矩形、文本的绘制,前景色和背景色的选取和设置,图片文件的打开与显示,以及橡皮擦功能,主要使用了 canvas 和 menu 组件,另外还用到了颜色选择对话框,同时还演示了鼠标事件处理函数的运用。

例 15-4 使用 tkinter 实现画图程序。

```
import tkinter
from PIL import Image

app=tkinter.Tk()
app.title('My Paint----by Dong Fuguo')
app['width']=800
app['height']=600

#控制是否允许画图的变量,1为允许,0为不允许
yesno=tkinter.IntVar(value=0)
#控制画图类型的变量,1为曲线,2为直线,3为矩形,4为文本,5为橡皮
what=tkinter.IntVar(value=1)
#记录鼠标位置的变量
X=tkinter.IntVar(value=0)
Y=tkinter.IntVar(value=0)
#前景色
foreColor='#000000'
backColor='#FFFFFF'
```

```python
#创建画布
image=tkinter.PhotoImage()
canvas=tkinter.Canvas(app, bg='white', width=800, height=600)
canvas.create_image(800, 600, image=image)
#单击,允许画图
def onLeftButtonDown(event):
    yesno.set(1)
    X.set(event.x)
    Y.set(event.y)
    if what.get()==4:
        #输出文本
        canvas.create_text(event.x, event.y, text=text)
canvas.bind('<Button-1>', onLeftButtonDown)

#记录最后绘制图形的id
lastDraw=0
#按住鼠标左键移动,画图
def onLeftButtonMove(event):
    if yesno.get()==0:
        return
    if what.get()==1:
        #使用当前选择的前景色绘制曲线
        canvas.create_line(X.get(), Y.get(), event.x, event.y, fill=foreColor)
        X.set(event.x)
        Y.set(event.y)
    elif what.get()==2:
        #绘制直线,先删除刚刚画过的直线,再画一条新的直线
        global lastDraw
        try:
            canvas.delete(lastDraw)
        except Exception as e:
            pass
        lastDraw=canvas.create_line(X.get(), Y.get(), event.x, event.y,
                                    fill=foreColor)
    elif what.get()==3:
        #绘制矩形,先删除刚刚画过的矩形,再画一个新的矩形
        global lastDraw
        try:
            canvas.delete(lastDraw)
        except Exception as e:
            pass
        lastDraw=canvas.create_rectangle(X.get(), Y.get(), event.x, event.y,
```

```
                                    fill=backColor, outline=foreColor)
        elif what.get()==5:
            #橡皮,使用背景色填充 10×10 的矩形区域
            canvas.create_rectangle(event.x-5, event.y-5, event.x+5, event.y+5,
                                    outline=backColor, fill=backColor)
canvas.bind('<B1-Motion>', onLeftButtonMove)

#鼠标左键抬起,不允许画图
def onLeftButtonUp(event):
    if what.get()==2:
        #绘制直线
        canvas.create_line(X.get(), Y.get(), event.x, event.y, fill=foreColor)
    elif what.get()==3:
        #绘制矩形
        canvas.create_rectangle(X.get(), Y.get(), event.x, event.y,
                                fill=backColor, outline=foreColor)
    yesno.set(0)
    global lastDraw
    lastDraw=0
canvas.bind('<ButtonRelease-1>', onLeftButtonUp)

#创建菜单
menu=tkinter.Menu(app, tearoff=0)
#打开图像文件
def Open():
    filename=tkinter.filedialog.askopenfilename(title='Open Image',
                                                filetypes=[('image', '*.jpg *.png *.gif')])
    if filename:
        global image
        image=tkinter.PhotoImage(file=filename)
        canvas.create_image(80, 80, image=image)
menu.add_command(label='Open', command=Open)
#添加菜单,清除绘制的所有图形
def Clear():
    for item in canvas.find_all():
        canvas.delete(item)
menu.add_command(label='Clear', command=Clear)
#添加分割线
menu.add_separator()
#创建子菜单,用来选择绘图类型
menuType=tkinter.Menu(menu, tearoff=0)
def drawCurve():
```

```python
        what.set(1)
menuType.add_command(label='Curve', command=drawCurve)
def drawLine():
        what.set(2)
menuType.add_command(label='Line', command=drawLine)
def drawRectangle():
        what.set(3)
menuType.add_command(label='Rectangle', command=drawRectangle)
def drawText():
        global text
        text=tkinter.simpledialog.askstring(title='Input what you want to draw',
                                        prompt='')
        what.set(4)
menuType.add_command(label='Text', command=drawText)
menuType.add_separator()
#选择前景色
def chooseForeColor():
        global foreColor
        foreColor=tkinter.colorchooser.askcolor()[1]
menuType.add_command(label='Choose Foreground Color', command=chooseForeColor)
#选择背景色
def chooseBackColor():
        global backColor
        backColor=tkinter.colorchooser.askcolor()[1]
menuType.add_command(label='Choose Background Color', command=chooseBackColor)
#橡皮
def onErase():
        what.set(5)
menuType.add_command(label='Erase', command=onErase)
menu.add_cascade(label='Type', menu=menuType)

#鼠标右键抬起,在鼠标位置弹出菜单
def onRightButtonUp(event):
        menu.post(event.x_root, event.y_root)
canvas.bind('<ButtonRelease-3>', onRightButtonUp)
canvas.pack(fill=tkinter.BOTH, expand=tkinter.YES)

app.mainloop()
```

程序运行结果如图 15-6 所示。

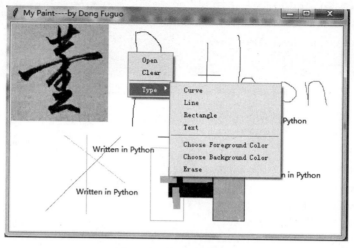

图 15-6 简单画图程序

15.5 电子时钟

下面的案例实现了电子时钟,使用 Label 组件实时显示当前日期和时间,涉及的知识主要有多线程、无标题栏、半透明、顶端显示、可拖动窗体的设计。

例 15-5 使用 tkinter 实现电子时钟。

```
import tkinter
import threading
import datetime
import time

app=tkinter.Tk()
#不显示标题栏
app.overrideredirect(True)
#半透明窗体
app.attributes('-alpha', 0.9)
#窗口总是在顶端显示
app.attributes('-topmost', 1)
#设置初始大小与位置
app.geometry('110×25+100+100')
labelDateTime=tkinter.Label(app)
labelDateTime.pack(fill=tkinter.BOTH, expand=tkinter.YES)
labelDateTime.configure(bg='gray')
#变量 X 和 Y 用来记录鼠标左键按下的位置
X=tkinter.IntVar(value=0)
Y=tkinter.IntVar(value=0)
#表示窗口是否可拖动的变量
```

```python
canMove=tkinter.IntVar(value=0)
#表示是否仍在运行的变量
still=tkinter.IntVar(value=1)

def onLeftButtonDown(event):
    #开始拖动时增加透明度
    app.attributes('-alpha', 0.4)
    #鼠标左键按下,记录当前位置
    X.set(event.x)
    Y.set(event.y)
    #标记窗口可拖动
    canMove.set(1)
#绑定鼠标左键单击事件处理函数
labelDateTime.bind('<Button-1>', onLeftButtonDown)

def onLeftButtonUp(event):
    #停止拖动时恢复透明度
    app.attributes('-alpha', 0.9)
    #鼠标左键抬起,标记窗口不可拖动
    canMove.set(0)
#绑定鼠标左键抬起事件处理函数
labelDateTime.bind('<ButtonRelease-1>', onLeftButtonUp)

def onLeftButtonMove(event):
    if canMove.get()==0:
        return
    #重新计算并修改窗口的新位置
    newX=app.winfo_x()+(event.x-X.get())
    newY=app.winfo_y()+(event.y-Y.get())
    g='110x25+'+str(newX)+'+'+str(newY)
    app.geometry(g)
#绑定鼠标左键移动事件处理函数
labelDateTime.bind('<B1-Motion>', onLeftButtonMove)

def onRightButtonDown(event):
    still.set(0)
    t.join(0.2)
    #关闭窗口
    app.destroy()
#绑定鼠标右键单击事件处理函数
labelDateTime.bind('<Button-3>', onRightButtonDown)
#显示当前时间的线程函数
def nowDateTime():
    while still.get()==1:
```

```
            now=datetime.datetime.now()
            s=str(now.year)+'-'+str(now.month)+'-'+str(now.day)+' '
            s=s+str(now.hour)+':'+str(now.minute)+':'+str(now.second)
            #显示当前时间
            labelDateTime['text']=s
            time.sleep(0.2)
#创建线程
t=threading.Thread(target=nowDateTime)
t.daemon=True
t.start()

app.mainloop()
```

程序运行界面如图 15-7 所示，电子时钟总是在顶端显示，用鼠标左键按住电子时钟可以拖动，并且拖动时窗口的透明度会发生改变，右击可以关闭电子时钟程序。

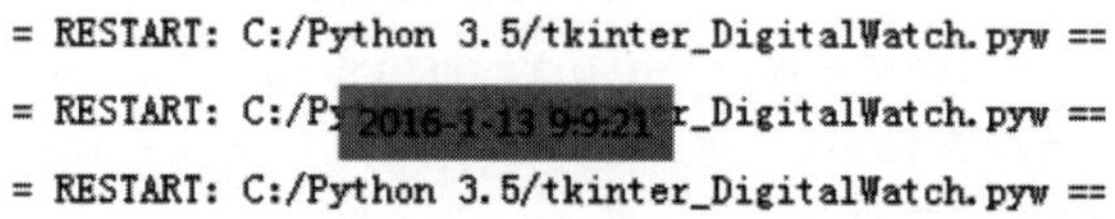

```
== RESTART: C:/Python 3.5/tkinter_DigitalWatch.pyw ==
== RESTART: C:/P 2016-1-13 9:9:21 r_DigitalWatch.pyw ==
== RESTART: C:/Python 3.5/tkinter_DigitalWatch.pyw ==
```

图 15-7　电子时钟运行截图

15.6　简单动画

tkinter 的 canvas 对象提供了 move()方法，可以将画布上的特定图形沿着 x 或 y 方向移动，利用这个方法可以制作简单的动画。

例 15-6　使用 tkinter 编写动画。

```
import tkinter
import time

app=tkinter.Tk()
app.title('tkinter animation')
app['width']=800
app['height']=600
canvas=tkinter.Canvas(app, bg='white', width=800, height=600)
#打开并加载图片
image=tkinter.PhotoImage(file='yingtaoxiaowanzi.png')
#记下图片的编号
id_actor=canvas.create_image(80, 80, image=image)
#控制是否自动运动的变量
flag=False
#单击，角色自动运动
def onLeftButtonDown(event):
```

```
        global flag
        flag=True
        while flag:
            #id_actor表示要运动的图形编号
            #第二个参数表示x方向的移动距离,5表示向右移动5个像素
            #第三个参数表示y方向的移动距离,0表示不移动
            canvas.move(id_actor, 5, 0)
            canvas.update()
            time.sleep(0.05)
canvas.bind('<Button-1>', onLeftButtonDown)

def onRightButtonUp(event):
    global flag
    flag=False
canvas.bind('<ButtonRelease-3>', onRightButtonUp)

#支持使用键盘上的4个方向键来控制图片的运动方向
def keyControl(event):
    if event.keysym =='Up':
        canvas.move(id_actor, 0, -5)
        canvas.update()
    elif event.keysym =='Down':
        canvas.move(id_actor, 0, 5)
        canvas.update()
    elif event.keysym =='Left':
        canvas.move(id_actor, -5, 0)
        canvas.update()
    elif event.keysym =='Right':
        canvas.move(id_actor, 5, 0)
        canvas.update()
canvas.bind_all('<KeyPress-Up>', keyControl)
canvas.bind_all('<KeyPress-Down>', keyControl)
canvas.bind_all('<KeyPress-Left>', keyControl)
canvas.bind_all('<KeyPress-Right>', keyControl)

canvas.pack(fill=tkinter.BOTH, expand=tkinter.YES)
canvas.focus()
#启动主循环
app.mainloop()
```

运行程序以后,单击图片自动向右移动,右击停止运动,通过键盘上的4个方向键可以灵活控制图片的运动方向,某个时刻的图片位置如图15-8所示。

拓展知识：Python标准库turtle也提供了很多绘图功能。下面的代码绘制结果如图15-9所示。程序运行后可以看到绘图的完整过程,就是绘图比较慢,绘制复杂图形

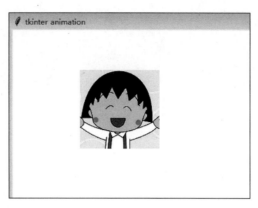

图 15-8 动画截图

时需要有足够的耐心等待画完。

```
import turtle

t=turtle.Pen()                    #使用钢笔
t.color(1, 0, 0)                  #设置钢笔颜色
t.up()                            #抬起,移动时不画
t.backward(280)                   #后退 280 像素
t.left(90)                        #左转 90°
t.forward(100)                    #前进 100 像素
t.right(90)                       #右转 90°
t.down()                          #落笔,开始画
for i in range(4):                #绘制矩形
    t.forward(150)
    t.left(90)

t.color(0, 0, 0)
t.up()
t.forward(200)
t.down()
for i in range(3):                #绘制等边三角形
    t.forward(200)
    t.left(120)

t.up()
t.forward(100)
t.down()
t.fillcolor(1, 0.6, 0.3)          #设置填充色
t.begin_fill()
t.circle(50)                      #绘制有填充色的圆,半径 50 像素
t.end_fill()
```

```
t.up()
t.forward(120)
t.left(90)
t.forward(90)
t.right(90)
t.down()
t.width(3)                        #设置钢笔粗细
t.fillcolor(0, 0.6, 0.8)          #设置填充色
t.begin_fill()
for i in range(5):                #绘制五角星
    t.forward(150)
    t.right(144)
t.end_fill()

t.up()
t.backward(270)
t.right(90)
t.forward(150)
t.write('Created using turtle, by 董付国', font=('隶书', 16, 'normal'))
t.forward(10)
t.left(90)
t.width(1)
t.down()
t.forward(350)
```

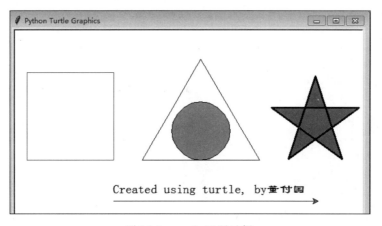

图 15-9 turtle 画图示例

15.7 多窗口编程

在大型软件中,不可能所有操作都在一个窗口中实现,需要根据功能进行分类并组织到不同的窗口中实现,下面的案例演示了如何自定义窗口并根据需要弹出不同的窗口。

例 15-7 弹出新窗口。

```python
import tkinter

#自定义窗口类
class myWindow:
    #构造函数
    def __init__(self, root, myTitle, flag):
        #创建窗口
        self.top=tkinter.Toplevel(root, width=300, height=200)
        #设置窗口标题
        self.top.title(myTitle)
        #设置顶端显示
        self.top.attributes('-topmost', 1)
        #根据不同情况在窗口上放置不同的组件
        if flag==1:
            label=tkinter.Label(self.top, text=myTitle)
            label.place(x=50, y=50)
        elif flag==2:
            def buttonOK():
                #弹出消息提示框
                tkinter.messagebox.showinfo(title='Python V5',
                                            message='I am Dong Fuguo')
            button=tkinter.Button(self.top, text=myTitle, command=buttonOK)
            button.place(x=50, y=50)
#创建应用程序主窗口
root=tkinter.Tk()
#设置主窗口大小
root.config(width=400)
root.config(height=200)
#设置主窗口标题
root.title('Multiple Windows Demo------Dong Fuguo')
window1=tkinter.IntVar(root, value=0)
window2=tkinter.IntVar(root, value=0)
#单击按钮 1,创建并弹出新窗口
def buttonClick1():
    if window1.get()==0:
        window1.set(1)
        w1=myWindow(root, 'First Window', 1)
        button1.wait_window(w1.top)
        window1.set(0)
button1=tkinter.Button(root, text='First Window', command=buttonClick1)
button1.place(x=70, y=40, height=40, width=200)
```

```python
def buttonClick2():
    if window2.get()==0:
        window2.set(1)
        w1=myWindow(root, 'Second Window', 2)
        button2.wait_window(w1.top)
        window2.set(0)
button2=tkinter.Button(root, text='Second Window', command=buttonClick2)
button2.place(x=70, y=100, height=40, width=200)
#启动消息主循环
root.mainloop()
```

15.8 屏幕任意区域截图

屏幕截图是一个很重要的功能,键盘上的 PrtScn 键实现了全屏幕截图的功能,QQ、微信电脑版实现了更加强大的区域截图功能,可以截取屏幕上任意区域的图像。一般来说,屏幕任意区域截图要依赖于 Windows API 函数来实现。但是如果稍微改变一下思路,使用 tkinter 的 Canvas 对象和 Python 扩展库 pillow 的 ImageGrab 模块也可以实现屏幕上任意区域截图的功能。本节案例的基本思路是:首先获取并显示全屏幕截图,然后在全屏幕截图上响应鼠标左键按下和抬起事件,最后进行二次截图。

例 15-8 使用 tkinter+pillow 实现屏幕任意区域截图。

```python
import tkinter
import tkinter.filedialog
import os
from PIL import ImageGrab
from time import sleep


root=tkinter.Tk()
root.geometry('100×40+400+300')
root.resizable(False, False)


class MyCapture:
    def __init__(self, png):
        #变量 X 和 Y 用来记录鼠标左键按下的位置
        self.X=tkinter.IntVar(value=0)
        self.Y=tkinter.IntVar(value=0)
        #屏幕尺寸
        screenWidth=root.winfo_screenwidth()
        screenHeight=root.winfo_screenheight()
        #创建顶级组件容器
        self.top=tkinter.Toplevel(root, width=screenWidth, height=screenHeight)
        #不显示最大化、最小化按钮
```

```python
        self.top.overrideredirect(True)
        self.canvas=tkinter.Canvas(self.top,bg='white', width=screenWidth,
                                   height=screenHeight)
        #显示全屏截图,在全屏截图上进行区域截图
        self.image=tkinter.PhotoImage(file=png)
        self.canvas.create_image(screenWidth//2, screenHeight//2,
                                 image=self.image)
        #鼠标左键按下的位置
        def onLeftButtonDown(event):
            self.X.set(event.x)
            self.Y.set(event.y)
            #开始截图
            self.sel=True
        self.canvas.bind('<Button-1>', onLeftButtonDown)
        #鼠标左键移动,显示选取的区域
        def onLeftButtonMove(event):
            if not self.sel:
                return
            global lastDraw
            try:
                #删除刚画完的图形,要不然鼠标移动的时候是黑乎乎的一片矩形
                self.canvas.delete(lastDraw)
            except Exception as e:
                pass
            lastDraw=self.canvas.create_rectangle(self.X.get(), self.Y.get(),
                                                 event.x, event.y, outline=
                                                 'black')
        self.canvas.bind('<B1-Motion>', onLeftButtonMove)
        #获取鼠标左键抬起的位置,保存区域截图
        def onLeftButtonUp(event):
            self.sel=False
            try:
                self.canvas.delete(lastDraw)
            except Exception as e:
                pass
            sleep(0.1)
            #考虑鼠标左键从右下方按下而从左上方抬起的截图
            left, right=sorted([self.X.get(), event.x])
            top, bottom=sorted([self.Y.get(), event.y])
            pic=ImageGrab.grab((left+1, top+1, right, bottom))
            #弹出保存截图对话框
            fileName=tkinter.filedialog.asksaveasfilename(title='保存截图',
                    filetypes=[('image', '*.jpg *.png')])
            if fileName:
```

```
                pic.save(fileName)
            #关闭当前窗口
            self.top.destroy()
        self.canvas.bind('<ButtonRelease-1>', onLeftButtonUp)
        self.canvas.pack(fill=tkinter.BOTH, expand=tkinter.YES)
#开始截图
def buttonCaptureClick():
    #最小化主窗口
    root.state('icon')
    sleep(0.2)

    filename='temp.png'
    im=ImageGrab.grab()
    im.save(filename)
    im.close()
    #显示全屏幕截图
    w=MyCapture(filename)
    buttonCapture.wait_window(w.top)
    #截图结束,恢复主窗口,并删除临时的全屏幕截图文件
    root.state('normal')
    os.remove(filename)
buttonCapture=tkinter.Button(root, text='截图', command=buttonCaptureClick)
buttonCapture.place(x=10, y=10, width=80, height=20)
#启动消息主循环
root.mainloop()
```

15.9 音乐播放器

Python 扩展库 pygame 提供了开发游戏所需要的所有功能,表 15-2 列出了其中一部分。其中,mixer 模块提供了 mp3、wav、ogg 等格式音乐文件播放的功能,主要功能见表 15-3。另外,跨平台音频/视频播放支持库 Phonon 也提供了播放音频和视频文件的功能,或者也可以使用 DirectSound 或 WMPlayer.ocx 或其他控件进行音乐文件播放。

表 15-2 pygame 的主要模块

模 块	说 明
cursors	控制鼠标指针
display	屏幕显示
draw	画图形
event	事件处理
font	使用字体
image	图像处理

续表

模 块	说 明
key	读取键盘按键
mask	图片遮罩
math	提供向量类
mixer	混音器有关功能
mouse	鼠标消息处理
movie	视频文件播放,需要安装 PyMedia
scrap	剪贴板支持
surface	绘制屏幕
time	时间控制
transform	修改和移动图像

表 15-3　mixer 模块的主要方法

方 法	说 明
pygame.mixer.init()	初始化,必须最先调用
pygame.mixer.music.load(filename)	打开音乐文件
pygame.mixer.music.play(count,start)	播放音乐文件
pygame.mixer.music.stop()	停止播放
pygame.mixer.music.pause()	暂停播放
pygame.mixer.music.unpause()	继续播放
pygame.mixer.music.get_busy()	检测声卡是否正被占用

例 15-9　使用 pygame+threading 编写音乐播放器。

```
import os
import tkinter
import tkinter.filedialog
import random
import time
import threading
import pygame

folder=''

def play():
    #默认播放 D:\music 文件夹中的所有 mp3 文件
    global folder
```

```python
        musics=[folder+'\\'+music for music in os.listdir(folder) \
                if music.endswith(('.mp3', '.wav', '.ogg'))]
        total=len(musics)
        #初始化混音器设备
        pygame.mixer.init()
        while playing:
            if not pygame.mixer.music.get_busy():
                #随机播放一首歌曲
                nextMusic=random.choice(musics)
                pygame.mixer.music.load(nextMusic.encode())
                #播放一次
                pygame.mixer.music.play(1)
                musicName.set('playing…'+nextMusic)
            else:
                time.sleep(0.3)

root=tkinter.Tk()
root.title('音乐播放器 v1.0---董付国')
root.geometry('280x70+400+300')
root.resizable(False, False)

#关闭程序时执行的代码
def closeWindow():
    global playing
    playing=False
    try:
        pygame.mixer.music.stop()
        pygame.mixer.quit()
    except:
        pass
    root.destroy()
root.protocol('WM_DELETE_WINDOW', closeWindow)

pause_resume=tkinter.StringVar(root, value='NotSet')
playing=False

#播放按钮
def buttonPlayClick():
    global folder
    if not folder:
        folder=tkinter.filedialog.askdirectory()
    if not folder:
        return
    global playing
```

```
    playing=True
    #创建一个线程来播放音乐
    t=threading.Thread(target=play)
    t.start()
    #根据情况禁用和启用相应的按钮
    buttonPlay['state']='disabled'
    buttonStop['state']='normal'
    buttonPause['state']='normal'
    buttonNext['state']='normal'
    pause_resume.set('Pause')
buttonPlay=tkinter.Button(root, text='Play', command=buttonPlayClick)
buttonPlay.place(x=20, y=10, width=50, height=20)

#停止按钮
def buttonStopClick():
    global playing
    playing=False
    pygame.mixer.music.stop()
    musicName.set('暂时没有播放音乐')
    buttonPlay['state']='normal'
    buttonStop['state']='disabled'
    buttonPause['state']='disabled'
buttonStop=tkinter.Button(root, text='Stop', command=buttonStopClick)
buttonStop.place(x=80, y=10, width=50, height=20)
buttonStop['state']='disabled'

#暂停与恢复,两个功能共用一个按钮
def buttonPauseClick():
    global playing
    if pause_resume.get()=='Pause':
        #playing=False
        pygame.mixer.music.pause()
        pause_resume.set('Resume')
    elif pause_resume.get()=='Resume':
        #playing=True
        pygame.mixer.music.unpause()
        pause_resume.set('Pause')
buttonPause=tkinter.Button(root, textvariable=pause_resume, \
                        command=buttonPauseClick)
buttonPause.place(x=140, y=10, width=50, height=20)
buttonPause['state']='disabled'

#下一首音乐
def buttonNextClick():
```

```
        global playing
        playing=False
        pygame.mixer.music.stop()
        pygame.mixer.quit()
        buttonPlayClick()
buttonNext=tkinter.Button(root, text='Next', command=buttonNextClick)
buttonNext.place(x=200, y=10, width=50, height=20)
buttonNext['state']='disabled'

musicName=tkinter.StringVar(root, value='暂时没有播放音乐…')
labelName=tkinter.Label(root, textvariable=musicName)
labelName.place(x=0, y=40, width=270, height=20)

#启动消息循环
root.mainloop()
```

这款小软件的运行界面如图 15-10 所示。

图 15-10　MP3 播放器界面

15.10　远程桌面监控系统

现在几乎所有学校机房都安装了远程监控软件,可以在主控机或教师机上实时查看学生端屏幕,如果发现有同学在做与学习无关的事情还可以远程关机,有力震慑了上机课玩游戏或者网购的同学,提高了教学质量(最起码理论上是这样的,但是也有同学退出了机房管理软件的客户端,就没法管理了,这属于另一个问题,不在本书讨论范围之内)。下面的代码实现了远程桌面监控功能,可以实时截取远程计算机的桌面并传送到监控端进行显示。

例 15-10　使用 tkinter＋pillow＋socket 编写远程桌面监控软件。

(1) 监控端代码(tkinter_RemoteDesktopMonitor_Server.pyw):

```
import tkinter
import socket
import time
import threading
import struct
from PIL import Image, ImageTk


def updateCanvas(canvas):
```

```python
global imageId
sock=socket.socket(socket.AF_INET, socket.SOCK_STREAM)
sock.bind(('', 10600))
sock.listen(1)
while running.get()==1:
    #自适应当前监控窗口大小
    width=canvas.winfo_width()
    height=canvas.winfo_height()
    conn, addr=sock.accept()
    tempImageBytes=b''
    #图像字节数量
    len_head=struct.calcsize('I128sI')
    data=conn.recv(len_head)
    length, size ,sizeLength=struct.unpack('I128sI',data)
    length=int(length)
    rest=length
    bufferSize=1024 * 10
    size=eval(size[:int(sizeLength)])
    while running.get()==1:
        if rest>bufferSize:
            data=conn.recv(1024 * 10)
        else:
            data=conn.recv(rest)
        tempImageBytes +=data
        rest=rest-len(data)

        #远程桌面截图接收完成,显示图像
        if rest ==0:
            tempImage=Image.frombytes('RGB', size, tempImageBytes)
            tempImage=tempImage.resize((width,height))
            #tempImage.save('temp.png')
            tempImage=ImageTk.PhotoImage(tempImage)
            #清除上一个截图
            try:
                canvas.delete(imageId)
            except:
                pass
            imageId=canvas.create_image(width//2, height//2, image=tempImage)
            #canvas.update()
            #通知客户端可以发送下一个截图
            conn.send(b'ok')
            print('ok')
            break
```

```
        conn.close()

root=tkinter.Tk()
#主程序窗口位置和大小
root.geometry('640×480+400+300')
width=640
height=480
root.title('远程桌面监考系统 v1.0---董付国')

#用来表示监控软件是否运行的变量
running=tkinter.IntVar(root, 1)

#关闭监控窗口时触发的消息处理代码
def closeWindow():
    running.set(0)
    root.destroy()
root.protocol('WM_DELETE_WINDOW', closeWindow)


canvas=tkinter.Canvas(root, width=width, height=height)
canvas.pack(fill=tkinter.BOTH, expand=tkinter.YES)
#使用子线程刷新监控窗口
t=threading.Thread(target=updateCanvas, args=(canvas,))
#主线程关闭时强制关闭刷新窗口的子线程
t.daemon=True
t.start()
root.mainloop()
```

(2) 被监控端代码(tkinter_RemoteDesktopMonitor_Client.pyw)：

```
import socket
import struct
from time import sleep
from PIL import ImageGrab


while True:
    try:
        sock=socket.socket(socket.AF_INET, socket.SOCK_STREAM)
        #假设监控端主机 IP 地址为 10.2.1.2,并监听 10600 端口
        sock.connect(('10.2.1.2', 10600))
        #本地全屏幕截图
        im=ImageGrab.grab()
        size=im.size
        #发本地截图转换为字节串进行发送
        imageBytes=im.tobytes()
        #发送字节串总长度和图像大小
```

```python
            fhead=struct.pack('I128sI',len(imageBytes), str(size).encode(),
                    len(str(size).encode()))
            sock.send(fhead)
            rest=len(imageBytes)
            bufferSize=1024*10
            while True:
                if rest>bufferSize:
                    temp=imageBytes[:bufferSize]
                    imageBytes=imageBytes[bufferSize:]
                else:
                    temp=imageBytes[:]
                sock.send(temp)
                rest=rest-len(temp)
                #本次截图发送完成
                if rest ==0:
                    if sock.recv(100)==b'ok':
                        print('ok')
                        break
        sock.close()
    except:
        print('无法连接监控端')
```

拓展知识：对于例 15-10 的程序有几点需要补充说明一下：①通过 socket 进行网络通信时收发数据包的数量和大小最好是一致的，要不然会导致混乱而无法正常监控；②如果需要进行屏幕广播，需要使用 UDP，可以参考本书第 16 章介绍的服务器发现功能代码；③在上面程序的实现中，发送端把屏幕截图转换为字节串后发送给接收端，而接收端接收完整个屏幕截图后还原为图像进行显示，这些操作都是在内存中进行的，如果同时监控多个客户端，可能会占用监控端大量的内存，可以接收截图的内容后不存储在内存中而是写入磁盘文件来缓解监控端的内存压力；④每次收发数据包的大小不能太大或太小，如果太大可能会因为出错引起的频繁重传而降低效率，如果太小则会因为额外开销（在网络体系结构中，发送端数据包每往下一层就会加一层"皮"，而接收端每往上一层就会为数据包去掉一层"皮"，这些"皮"是有开销的）太大而降低效率。

第 16 章 课堂教学管理系统设计与实现

目前越来越多的学校对程序设计类课程采取了边讲边练的授课方式,比传统的"课堂授课+实验"的教学模式具有更好的教学效果。本章设计的"边讲边练类课堂教学管理系统"实现了在线点名、交作业、在线答疑、自我测试、在线考试等功能,更大程度地提高了课堂时间利用率,有效保证了教学进度。从技术角度来讲,除了 Python 基础语法和基本类型之外,还涉及文件操作、数据库操作、多线程编程、GUI 编程、正则表达式等。

16.1 功能简介

16.1.1 教师端功能

教师端或服务端是整个系统的核心,主界面如图 16-1 所示,主要提供了以下功能。

图 16-1 教师端主界面

(1) 信息导入:包括学生名单和题库的导入,其中学生名单支持 xls 文件格式,而题库导入支持 xlsx 和 docx 两种文件格式。

(2) 在线点名与离线点名:由任课教师控制点名时间段,学生在学生端输入学号和姓名进行在线点名,几百人可以瞬间完成点名,大幅度提高了课堂时间的利用率,并且可以有效防止有人替同学答到；离线点名可以根据情况用来为迟到或请假的同学补点名。

（3）随机提问：教师选择专业后由系统随机分配一个学生进行提问，学生回答结束后可以在系统中记录答题情况与得分。

（4）加分与减分：可以为认真听课并且作业完成特别好的同学额外加分，给不认真听课的学生减分。

（5）在线收作业：系统可以接收学生端计算机全屏幕截图作为作业，适合检查学生随堂练习或者小段代码的作业的完成情况；也可以接收学生端上传的 Python 程序文件或 rar/zip 压缩文件作为作业，适合需要较多代码的作业。由教师控制交作业时间段，可以有效利用课堂的宝贵时间，还可以保证教学进度，并且在有限的时间内激发学生潜能。

（6）在线答疑：学生在线提出疑问，教师及时解答，而对于提问较多的问题，可以再对所有同学重点讲一下，不用一一解答，可以节省宝贵时间。

（7）在线测试与考试：学生可以在老师允许的时间段内查看系统中提供的练习题，并且可以查看标准答案；在线考试功能可以用来完成单元测试、期中考试或者期末考试，由老师控制考试时间段，由系统为每个学生随机分配题目，座位相邻的学生在同一时间很难分配到同一道题，使得作弊的可能性几乎为 0。另外，考试模式下禁用学生端的主流文本编辑器和浏览器，进一步降低作弊的可能。

（8）信息查看：可以在线查看学生名单、出勤情况、提问情况、考试情况等各类信息。

（9）数据导出：这个功能用来把系统中的学生点名、随机提问、在线答疑、考试成绩等各类数据导出到 xlsx 文件，可以实现离线数据查看。

16.1.2　学生端功能

学生端主要实现在线点名、交作业、在线提问以及在线自测和考试等功能，主界面如图 16-2 所示。

（1）报到：输入学号和姓名之后，单击"报到"按钮可以进行在线点名，系统根据已导入的学生名单来识别学号和姓名是否正确。

（2）提问：单击"提问"按钮后弹出新窗口，输入完整问题后发送给教师端，任课教师看到后及时解答。

图 16-2　学生端主界面

（3）交作业：可以根据老师的作业要求选择"全屏幕截图交作业"或者"上传文件交作业"。

（4）在线自测与考试：单击"自我测试"或"考试"按钮，弹出新窗口查看题库中的题目。自我测试时题目按题库中的顺序依次出现，并且可以查看每道题的答案；而考试时题目随机出现，并且不允许查看答案，答完 100 道题之后系统自动进行评分，考生可以立刻得知自己的分数（对于没好好学习的学生好像是有点残忍）。

16.2 数据库设计

系统采用了 Python 内置的 SQLite 数据库,所用到的表结构如表 16-1 所示。

表 16-1 表结构

数据表	字段	字段类型
students	kecheng	TEXT
	xuehao	TEXT
	xingming	TEXT
	zhuanye	TEXT
tiku	id	INTEGER PRIMARY KEY
	kechengmingcheng	TEXT
	zhangjie	TEXT
	timu	TEXT
	daan	TEXT
dianming	xuehao	TEXT
	shijian	TEXT
tiwen	xuehao	TEXT
	shijian	TEXT
	defen	NUMERIC
xueshengtiwen	xuehao	TEXT
	wenti	TEXT
	shijian	TEXT
tikufangwenqingkuang	xuehao	TEXT
	xingming	TEXT
	shijian	TEXT
kaoshi	id	INTEGER PRIMARY KEY
	xuehao	TEXT
	xingming	TEXT
	shijian	TEXT
	timubianhao	NUMERIC
	xueshengdaan	TEXT
	biaozhundaan	TEXT
	shifouzhengque	TEXT

16.3 系统总框架与通用功能设计

（1）导入系统功能所需要的 Python 标准库和扩展库，并根据需要及时升级 pip 工具：

```
import datetime
import tkinter
import tkinter.scrolledtext
import tkinter.messagebox
import tkinter.filedialog
import tkinter.ttk
import socket
import sqlite3
import random
import threading
import time
import struct
import os
import sys
import string

#先把 pip 升级到最新版本
path='"'+os.path.dirname(sys.executable)+\
    '\\scripts\\pip" install --upgrade pip'
os.system(path)
#导入必要的扩展库
#docx 扩展库
try:
    import docx
except:
    path='"'+os.path.dirname(sys.executable)+\
        '\\scripts\\pip" install python-docx'
    os.system(path)
    import docx
#xlrd 扩展库
try:
    import xlrd
except:
    path='"'+os.path.dirname(sys.executable)+'\\scripts\\pip" install xlrd'
    os.system(path)
    import xlrd
#openpyxl 扩展库
try:
```

```
    import openpyxl
except:
    path='"'+os.path.dirname(sys.executable)+\
        '\\scripts\\pip" install openpyxl'
    os.system(path)
    import openpyxl
```

(2) 创建 tkinter 应用程序主窗口,并自定义关闭窗口的事件处理函数:

```
root=tkinter.Tk()
#定义窗口大小
#root.config(width=360)
#root.config(height=260)
root.geometry('360×340+400+300')
#不允许改变窗口大小
root.resizable(False, False)
root.title('边讲边练类课程教学管理系统 v1.0---董付国')
#关闭程序时,取消点名、收作业、接受提问以及接受客户端查询等状态
#避免端口一直占用
def closeWindow():
    #结束点名
    if int_canDianming.get()==1:
        int_canDianming.set(0)
    #结束收作业
    if int_zuoye.get()==1:
        int_zuoye.set(0)
    #结束学生主动提问
    if int_xueshengTiwen.get()==1:
        int_xueshengTiwen.set(0)
    #结束服务状态
    if int_server.get()==1:
        int_server.set(0)
    root.destroy()
#绑定事件处理函数
root.protocol('WM_DELETE_WINDOW', closeWindow)
```

(3) 定义一个通用类,封装系统中频繁使用的功能:

```
#通用功能类
class Common:
    #查询数据库,获取学生专业列表
    def getZhuanye():
        conn=sqlite3.connect('database.db')
        cur=conn.cursor()
        cur.execute('SELECT distinct(zhuanye) FROM students')
        temp=cur.fetchall()
```

```python
        conn.close()
        xueshengZhuanye=[]
        for line in temp:
            xueshengZhuanye.append(line[0])
        return xueshengZhuanye

    #获取指定专业的学生名单
    def getXuehaoXingming(zhuanye):
        conn=sqlite3.connect('database.db')
        cur=conn.cursor()
        cur.execute("SELECT xuehao,xingming FROM students WHERE zhuanye='"+
                    zhuanye+"' ORDER BY xuehao")
        temp=cur.fetchall()
        conn.close()
        xueshengXinxi=[]
        for line in temp:
            xueshengXinxi.append(line[0]+','+line[1])
        return xueshengXinxi

    #获取指定学号的出勤次数
    def getChuqinCishu(xuehao):
        conn=sqlite3.connect('database.db')
        cur=conn.cursor()
        cur.execute("SELECT count(xuehao) FROM dianming WHERE xuehao='"+xuehao
                    +"'")
        temp=cur.fetchall()
        conn.close()
        xueshengXinxi=[]
        for line in temp:
            xueshengXinxi.append(line[0])
        return xueshengXinxi

    #获取指定学号的学生提问总得分
    def getTiwenDefen(xuehao):
        conn=sqlite3.connect('database.db')
        cur=conn.cursor()
        cur.execute("SELECT sum(defen) FROM tiwen WHERE xuehao='"+xuehao+"'")
        temp=cur.fetchall()
        conn.close()
        xueshengXinxi=[]
        for line in temp:
            xueshengXinxi.append(line[0])
        return xueshengXinxi
```

```python
#获取指定学号的学生主动提问次数
def getZhudongTiwenCishu(xuehao):
    conn=sqlite3.connect('database.db')
    cur=conn.cursor()
    cur.execute("SELECT count(xuehao) FROM xueshengtiwen WHERE xuehao='"+
                xuehao+"' AND wenti NOT LIKE '老师回复%'")
    temp=cur.fetchall()
    conn.close()
    xueshengXinxi=[]
    for line in temp:
        xueshengXinxi.append(line[0])
    return xueshengXinxi

#查看学生在线考试得分
def getKaoshiDefen(xuehao):
    conn=sqlite3.connect('database.db')
    cur=conn.cursor()
    cur.execute("SELECT count(xuehao) FROM kaoshi WHERE xuehao='"+xuehao+
                "' AND shifouzhengque='Y'")
    temp=cur.fetchall()
    conn.close()
    xueshengkaoshi=[]
    for line in temp:
        xueshengkaoshi.append(line[0])
    return xueshengkaoshi

#获取指定SQL语句查询结果
def getDataBySQL(sql):
    conn=sqlite3.connect('database.db')
    cur=conn.cursor()
    cur.execute(sql)
    result=cur.fetchall()
    conn.close
    return result

#执行SQL语句
def doSQL(sql):
    conn=sqlite3.connect('database.db')
    cur=conn.cursor()
    cur.execute(sql)
    conn.commit()
    conn.close()

#当前日期时间,格式为"年-月-日 时:分:秒"
```

```
def getCurrentDateTime():
    return str(datetime.datetime.now())[:19]

#当前日期时间之前一个半小时前的时间,主要用来避免重复点名
def getStartDateTime():
    now=datetime.datetime.now()
    now=now+datetime.timedelta(minutes=-90)
    return str(now)[:19]
```

16.4 数据导入功能

16.4.1 学生名单导入

系统支持把 Excel 文件中的学生名单导入数据库中,主要代码如下:

```
#导入学生信息
def buttonImportXueshengXinxiClick():
    filename=tkinter.filedialog.askopenfilename(title='请选择 Excel 文件',
                                                filetypes=[('Excel Files','*.xls')])
    if filename:
        #读取数据并导入数据库
        workbook=xlrd.open_workbook(filename=filename)
        sheet1=workbook.sheet_by_index(0)
        #Excel 文件必须包含 4 列,分别为学号、姓名、专业年级、课程名称
        if sheet1.ncols !=4:
            tkinter.messagebox.showerror(title='抱歉',
                                         message='Excel 文件格式不对')
            return
        for rowIndex in range(1, sheet1.nrows):#遍历 Excel 文件每一行
            row=sheet1.row(rowIndex)
            sql="INSERT INTO students (xuehao, xingming, zhuanye, kecheng) VALUES ('"+str(row[0].value).strip()+"','"+str(row[1].value)+"','"+str(row[2].value)+"','"+str(row[3].value)+"')"
            Common.doSQL(sql)
        tkinter.messagebox.showinfo(title='恭喜', message='导入成功')
#创建按钮并放置到主窗口上
buttonImportXueshengXinxi=tkinter.Button(root, text='导入学生信息',
                                         command=buttonImportXueshengXinxiClick)
buttonImportXueshengXinxi.place(x=20, y=20, height=30, width=100)
```

16.4.2 题库导入

题库导入功能支持 xls 和 docx 两种格式的文件,可以读取题库内容并自动判断

题型。

```python
def buttonDaoruTikuClick():
    filename=tkinter.filedialog.askopenfilename(title='请选择 Excel 2003 或 Word 2007 版本的题库文件', filetypes=[('Excel Files','*.xls'),('Word 2007 Files','*.docx')])
    if filename:
        #读取数据并导入数据库
        #Excel 文件
        if filename.endswith('.xls'):
            workbook=xlrd.open_workbook(filename=filename)
            sheet1=workbook.sheet_by_index(0)
            #Excel 文件必须包含 4 列,分别为课程名称、章节、题目、答案
            if sheet1.ncols !=4:
                tkinter.messagebox.showerror(title='抱歉', message='题库格式不对')
                return
            for rowIndex in range(1, sheet1.nrows):#遍历 Excel 文件每一行
                row=sheet1.row(rowIndex)
                sql=" INSERT INTO tiku (kechengmingcheng, zhangjie, timu, daan) VALUES('"+str(row[0].value).strip()+"','"+str(row[1].value)+"','"+str(row[2].value)+"','"+str(row[3].value)+"')"
                Common.doSQL(sql)
            tkinter.messagebox.showinfo(title='恭喜', message='导入成功')
        #docx 文件
        elif filename.endswith('.docx'):
            #docx 文档题库包含很多段,每段一个题目
            #datase.db 中 tiku 表包含 kechengmingcheng、zhangjie、timu、daan 四个字段
            from docx import Document
            doc=Document(filename)
            #连接数据库
            conn=sqlite3.connect('database.db')
            cur=conn.cursor()
            #先清空原来的题,可选操作
            cur.execute('delete from tiku')
            conn.commit()

            for p in doc.paragraphs:
                text=p.text
                if '(' in text and ')' in text:
                    index=text.index('(')
                    #分离问题和答案
                    question=text[:index]
                    if '___' in question:
                        question='填空题:'+question
```

```
                else:
                    question='判断题:'+question
                answer=text[index+1:-1]
                #将数据写入数据库
                sql='INSERT INTO tiku(kechengmingcheng,zhangjie,timu,daan) 
                VALUES("Python 程序设计","未分类","'+question+'","'+answer
                +'")'
                cur.execute(sql)
        conn.commit()
        #关闭数据库连接
        conn.close()
        tkinter.messagebox.showinfo(title='恭喜', message='导入成功')
buttonDaoruTiku = tkinter. Button ( root, text = '导入题库', command =
buttonDaoruTikuClick)
buttonDaoruTiku.place(x=20, y=220, height=30, width=100)
```

16.5 点名与加分功能

16.5.1 在线点名

在线点名功能使用多线程技术实现,通过主窗口上的按钮来启动和停止负责点名的线程。学生端输入学号和姓名并单击"报到"按钮之后,系统首先判断学号和姓名是否与已导入的学生名单匹配,并且检测是否为重复点名,如果是正常点名则把学号和点名时间插入数据库。另外,系统还对学生端的 IP 地址进行识别和记录,限制每台计算机只能一个学生点名,有效防止了有同学代替别人签名报到的情况。

服务端主要代码如下:

```
#控制是否可以点名的变量,1 表示可以,0 表示不可以
int_canDianming=tkinter.IntVar(root, value=0)
def thread_Dianming():
    #开始监听
    global sockDianming
    sockDianming=socket.socket(socket.AF_INET, socket.SOCK_STREAM)
    sockDianming.bind(('', 30000))
    sockDianming.listen(200)
    while int_canDianming.get()==1:
        try:
            #接受一个客户端连接
            conn, addr=sockDianming.accept()
        except:
            return
        data=conn.recv(1024)
        data=data.decode()
```

```python
        #客户端发来的消息格式为"学号,姓名"
        xuehao,xingming=data.split(',')
        #首先检查学号与姓名是否匹配,并且与数据库中的学生信息一致
        sqlIfMatch="SELECT count(xuehao)FROM students WHERE xuehao='"+\
                   xuehao+"' AND xingming='"+xingming+"'"
        if Common.getDataBySQL(sqlIfMatch)[0][0] !=1:
            conn.sendall('notmatch'.encode())
            conn.close()
        else:
            #记录该学生点名信息:学号、姓名、时间
            #并反馈给客户端点名成功,然后客户端关闭连接
            currentTime=Common.getCurrentDateTime()
            #获取一个半小时之前的时间
            startTime=Common.getStartDateTime()
            #查看是否已经点名过,避免一个半小时内重复点名
            sqlShifouChongfuDianming="SELECT count(xuehao) FROM dianming WHERE
            xuehao='"+xuehao+"' AND shijian >='"+startTime+"'"
            if Common.getDataBySQL(sqlShifouChongfuDianming)[0][0] !=0:
                conn.sendall('repeat'.encode())
                conn.close()
            else:
                #检查是否代替点名,根据学生端IP地址识别
                sqlShifouDaiDianming="SELECT count(ip)FROM dianming WHERE ip='"
                                     +addr[0]+"' AND shijian >='"+startTime+"'"
                if Common.getDataBySQL(sqlShifouDaiDianming)[0][0] !=0:
                    conn.sendall('daidianming'.encode())
                    conn.close()
                else:
                    #点名
                    sqlDianming="INSERT INTO dianming(xuehao,shijian,ip)VALUES
                    ('"+xuehao+"','"+currentTime+"','"+addr[0]+"')"
                    Common.doSQL(sqlDianming)
                    conn.sendall('ok'.encode())
                    conn.close()
    sockDianming.close()
    sockDianming=None
```

学生端代码相对简单,获取学生输入的姓名和学号,连接服务器,通过 Socket 把信息发送到教师端,然后根据教师端的反馈进行相应的提示。核心代码如下:

```python
#登录按钮事件处理函数
def buttonOKClick():
    #获取学号和姓名
    xuehao=entryXuehao.get()
```

```python
xingming=entryXingming.get()
serverIP=entryServerIP.get()
if not re.match('^\d{1,3}\.\d{1,3}\.\d{1,3}\.\d{1,3}$', serverIP):
    tkinter.messagebox.showerror('很抱歉','服务器 IP 地址不合法')
    return
#创建 Socket,并连接教师端
sock=socket.socket(socket.AF_INET, socket.SOCK_STREAM)
try:
    sock.connect((serverIP, 30300))
except Exception as e:
    tkinter.messagebox.showerror('很抱歉','现在不是点名时间')
    return
#向教师端发送学号和姓名,然后等待教师端反馈信息
sock.sendall((xuehao+','+xingming).encode())
data=sock.recv(1024)
data=data.decode()
#根据教师端的反馈信息进行相应的提示
if data.lower()=='ok':
    #点名成功
    tkinter.messagebox.showinfo('恭喜', xuehao+','+xingming+'  报到点名成功')
    sock.close()
    return
elif data.lower()=='repeat':
    tkinter.messagebox.showerror('很抱歉','不允许重复报到')
    sock.close()
    return
elif data.lower()=='notmatch':
    tkinter.messagebox.showerror('很抱歉','学号与姓名不匹配')
    sock.close()
    return
elif data.lower()=='daidianming':
    tkinter.messagebox.showerror('很抱歉','不允许替别人点名,警告一次')
    sock.close()
    return
#在学生端应用程序主界面上创建按钮并设置单击事件处理函数
buttonOk=tkinter.Button(root, text='报到', command=buttonOKClick)
buttonOk.place(x=30, y=90, width=80, height=20)
```

16.5.2 离线点名与加分

离线点名功能主要用来给迟到或请假的同学点名,加分功能用来为上课时听课非常认真的同学进行加分,减分功能用来为上课时听课非常不认真的同学进行减分,这 3 个功能由教师在软件教师端操作完成,如图 16-3 所示。该窗口主要代码如下:

```python
class windowChakanXueshengXinxi:
    def __init__(self, root, myTitle):
        self.top=tkinter.Toplevel(root, width=350, height=400)
        self.top.title(myTitle)
        self.top.attributes('-topmost', 1)

        #用组合框来显示学生的专业
        #调用通用功能类中的方法获取学生的专业
        xueshengZhuanye=Common.getZhuanye()
        comboboxZhuanye=tkinter.ttk.Combobox(self.top, values=
xueshengZhuanye)
        comboboxZhuanye.place(x=20, y=20, height=20, width=100)
        def buttonChakanClick():
            zhuanye=comboboxZhuanye.get()
            if not zhuanye:
                tkinter.messagebox.showerror(title='抱歉', message='请选择专业')
                return
            #根据选择的专业,获取该专业所有学生名单,格式为"学号,姓名"
            temp=Common.getXuehaoXingming(zhuanye)
            for row in treeXueshengMingdan.get_children():    #删除原有的所有行
                treeXueshengMingdan.delete(row)
            iii=0
            for student in temp:
                student=student.split(',')                    #分隔学号和姓名
                treeXueshengMingdan.insert('', iii, values=(student[0], student[1]))
                iii=iii+1
        #创建表格,设置表头,show="headings"用来隐藏树形控件的默认首列
        self.frame=tkinter.Frame(self.top)
        self.frame.place(x=20, y=50, width=200, height=280)
        #垂直滚动条
        scrollBar=tkinter.Scrollbar(self.frame)
        scrollBar.pack(side=tkinter.RIGHT, fill=tkinter.Y)
        #使用树形控件实现表格
        treeXueshengMingdan=tkinter.ttk.Treeview(self.frame, columns=('col1',
'col2'), show="headings", yscrollcommand=scrollBar.set)
        treeXueshengMingdan.column('col1', width=90, anchor='center')
        treeXueshengMingdan.column('col2', width=90, anchor='center')
        treeXueshengMingdan.heading('col1', text='学号')
        treeXueshengMingdan.heading('col2', text='姓名')
        #双击某行学生信息,双击实现离线点名
        def onDBClick(event):
            if not treeXueshengMingdan.selection():
                tkinter.messagebox.showerror('很抱歉', '请选择学生')
                return
```

```python
            item=treeXueshengMingdan.selection()[0]
            xuehaoDianming=treeXueshengMingdan.item(item, 'values')[0]
            xingmingDianming=treeXueshengMingdan.item(item, 'values')[1]
            currentTime=Common.getCurrentDateTime()
            #获取一个半小时之前的时间
            startTime=Common.getStartDateTime()
            #查看是否已经点名过,避免一个半小时内重复点名
            sqlShifouChongfuDianming="SELECT count (xuehao) FROM dianming WHERE xuehao='"+xuehaoDianming+"' AND shijian >='"+startTime+"'"
            if Common.getDataBySQL(sqlShifouChongfuDianming)[0][0] !=0:
                tkinter.messagebox.showerror('很抱歉', xuehaoDianming+','+
                                             xingmingDianming+'重复点名')
                return
            #点名
            sqlDianming=" INSERT INTO dianming (xuehao, shijian) VALUES ('" +
                    xuehaoDianming+"','"+currentTime+"')"
            Common.doSQL(sqlDianming)
            tkinter.messagebox.showinfo('恭喜',
                        xuehaoDianming+','+xingmingDianming+'  点名成功')
        treeXueshengMingdan.bind("<Double-1>", onDBClick)
        treeXueshengMingdan.pack(side=tkinter.LEFT, fill=tkinter.Y)
        #树形控件与垂直滚动条结合
        scrollBar.config(command=treeXueshengMingdan.yview)
        buttonChakan=tkinter.Button(self.top, text='查看', command=
                            buttonChakanClick)
        buttonChakan.place(x=130, y=20, height=20, width=40)
        def buttonJiafenClick():
            #首先选择一个学生,然后加分
            #为该学生添加一个提问得分,5分
            if not treeXueshengMingdan.selection():
                tkinter.messagebox.showerror('很抱歉','请选择学生')
                return
            item=treeXueshengMingdan.selection()[0]
            xuehaoJiafen=treeXueshengMingdan.item(item, 'values')[0]
            sqlJiafen="INSERT INTO tiwen (xuehao, shijian, defen) VALUES ('" +
                    xuehaoJiafen +"','" +Common.getCurrentDateTime()+"',5)"
            Common.doSQL(sqlJiafen)
            tkinter.messagebox.showinfo('恭喜','加分成功')
        buttonJiafen=tkinter.Button(self.top, text='听课认真加分',
                            command=buttonJiafenClick)
        buttonJiafen.place(x=30, y=350, height=20, width=100)
        def buttonJianfenClick():
            #首先选择一个学生,然后减分
            #为该学生添加一个提问得分,-5分
```

```
        if not treeXueshengMingdan.selection():
            tkinter.messagebox.showerror('很抱歉','请选择学生')
            return
        item=treeXueshengMingdan.selection()[0]
        xuehaoJiafen=treeXueshengMingdan.item(item,'values')[0]
        sqlJiafen="INSERT INTO tiwen(xuehao,shijian,defen)VALUES
                ('"+xuehaoJiafen+"','"+Common.getCurrentDateTime()+"',-5)"
        Common.doSQL(sqlJiafen)
        tkinter.messagebox.showinfo('恭喜','减分成功')
    buttonJianfen=tkinter.Button(self.top, text='听课不认真减分',
                                command=buttonJianfenClick)
    buttonJianfen.place(x=140, y=350, height=20, width=100)
    labelTishi=tkinter.Label(self.top, text='温馨提示:双击表格中某个学生可以离线点名,或者补点名。', fg='red')
    labelTishi.place(x=10, y=380, height=20)
```

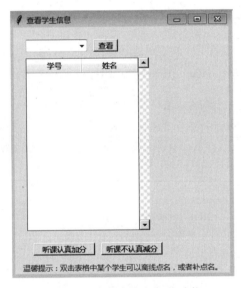

图 16-3　离线点名与加分功能

16.6　随机提问功能

随机提问功能用来从指定专业的学生名单中随机抽取一名同学,然后由教师提问题并记录回答情况。这样的功能对所有同学都是公平的,可以避免老师总是习惯性地提问学习较好的同学,真是替那些不爱学习的孩子们高兴啊。

```
class windowTiwen:
    def __init__(self, root, myTitle):
        self.top=tkinter.Toplevel(root, width=300, height=150)
        self.top.title(myTitle)
```

```python
self.top.attributes('-topmost', 1)
#学生专业
xueshengZhuanye=Common.getZhuanye()
comboboxZhuanye=tkinter.ttk.Combobox(self.top, values=xueshengZhuanye)
comboboxZhuanye.place(x=20, y=20, height=20, width=100)
#被提问到的学生学号
xueshengXuehao=tkinter.StringVar(self.top, value='')
def buttonTiwenClick():
    zhuanye=comboboxZhuanye.get()
    if not zhuanye:
        tkinter.messagebox.showerror(title='很抱歉', message='请选择专业')
        return
    #获取该专业的学生名单
    conn=sqlite3.connect('database.db')
    cur=conn.cursor()
    cur.execute("SELECT xuehao,xingming FROM students WHERE zhuanye='"
                +zhuanye+"'")
    temp=cur.fetchall()
    conn.close
    #从该专业所有学生名单中随机选择一个
    temp=random.choice(temp)
    xuehao=temp[0]
    xueshengXuehao.set(xuehao)
    tkinter.messagebox.showinfo(title='恭喜', message='本次中奖同学为'
                    +str((temp[0],temp[1])))

buttonTiwen=tkinter.Button(self.top, text='看看谁最幸运',
                    command=buttonTiwenClick)
buttonTiwen.place(x=130, y=20, height=20, width=80)
#根据学生答题情况进行加分或减分
comboboxDefen=tkinter.ttk.Combobox(self.top,
                    values=['-2', '-1', '0', '1', '2', '3', '4', '5'])
comboboxDefen.place(x=20, y=50, height=20, width=100)
def buttonDefenClick():
    if xueshengXuehao.get()=='':
        tkinter.messagebox.showerror(title='很抱歉',message='请先选择'
                '同学')
        return
    defen=comboboxDefen.get()
    if not defen:
        tkinter.messagebox.showerror(title='很抱歉', message='请选择得分')
    else:
        #记录该学生得分
        sql=" INSERT INTO tiwen (xuehao, shijian, defen) VALUES ('" +
```

```
                    xueshengXuehao.get()+"','"+Common.getCurrentDateTime()+"','"+
                    defen+")"
                    Common.doSQL(sql)
                    tkinter.messagebox.showinfo(title='恭喜',
                                                message='添加提问情况成功!')
                buttonDefen=tkinter.Button(self.top, text='确认得分',
                                           command=buttonDefenClick)
                buttonDefen.place(x=130, y=50, height=20, width=80)
```

随机提问界面如图 16-4 所示。

图 16-4　随机提问界面

16.7　在线收作业功能

在这个设计中,支持学生端进行全屏幕截图和上传文件这两种交作业的方式。其中,全屏幕截图是把学生端计算机整个屏幕的内容进行截图然后以图像文件的形式发送到教师端,而上传文件是学生端选择 Python 源程序文件或者 RAR 压缩包发送到教师端。通过 Socket 发送文件的代码是相似的,这里只介绍截图交作业的代码,上传文件交作业的代码可以查看本书配套资源。

16.7.1　学生端

学生端使用 pillow 中 ImageGrab 模块的 grab()方法进行全屏幕截图,把截图文件保存到当前文件夹中,然后把这个图像文件通过 Socket 发送给教师端。

```
def buttonZuoyeClick():
    #获取学号和姓名
    xuehao=entryXuehao.get()
    xingming=entryXingming.get()
    #获取输入的服务器 IP 地址,并检查是否合法
    serverIP=entryServerIP.get()
    if not re.match('^\d{1,3}\.\d{1,3}\.\d{1,3}\.\d{1,3}$', serverIP):
        tkinter.messagebox.showerror('很抱歉','服务器 IP 地址不合法')
        return
    #创建 Socket,并连接教师端
    sock=socket.socket(socket.AF_INET, socket.SOCK_STREAM)
    try:
```

```
        sock.connect((serverIP,30300))
    except Exception as e:
        tkinter.messagebox.showerror('很抱歉','现在不是交作业时间')
        return
    #截图并保存成文件
    filename=xuehao+'_'+xingming+'.png'
    im=ImageGrab.grab()
    im.save(filename)
    im.close()


    BUFSIZE=1024
    FILEINFO_SIZE=struct.calcsize('I128sI')

    #发送截图文件的基本信息
    fhead=struct.pack('I128sI',len(filename),filename.encode(),
                      os.stat(filename).st_size)
    #发送文件名和大小等信息,接收服务器反馈
    sock.send(fhead)
    data=sock.recv(1024)
    data=data.decode()
    if data.lower()=='notmatch':
        tkinter.messagebox.showerror('很抱歉','学号与姓名不匹配')
        sock.close()
        return
    #发送文件,发送结束后关闭Socket
    fp=open(filename,'rb')
    while True:
        filedata=fp.read(BUFSIZE)
        if not filedata:
            break
        sock.send(filedata)
    fp.close()
    sock.close()
    tkinter.messagebox.showinfo('恭喜','交作业成功')
#在学生端主窗口上添加按钮
buttonZuoye=tkinter.Button(root, text='全屏截图交作业', command=buttonZuoyeClick)
buttonZuoye.place(x=120, y=90, width=100, height=20)
```

16.7.2 教师端

教师端的功能要复杂很多,因为要同时接收很多学生上传的图片,所以使用了多线程,单击"开始接收截图作业"按钮之后,启动一个线程来监听端口,每当有学生端建立连接提交作业时,针对每个连接再创建一个负责接收图片文件的线程。

首先添加一个用来控制系统状态的变量 int_zuoye,0 表示停止收作业,1 表示开始收作业:

```python
int_zuoye=tkinter.IntVar(root, value=0)
```

负责监听端口的线程函数代码如下:

```python
def thread_ZuoyeMain():
    today=Common.getCurrentDateTime().split()[0]
    if not os.path.exists(today):
        os.mkdir(today)
    #创建Socket,开始监听端口
    global sockShouzuoye
    sockShouzuoye=socket.socket(socket.AF_INET, socket.SOCK_STREAM)
    sockShouzuoye.bind(('', 30300))
    sockShouzuoye.listen(200)
    while int_zuoye.get()==1:
        time.sleep(0.05)
        try:
            #接受学生端连接
            conn, addr=sockShouzuoye.accept()
        except:
            return
        #针对每个学生端的连接再创建一个负责接收图片文件的线程
        t=threading.Thread(target=thread_ShouZuoye, args=(conn,today))
        t.start()
    sock.close()
```

负责接收图片文件的线程函数代码为

```python
def thread_ShouZuoye(conn, today):
    #最终把客户端发来的截图保存为"学号_姓名.jpg"
    #开始接收客户端发来的截图
    BUFSIZE=1024
    FILEINFO_SIZE=struct.calcsize('I128sI')
    fhead=conn.recv(FILEINFO_SIZE)
    filenamelength, filename,filesize=struct.unpack('I128sI',fhead)
    filename=filename.decode()
    filename=filename[:filenamelength]
    ttt=filename.split('.')[0]
    #如果学号和姓名不匹配,拒绝接收作业图片
    xuehao, xingming=ttt.split('_')
    sql="SELECT count(xuehao)FROM students WHERE xuehao='"+
        xuehao.strip()+"' AND xingming='"+xingming.strip()+"'"
    t=Common.getDataBySQL(sql)[0][0]
    if t !=1:
        conn.sendall('notmatch'.encode())
        conn.close()
        return
```

```python
    else:
        conn.sendall('ok'.encode())
#接收截图文件基本信息
filename=filename[:-4]+'_'.join(Common.getCurrentDateTime().split())+ \
         filename[-4:]
filename=filename.replace('-', '_')
filename=filename.replace(':', '_')
filename=today+'\\'+filename

#首先删除本次作业期间之前上交的作业,只保留最后一次的作业
for f in os.listdir(today):
    if f.startswith(ttt):
        os.remove(today+'\\'+f)

#接收本次作业
fp=open(filename,'wb')
restsize=filesize
while True:
    if restsize>BUFSIZE:
        filedata=conn.recv(BUFSIZE)
    else:
        filedata=conn.recv(restsize)
    if not filedata:
        break
    fp.write(filedata)
    restsize=restsize-len(filedata)
    if restsize ==0:
        break
fp.close()
conn.close()
```

16.8　在线自测与在线考试功能

利用已导入的题库,在线自测功能可以让学生在规定的时间段内检验自己对所学知识掌握的程度,可以自由选择题号进行查看,也可以查看标准答案。而在线考试功能则由系统从题库中为每个学生随机抽取 100 道题发送给学生端,每个学生在同一时间所答题目不一样,完全避免了作弊的可能,答完题后系统自动阅卷并计算得分。在线自测功能的代码与在线考试的功能类似,都是监听端口和启动线程然后通过 Socket 发送题目,区别在于在线自测功能可以自由选择题目并且可以查看答案,而在线考试功能是由系统随机分配题目并且不允许查看答案。下面重点介绍在线考试功能,自测功能代码可以查看配套资源。

16.8.1 学生端

学生端输入学号和姓名之后,系统首先根据已导入的学生名单进行检验,如果学号和姓名正确则进入考试界面,系统随机分配题目,每道题必须答完才能进入下一题,不允许跳过任何一道题。这部分功能的主要代码如下:

```python
class windowKaoshi:
    def __init__(self, root, conn, xuehaoxingming):
        #创建面板容器,用于放置其他控件
        self.top=tkinter.Toplevel(root, width=300, height=220)
        self.top.title('学生自测---'+xuehaoxingming)
        self.top.attributes('-topmost', 1)
        #不允许改变窗口大小
        self.top.resizable(False, False)
        
        #自定义关闭窗口事件
        def closeWindow():
            if int_windowZice.get()==1:
                int_windowZice.set(0)
                conn.sendall('xxxx'.encode())
                conn.close()
            self.top.destroy()
        self.top.protocol('WM_DELETE_WINDOW', closeWindow)
        
        #从服务器接收课程名称清单,用组合框显示
        data=conn.recv(1024)
        data=data.decode()
        kechengQingdan=data.split(',')
        labelKechengmingcheng=tkinter.Label(self.top, text='请选择课程名称:')
        labelKechengmingcheng.place(x=10, y=10, height=20, width=100)
        combobox Kechengmingcheng=tkinter.ttk.Combobox(self.top,
                                                values=kechengQingdan)
        comboboxKechengmingcheng.place(x=120, y=10, height=20, width=130)
        #每次改变课程名称时,把 self.currentID 重新设置为 0
        def comboxboxKechengmingChanged(event):
            self.currentID=0
        comboboxKechengmingcheng.bind('<<ComboboxSelected>>',
                                      comboxboxKechengmingChanged)
        
        #使用标签组件显示课程名称
        string_Kecheng=tkinter.StringVar(self.top, value='')
        labelKecheng=tkinter.Label(self.top, text='', textvariable=string_
Kecheng)
```

```python
labelKecheng.place(x=10, y=40, height=20, width=100)
#使用可滚动的文本框显示题目内容
entryMessage=tkinter.scrolledtext.ScrolledText(self.top,
                                                wrap= tkinter.WORD)
entryMessage.place(x=10, y=70, width=280, height=70)

#下一题按钮
def buttonNextClick():
    #获取当前选择的课程
    kechengmingchengSelected=comboboxKechengmingcheng.get()
    if not kechengmingchengSelected:
        tkinter.messagebox.showerror('很抱歉','请选择课程名称')
        return

    #必须做系统随机分配的每一道题
    if entryMessage.get(0.0).strip()!='' and entryDaan.get().strip()=='':
        tkinter.messagebox.showinfo('很抱歉','必须做这个题')
        return

    #检查答案长度,禁止向服务器发送太长的内容
    if len(entryDaan.get())>=200:
        tkinter.messagebox.showerror('很抱歉','答案太长')
        return

    #提交答案,同时获取下一题
    message = (kechengmingchengSelected+'xx'+str(self.currentID)+'xx'+
               entryDaan.get()+'xxnext')
    conn.sendall(message.encode())
    data=conn.recv(1024)
    data=data.decode()
    if data.startswith('no,'):
        fenshu=data.split(',')[1]
        tkinter.messagebox.showinfo(title='恭喜', message='得分:'+
                                                   fenshu)
        #禁用"下一题"按钮
        buttonNext['state']='disabled'
        return
    kechengmingcheng, zhangjie, timu, self.currentID=data.split('xx')
    #删除原来的题目内容,显示新题目内容
    entryMessage.delete(0.0, tkinter.END)
    entryMessage.insert(tkinter.INSERT, timu)
    string_Kecheng.set(kechengmingcheng)
    #删除上一题学生输入的答案
    entryDaan.delete(0, tkinter.END)
```

```python
        buttonNext=tkinter.Button(self.top, text='下一题', command=
buttonNextClick)
        buttonNext.place(x=10, y=150, width=60, height=20)

        #填写答案的文本框
        entryDaan=tkinter.Entry(self.top,)
        entryDaan.place(x=10,y=180, width=270, height=20)

        #开始答题
        buttonNextClick()
```

16.8.2 教师端

教师端通过两个按钮来控制考试时间,开始考试后,系统启用一个线程监听端口并接受学生端的连接,针对每个学生连接再创建一个线程来随机分配题目并接收学生答案。负责监听端口的线程函数代码如下:

```python
def thread_xueshengKaoshiMain():
    global sockKaoshi
    #创建 Socket,监听 18000 端口
    sockKaoshi=socket.socket(socket.AF_INET , socket.SOCK_STREAM)
    sockKaoshi.bind(('', 18000))
    #最多允许同时 200 个学生考试
    sockKaoshi.listen(200)
    #int_xueshengKaoshi 是表示系统状态的变量,1 表示正在考试
    while int_xueshengKaoshi.get()==1:
        try:
            conn, addr=sockKaoshi.accept()                #接受一个连接
        except:
            return
        #创建并启动考试线程
        t_Kaoshi=threading.Thread(target=thread_xueshengKaoshi, args=(conn,))
        t_Kaoshi.start()
    sockKaoshi.close()
```

负责随机分配题目和接收学生答案的线程函数代码如下:

```python
def thread_xueshengKaoshi(conn):
    #接收学号和姓名
    data=conn.recv(1024)
    data=data.decode()
    xuehao, xingming=data.split(',')
    #检查学号、姓名是否匹配和正确
    sql = ("SELECT count(xuehao) FROM students WHERE xuehao='"+xuehao.strip()+
           "' AND xingming='"+xingming.strip()+"'")
```

```python
        t=Common.getDataBySQL(sql)[0][0]
        if t !=1:
            conn.sendall('notmatch'.encode())
            conn.close()
            return
        else:
            conn.sendall('ok'.encode())

        #获取并向客户端发送题库中的课程名称,以逗号隔开
        sqlKechengmingcheng="SELECT distinct(kechengmingcheng)FROM tiku"
        kechengQingdan=[]
        for kecheng in Common.getDataBySQL(sqlKechengmingcheng):
            kechengQingdan.append(str(kecheng[0]))
        kechengQingdan=','.join(kechengQingdan)
        conn.sendall(kechengQingdan.encode())
        #开始考试,如果客户端发来 xxxx,表示结束考试
        while int_xueshengKaoshi.get()==1:
            data=conn.recv(1024)
            data=data.decode()
            if data =='xxxx':
                conn.recv(1024)
                break
            #接收学生提交的答案
            kechengmingcheng, currentID, daan, pre_next=data.split('xx')
            if pre_next =='next':
                #把本题学生答案记录,不记录 0 号题
                if currentID !='0':
                    #自动阅卷
                    if daan ==biaozhundaan:
                        shifouzhengque='Y'
                    #如果标准答案中有空格,重新修正学生答案
                    #考虑学生输入多个连续空格的情况
                    #例如 pip list 等价于 pip    list
                    elif ' '.join(biaozhundaan.split())==' '.join(daan.split()):
                        shifouzhengque='Y'
                    #填空题,help 和 help()两种形式的答案都算对
                    elif biaozhundaan.endswith('()')and daan==biaozhundaan[:-2]:
                        shifouzhengque='Y'
                    #考虑学生没有输入空格的情况,例如[1,2,3]等价于[1, 2, 3]
                    elif ''.join(biaozhundaan.split())==''.join(daan.split()):
                        shifouzhengque='Y'
                    else:
                        shifouzhengque='N'
                    sql = (' INSERT INTO kaoshi (xuehao, xingming, timubianhao,
```

```
                xueshengdaan, biaozhundaan, shifouzhengque, shijian) VALUES("'+
                xuehao+ '","'+ xingming+ '", '+currentID+ ',"'+ daan+ '","'+
                biaozhundaan + '"," ' + shifouzhengque + '"," ' + Common.
                getCurrentDateTime()+'")')
                Common.doSQL(sql)

                #判断学生是否已答100题,若是则不允许继续答题,同时发送考试得分
                sql="SELECT count(xuehao)FROM kaoshi WHERE xuehao='"+xuehao+"'"
                total=Common.getDataBySQL(sql)[0]
                if total[0] >=100:
                    conn.sendall(('no,'+str(Common.getKaoshiDefen(xuehao))).encode())
                    break

                #发送下一题
                sqlHasMore=("SELECT kechengmingcheng, zhangjie, timu, id, daan FROM
                tiku WHERE kechengmingcheng='"+kechengmingcheng+"' AND id NOT IN
                (SELECT timubianhao FROM kaoshi WHERE xuehao='"+xuehao+"') ORDER BY
                random()limit 1")
                ttt=Common.getDataBySQL(sqlHasMore)
                if ttt:
                    tttt=ttt[0]
                    message=tttt[0]+'xx'+tttt[1]+'xx'+tttt[2]+'xx'+str(tttt[3])
                    #记录本题答案,评分用
                    biaozhundaan=str(tttt[4])
                    conn.sendall(message.encode())
                else:
                    conn.sendall(('no,'+str(Common.getKaoshiDefen(xuehao))).encode())
        conn.close()
```

16.9 信息查看功能

系统支持查看学生出勤情况、提问情况、考试得分等各类数据,为老师计算学生的平时成绩和最终成绩提供重要参考和依据。信息查看界面主要代码如下:

```
class windowChakanTongjiQingkuang:
    def __init__(self, root, myTitle):
        self.top=tkinter.Toplevel(root, width=600, height=380)
        self.top.title(myTitle)
        self.top.attributes('-topmost', 1)
        #获取学生专业列表,使用组合框显示
        xueshengZhuanye=Common.getZhuanye()
        comboboxZhuanye=tkinter.ttk.Combobox(self.top, values=xueshengZhuanye)
        comboboxZhuanye.place(x=20, y=20, height=20, width=120)
        #查看指定专业所有同学的提问情况
```

```python
def chakanZhuanye():
    zhuanye=comboboxZhuanye.get()
    if not zhuanye:
        tkinter.messagebox.showerror('很抱歉','请选择专业')
        return
    else:
        xuehaoXingmings=Common.getXuehaoXingming(zhuanye)
        xuehaos=[xingming.split(',')[0] for xingming in xuehaoXingmings]
        xingmings=[xingming.split(',')[1] for xingming in xuehaoXingmings]
        #获取每个同学的出勤次数,缺勤算 0,没提问到也算 0
        chuqinCishu=[Common.getChuqinCishu(xuehao) for xuehao in
                     xuehaos]
        #获取每个学生的提问得分和主动提问次数
        tiwenDefen=[Common.getTiwenDefen(xuehao) for xuehao in xuehaos]
        zhudongTiwenCishu=[Common.getZhudongTiwenCishu(xuehao) for
                           xuehao in xuehaos]
        #获取每个学生的考试得分
        kaoshidefen=[Common.getKaoshiDefen(xuehao) for xuehao in xuehaos]
        for row in treeXueshengMingdan.get_children():#删除原有的所有行
            treeXueshengMingdan.delete(row)
        iii=0
        #各类信息汇总
        tj=zip(xuehaos, xingmings, chuqinCishu, tiwenDefen,
               zhudongTiwenCishu, kaoshidefen)
        for xuehao, xingming, chuqin, tiwen, zhudongtiwen, kaoshidefen in tj:
            #把信息插入表格
            treeXueshengMingdan.insert('', iii, values=(xuehao, xingming,
                                       chuqin, tiwen, zhudongtiwen,
                                       kaoshidefen))
            iii=iii+1
buttonZhuanye=tkinter.Button(self.top, text='查看', command=
                             chakanZhuanye)
buttonZhuanye.place(x=150, y=20, height=20, width=80)

self.frame=tkinter.Frame(self.top)
self.frame.place(x=20, y=50, width=560, height=320)
#垂直滚动条
scrollBar=tkinter.Scrollbar(self.frame)
scrollBar.pack(side=tkinter.RIGHT, fill=tkinter.Y)
#使用树形控件实现表格,show="headings"用来隐藏树形控件的默认首列
treeXueshengMingdan=tkinter.ttk.Treeview(self.frame, columns=('col1',
'col2', 'col3', 'col4', 'col5', 'col6'),
                                         show="headings",
                                         yscrollcommand=scrollBar.set)
```

```
#设置表头
treeXueshengMingdan.column('col1', width=70, anchor='center')
treeXueshengMingdan.column('col2', width=50, anchor='center')
treeXueshengMingdan.column('col3', width=120, anchor='center')
treeXueshengMingdan.column('col4', width=120, anchor='center')
treeXueshengMingdan.column('col5', width=80, anchor='center')
treeXueshengMingdan.column('col6', width=80, anchor='center')
treeXueshengMingdan.heading('col1', text='学号')
treeXueshengMingdan.heading('col2', text='姓名')
treeXueshengMingdan.heading('col3', text='出勤次数')
treeXueshengMingdan.heading('col4', text='老师提问得分')
treeXueshengMingdan.heading('col5', text='主动提问次数')
treeXueshengMingdan.heading('col6', text='考试得分')
treeXueshengMingdan.pack(side=tkinter.LEFT, fill=tkinter.Y)
#树形控件与垂直滚动条结合
scrollBar.config(command=treeXueshengMingdan.yview)
```

16.10 数据导出功能

系统的数据导出功能用来把详细的学生出勤记录、提问记录、题库访问记录和考试情况导出到 xlsx 文件,支持数据的离线查看。下面给出了学生点名记录的代码,提问记录、题库访问记录和考试情况的代码类似,完整代码请查看配套资源。

```
def buttonDaochuClick():
    try:
        import openpyxl
        from openpyxl import Workbook
    except:
        tkinter.messagebox.showerror('抱歉', '您需要安装 openpyxl 扩展库')
    #创建 Workbook 对象
    wb=Workbook()
    #删除默认的 worksheet
    wb.remove_sheet(wb.worksheets[0])
    #创建新的 worksheet,导出点名记录
    ws=wb.create_sheet(title='在线点名情况')
    ws.append(['学号', '姓名', '点名时间'])
    sql=' SELECT students.xuehao, students.xingming, shijian FROM students,
dianming WHERE students.xuehao=dianming.xuehao ORDER BY students.xuehao'
    data=Common.getDataBySQL(sql)
    #把数据写入 xlsx 文件
    for d in data:
        ws.append([d[0], d[1], d[2]])
    #保存文件
    wb.save('数据导出.xlsx')
```

```
            tkinter.messagebox.showinfo('恭喜','导出成功,请查看"数据导出.xlsx"文件')
#在系统主界面上添加按钮
buttonDaochu=tkinter.Button(root, text='数据导出', command=buttonDaochuClick)
buttonDaochu.place(x=240, y=260, height=30, width=100)
```

16.11 其他辅助功能

考场的作弊和防作弊是个永恒的话题(严格来说前面提到并有效防范的同学代替别人签到也算是一种作弊),可谓道高一尺魔高一丈,作弊手法千变万化,这里只是给出一个在线考试防作弊的思路。另外,为了自适应不同的机房,不需要为客户端手工设置服务器IP地址,这个系统采用了UDP广播的技术实现了这一点,极大地方便了用户使用。最后,系统还提供了试卷生成功能支持离线考试。

16.11.1 防作弊功能

由于本系统提供了在线自测功能,学生在平时练习时可能会整理题库并保存成Word文档,这样的话考试的时候就可以用得上了。不幸的是,这种形式的作弊手法已经被我预料到并且有效地避免了,那就是考试时禁用学生机上的Word、WPS和记事本等文本编辑器进程。核心代码如下:

```
def funcJinyong():
    import threading
    import psutil                              #导入扩展库psutil,需要先安装
    from os.path import basename
    while True:
        for id in psutil.pids():               #列出当前所有进程ID
            try:
                p=psutil.Process(id)           #获取进程,判断对应程序的扩展名
                if basename(p.exe()).lower()in('notepad.exe', 'winword.exe',
                                    'wps.exe'):
                    p.kill()                   #结束进程
            except:
                pass
        time.sleep(3)                          #暂停3s
t_jinyong=threading.Thread(target=funcJinyong)
                                               #创建线程
t_jinyong.start()                              #启动线程
```

16.11.2 服务器自动发现功能

在整个系统的设计中,学生端和教师端的所有通信都是通过Socket实现的,这就要求学生端必须清楚地知道教师端计算机的IP地址。这个系统刚投入使用时都是教师先

在系统中查看一下本机 IP 地址,把这个 IP 地址告诉学生,再由学生手工输入到服务器 IP 地址文本框里,虽然不是很复杂的操作,但是也比较麻烦。后来给学生上课讲到网络编程那一章时,突然想到了一个思路,那就是教师端使用 UDP 定期广播本机 IP 地址,而学生端定时接收这样的广播信息并修正最新的服务器 IP 地址。于是,服务器自动发现功能就这样诞生了,果然是教学相长啊。

(1)教师端使用 UDP 定期广播的核心代码如下:

```
def sendServerIP():
    sock=socket.socket(socket.AF_INET, socket.SOCK_DGRAM)      #创建 Socket 对象
    while True:
        IP=socket.gethostbyname(socket.gethostname())          #获取本机 IP
        IP=IP[:IP.rindex('.')]+'.255'                          #255 表示广播地址
        sock.sendto('ServerIP'.encode(),(IP, 5000))            #发送信息
        time.sleep(3)                                          #暂停 3s
thread_sendServerIP=threading.Thread(target=sendServerIP)     #创建线程
thread_sendServerIP.start()                                    #启动线程
```

(2)学生端定期接收广播信息的核心代码如下:

```
def findServer():
    sock=socket.socket(socket.AF_INET, socket.SOCK_DGRAM)      #创建 Socket 对象
    sock.bind(('', 5000))                                      #绑定 Socket
    while int_searchServer.get()==1:
        data, addr=sock.recvfrom(1024)                         #接收信息
        if data.decode()=='ServerIP':                          #服务器广播信息
            server_IP.set(addr[0])                             #修正服务器 IP
        time.sleep(3)
thread_findServer=threading.Thread(target=findServer)         #创建线程
thread_findServer.start()                                      #启动线程
```

16.11.3 Word 版试卷生成功能

虽然现在很多学校的程序设计课是在机房里以边讲边练的形式授课,但也有的时候需要(也或者是任课老师的个人喜好)出题以传统的纸质试卷的形式进行考试。本系统提供了试卷生成功能,可以从题库中随机抽取 100 道题并生成 docx 文件。为了使用这个功能,需要先使用 pip install python-docx 命令安装 docx 库才行,主要代码如下:

```
def buttonGenerateShijuanClick():
    num=tkinter.simpledialog.askinteger('请输入题目数量','题目数量')
    if not num:
        return
    conn=sqlite3.connect('database.db')
    cur=conn.cursor()
    cur.execute('SELECT timu,daan FROM tiku')
```

```python
            temp=cur.fetchall()
        conn.close
        temp=random.sample(temp, num)
        yesno=tkinter.messagebox.askyesno('按题型排序吗?')
        if yesno:
            #对题目类型排序,填空题在前,判断题在后
            temp.sort(reverse=True)
        from docx import Document
        document=Document()
        document.add_paragraph('试题')
        for i, t in enumerate(temp):
            document.add_paragraph(str(i+1)+'、'+t[0])
        document.add_page_break()
        document.add_paragraph('答案')
        for i, t in enumerate(temp):
            document.add_paragraph(str(i+1)+'、'+t[1])
        document.save('试卷_答案.docx')
        tkinter.messagebox.showinfo('恭喜', '生成试卷成功')
        os.startfile('试卷_答案.docx')
buttonGenerateShijuan=tkinter.Button(root, text='生成 Word 试卷',
                                     command=buttonGenerateShijuanClick)
buttonGenerateShijuan.place(x=20, y=340, height=30,width=100)
```

结　束　语

当看到这一页内容的时候,您应该是松了一口气:总算把这本厚厚的书看完了。在阅读和学习本书的过程中,您或许遇到过不少困难,甚至有可能想过放弃。但看完以后,您应该会庆幸自己的坚持。仔细想想,其实学习过程中更大的收获,应该是学会一个又一个知识点之后的快乐,这种快乐只有全身心投入其中的人才更能真切体会。

镜头回到1999年春天,当时还在山东师范大学物理系读大三的我正在积极备战计算机等级四级考试(后来去拿证书的时候据考试中心的老师说,当时济南考点有800多人参加四级考试,共17个人通过,只有2个人是优秀,而我就是其中之一)。当时有一道模拟题耗费了我整整一周的时间还没做对,每天除了吃饭和睡觉之外都在冥思苦想,一天差不多有16个小时泡在机房里(当时我利用课余时间在机房帮忙值班近2年,随时可以使用计算机)。第八天中午,我去山东师范大学北街买了25个肉包子(不是我饭量大,而是脑力劳动消耗太多,我这么说您肯定相信),回机房的路上突然有了一个思路,然后一口气冲到机房里,包子也顾不上吃,一鼓作气把思路实现了,程序运行结果非常完美。当时我一个人在机房里激动地走来走去,每隔几分钟就开门看看有没有人来上机,后来终于等来了一个上机的同学。我和他商量了一下,把我刚想出来的思路和实现的代码讲给他听,他只要分享我的快乐就可以免费上机一下午。17年过去了,我至今还清晰地记得当时的那道题和我解决问题的思路,也还记得当时那种迫切地想找个人和我分享快乐的激动心情。我希望,在阅读本书的过程中,在不断提高自己的过程中,您也能体验到类似的快乐!一分耕耘,一分收获,这是永恒的道理。只要努力了就会有收获,学习知识更是如此。

欲穷千里目,更上一层楼。您可能已经发现了,目前国内市面上像本书涉及面这么广泛并且深入的Python类图书应不多见。尽管如此,我们必须认识到,在通往Python之巅的道路上我们也只是才走了很少一小段,后面的路还很长。问渠那得清如许,为有源头活水来。只有坚持不懈地学习才能一直保持前进的步伐。浏览Python社区,阅读Python官方文档,阅读Python标准库和扩展库的源代码,反复优化自己写过的代码,找个实际的项目做做……这都是不断提高自己的有效方式。

莫愁前路无知己,天下谁人不识君。在本书的最后,把《别董大》里的这句名言赠予各位读者朋友。在茫茫Python社区,希望每个人都能遇到几个志同道合的朋友,一起朝着远方出发!祝您早日成为Python高手,也祝家庭幸福美满!在学习Python之余,记得要注意锻炼身体(例如像我一样打打太极拳),更别忘了多陪陪父母、对象、孩子或者身边的朋友。生活不止是眼前的Python,还有亲情和友情!

董付国

附录 A 本书中例题清单

(1) 例 3-1　面试资格确认。
(2) 例 3-2　用户输入若干个成绩,求所有成绩的平均分。每输入一个成绩后询问是否继续输入下一个成绩,回答 yes 就继续输入下一个成绩,回答 no 就停止输入成绩。
(3) 例 3-3　编写程序,判断今天是今年的第几天。
(4) 例 3-4　输出序列中的元素。
(5) 例 3-5　求 1~100 之间能被 7 整除,但不能同时被 5 整除的所有整数。
(6) 例 3-6　输出"水仙花数"。所谓水仙花数是指一个 3 位的十进制数,其各位数字的立方和恰好等于该数本身。例如,153 是水仙花数,因为 $153=1^3+5^3+3^3$。
(7) 例 3-7　求平均分。
(8) 例 3-8　打印九九乘法表。
(9) 例 3-9　求 200 以内能被 17 整除的最大正整数。
(10) 例 3-10　判断一个数是否为素数。
(11) 例 3-11　鸡兔同笼问题。假设共有鸡、兔 30 只,脚 90 只,求鸡、兔各有多少只?
(12) 例 3-12　编写程序,输出由 1、2、3、4 这 4 个数字组成的每位数都不相同的所有三位数。
(13) 例 3-13　编写程序,计算组合数 $C(n,i)$,即从 n 个元素中任选 i 个,有多少种选法?
(14) 例 3-14　编写程序,计算理财产品收益,假设利息和本金一起滚动。
(15) 例 3-15　编写函数计算圆的面积。
(16) 例 3-16　编写函数,接收任意多个实数,返回一个元组,其中第一个元素为所有参数的平均值,其他元素为所有参数中大于平均值的实数。
(17) 例 3-17　编写函数,接收字符串参数,返回一个元组,其中第一个元素为大写字母的个数,第二个元素为小写字母的个数。
(18) 例 3-18　编写函数,接收包含 20 个整数的列表 lst 和一个整数 k 作为参数,返回新列表。处理规则:将列表 lst 中下标 k 之前的元素逆序,下标 k 之后的元素逆序,然后将整个列表 lst 中的所有元素逆序。
(19) 例 3-19　编写函数,接收整数参数 t,返回斐波那契数列中大于 t 的第一个数。
(20) 例 3-20　编写函数,接收一个包含若干整数的列表参数 lst,返回一个元组,其中第一个元素为列表 lst 中的最小值,其余元素为最小值在列表 lst 中的下标。

(21) 例 3-21　编写函数,接收一个整数 t 为参数,打印杨辉三角前 t 行。

(22) 例 3-22　编写函数,接收一个正偶数为参数,输出两个素数,并且这两个素数之和等于原来的正偶数。如果存在多组符合条件的素数,则全部输出。

(23) 例 3-23　编写函数,接收两个正整数作为参数,返回一个元组,其中第一个元素为最大公约数,第二个元素为最小公倍数。

(24) 例 3-24　编写函数,接收一个所有元素值都不相等的整数列表 x 和一个整数 n,要求将值为 n 的元素作为支点,将列表中所有值小于 n 的元素全部放到 n 的前面,所有值大于 n 的元素放到 n 的后面。

(25) 例 3-25　编写函数,计算字符串匹配的准确率。

(26) 例 3-26　编写函数,对整数进行因数分解。

(27) 例 3-27　韩信点兵。

(28) 例 3-28　模拟发红包算法。

(29) 例 3-29　编写函数,将 YYYY-MM-DD 的日期形式转换为 YYYYQ 的形式,其中 Q 表示季度。

(30) 例 3-30　模拟一维信号卷积,并模拟整数乘法。

(31) 例 3-31　猜数游戏。系统随机产生一个数,玩家最多可以猜 5 次,系统会根据玩家的猜测进行提示,玩家则可以根据系统的提示对下一次的猜测进行适当调整。

(32) 例 3-32　计算形式如 a+aa+aaa+aaaa+…+aaa…aaa 的表达式的值,其中 a 为小于 10 的自然数。

(33) 例 3-33　有 n 个人围成一圈,顺序排号。从第一个人开始从 1 到 k(假设 k=3)报数,报到 k 的人退出圈子,然后圈子缩小,从下一个人继续游戏,问最后留下的是原来的第几号?

(34) 例 3-34　汉诺塔问题。

(35) 例 3-35　编写函数计算任意位数的黑洞数。黑洞数是指这样的整数:由这个数字每位数字组成的最大数减去每位数字组成的最小数仍然得到这个数自身。例如,3 位黑洞数是 495,因为 954−459=495,4 位数字是 6174,因为 7641−1467=6174。

(36) 例 3-36　24 点游戏是指随机选取 4 张扑克牌(不包括大小王),然后通过四则运算来构造表达式,如果表达式的值恰好等于 24 就赢一次。下面的代码定义了一个函数用来测试随机给定的 4 个数是否符合 24 点游戏规则,如果符合就输出所有可能的表达式。

(37) 例 3-37　双色球是一种比较常见的彩票玩法,每一注彩票由 6 个介于 1 到 33 之间的不重复数字和 1 个介于 1 到 16 之间的数字组成。下面的代码用来随机生成一注双色球彩票,结果是完全随机的。

(38) 例 3-38　八皇后问题。八皇后问题是高斯先生(就是小时候就把 1+2+3+…+100 转换成(1+100)×50 的那个数学家)在 60 多年以前提出来的,是一个经典的回溯算法问题,其核心为:在国际象棋棋盘(8 行 8 列)上摆放 8 个皇后,要求 8 个皇后中任意两个都不能位于同一行、同一列或同一斜线上。

(39) 例 4-1　设计 Person 类,并根据 Person 派生 Teacher 类,分别创建 Person 类与

Teacher 类的对象。

（40）例 4-2 自定义一个数组类,支持数组与数字之间的四则运算,数组之间的加法运算、内积运算和大小比较,数组元素访问和修改,以及成员测试等功能。

（41）例 4-3 模拟矩阵运算,支持矩阵转置,修改矩阵大小,矩阵与数字的加、减、乘运算,以及矩阵与矩阵的加、减、乘运算。

（42）例 4-4 设计自定义队列类,模拟入队、出队等基本操作。

（43）例 4-5 设计自定义栈类,模拟入栈、出栈、判断栈是否为空、是否已满以及改变栈大小等操作。

（44）例 4-6 设计二叉树类,模拟二叉树创建、插入子节点以及前序遍历、中序遍历和后序遍历等遍历方式,同时还支持二叉树中任意子树的节点遍历。

（45）例 4-7 设计有向图类,模拟有向图的创建和路径搜索功能。

（46）例 4-8 自定义集合类。

（47）例 5-1 编写函数实现字符串加密和解密,循环使用指定密钥,采用简单的异或算法。

（48）例 5-2 编写程序,生成大量随机信息。

（49）例 5-3 使用正则表达式提取字符串中的电话号码。

（50）例 5-4 使用正则表达式提取 Python 程序中的类名、函数名以及变量名等标识符。

（51）例 5-5 使用正则表达式检查 Python 程序的代码风格是否符合规范。

（52）例 6-1 向文本文件中写入内容。

（53）例 6-2 读取文本文件内容。

（54）例 6-3 读取并显示文本文件的所有行。

（55）例 6-4 移动文件指针。假设文件 sample.txt 中的内容原为"Hello world\n 文本文件的读取方法\n 文本文件的写入方法"。

（56）例 6-5 假设文件 data.txt 中有若干整数,整数之间使用英文逗号分隔,编写程序读取所有整数,将其按升序排序后再写入文本文件 data_asc.txt 中。

（57）例 6-6 编写程序,保存为 demo.py,运行后生成文件 demo_new.py,其中的内容与 demo.py 一致,但是在每行的行尾加上了行号。

（58）例 6-7 计算文本文件中最长行的长度和该行的内容。

（59）例 6-8 Python 程序代码复用度检查。

（60）例 6-9 使用 pickle 模块写入二进制文件。

（61）例 6-10 使用 pickle 模块读取例 6-9 中写入二进制文件的内容。

（62）例 6-11 使用 struct 模块写入二进制文件。

（63）例 6-12 使用 struct 模块读取例 6-11 中二进制文件的内容。

（64）例 6-13 将当前目录的所有扩展名为 html 的文件重命名为扩展名为 htm 的文件。

（65）例 6-14 计算文件的 CRC32 值。

（66）例 6-15 判断一个文件是否为 GIF 图像文件。任何一种文件都具有专门的文件头结构,在文件头中存放了大量的信息,其中就包括该文件的类型。通过文件头信息来判断文件类型的方法可以得到更加准确的信息,而不依赖于文件扩展名。

(67) 例 6-16　使用 xlwt 模块写入 Excel 文件。

(68) 例 6-17　使用 xlrd 模块读取 Excel 文件。

(69) 例 6-18　使用 Pywin32 操作 Excel 文件。

(70) 例 6-19　检查 Word 文档的连续重复字。在 Word 文档中,经常会由于键盘操作不小心而使得文档中出现连续的重复字,例如,"用户的资料"或"需要需要用户输入"之类的情况。本例使用 Pywin32 模块中 win32com 对 Word 文档进行检查并提示类似的重复汉字。

(71) 例 6-20　编写程序,进行文件夹增量备份。

(72) 例 6-21　编写程序,统计指定文件夹大小以及文件和子文件夹数量。本例也属于系统运维范畴,可用于磁盘配额的计算,例如 E-mail、博客、FTP、快盘等系统中每个账号所占空间大小的统计。

(73) 例 6-22　编写程序,统计指定目录所有 C++ 源程序文件中不重复代码行数。

(74) 例 6-23　编写程序,递归删除指定文件夹中指定类型的文件。

(75) 例 6-24　使用扩展库 openpyxl 读写 Excel 2007 及更高版本的 Excel 文件。

(76) 例 6-25　编写代码,查看指定 ZIP 和 RAR 压缩文件中的文件列表。

(77) 例 6-26　小学口算题库生成器。

(78) 例 6-27　将 docx 文档中的题库导入 SQLite 数据库。

(79) 例 6-28　提取 docx 文档中例题、插图和表格清单。

(80) 例 6-29　将指定文件夹中的文件压缩至已有压缩包。

(81) 例 6-30　使用密码字典暴力破解 RAR 或 ZIP 文件密码。

(82) 例 6-31　把 Excel 2007$^+$ 文件中的多个同结构 worksheet 的内容合并到新文件中的一个 worksheet 中。

(83) 例 6-32　把记事本文件 test.txt 转换成 Excel 2007$^+$ 文件。假设 test.txt 文件中第一行为表头,从第二行开始是实际数据,并且表头和数据行中的不同字段信息都是用逗号分隔。

(84) 例 7-1　使用 doctest 模块测试 Python 代码。

(85) 例 7-2　编写单元测试程序。

(86) 例 7-3　使用 IDLE 调试 Python 程序。

(87) 例 9-1　UDP 通信程序。

(88) 例 9-2　会聊天的小机器人。

(89) 例 9-3　网络嗅探器程序。

(90) 例 9-4　端口扫描器程序。

(91) 例 9-5　网页爬虫程序。

(92) 例 9-6　使用 Flask 框架编写网站程序。

(93) 例 9-7　Python+Flask+Flask-email 发送带附件的电子邮件。

(94) 例 9-8　创建第一个 django Web 应用。

(95) 例 9-9　素数判断。
(96) 例 9-10　使用网页模板。
(97) 例 10-1　线程对象的 join() 方法。
(98) 例 10-2　线程状态检测。
(99) 例 10-3　线程对象的 daemon 属性。
(100) 例 10-4　使用 Lock/RLock 对象实现线程同步。
(101) 例 10-5　使用 Condition 对象实现线程同步。
(102) 例 10-6　使用 queue 对象实现线程同步。
(103) 例 10-7　使用 Event 对象实现线程同步。
(104) 例 10-8　进程创建与启动。
(105) 例 10-9　使用 Queue 对象在进程间交换数据，一个进程把数据放入 Queue 对象，另一个进程从 Queue 对象中获取数据。
(106) 例 10-10　使用管道实现进程间数据交换。管道有两个端，一个接收端和一个发送端，相当于在两个进程之间建立了一个用于传输数据的通道。
(107) 例 10-11　使用共享内存实现进程间数据传递，比较适合大量数据的场合。
(108) 例 10-12　使用 Manager 对象实现进程间数据交换。
(109) 例 10-13　使用 Lock 对象实现进程同步。
(110) 例 10-14　使用 Event 对象实现进程同步。
(111) 例 12-1　绘制三维直线、二维三角形和圆。
(112) 例 12-2　纹理映射与计算机动画。
(113) 例 12-3　法向量与光照模型。
(114) 例 12-4　计算椭圆中心。
(115) 例 12-5　动态生成比例分配图。
(116) 例 12-6　生成验证码图片。
(117) 例 12-7　GIF 动态图像分离与生成。
(118) 例 12-8　把固定大小的图片进行缩放并映射到任意大小物体表面。
(119) 例 12-9　使用 pillow 进行图像空域融合。
(120) 例 12-10　使用 pillow 生成国际象棋棋盘纹理。
(121) 例 14-1　恺撒密码算法。
(122) 例 14-2　维吉尼亚密码。
(123) 例 14-3　换位密码算法。
(124) 例 14-4　计算文件的 MD5 值。
(125) 例 14-5　使用 Python 扩展库 pycrypto 提供的 AES 算法实现消息加密和解密。
(126) 例 14-6　使用 rsa 模块来实现消息加密和解密。
(127) 例 15-1　tkinter 实现用户登录界面。
(128) 例 15-2　tkinter 单选按钮、复选框、组合框、列表框综合运用案例。
(129) 例 15-3　使用 tkinter 实现文本编辑器。

(130) 例 15-4　使用 tkinter 实现画图程序。
(131) 例 15-5　使用 tkinter 实现电子时钟。
(132) 例 15-6　使用 tkinter 编写动画。
(133) 例 15-7　弹出新窗口。
(134) 例 15-8　使用 tkinter＋pillow 实现屏幕任意区域截图。
(135) 例 15-9　使用 pygame＋threading 编写音乐播放器。
(136) 例 15-10　使用 tkinter＋pillow＋socket 编写远程桌面监控软件。

附录 B 本书中插图清单

(1) 图 1-1　Python 3.5.1 IDLE 的界面
(2) 图 1-2　wingIDE 的运行界面
(3) 图 1-3　PyCharm 的运行界面
(4) 图 1-4　Eclipse＋PyDev 的运行界面
(5) 图 1-5　Python 官方网站提供的 Interactive Shell 入口
(6) 图 1-6　Python 官方网站提供的 Interactive Shell 界面
(7) 图 1-7　Windows 7 环境中系统 Path 变量的修改方法
(8) 图 1-8　在 IDLE 中运行程序
(9) 图 1-9　在命令提示符中运行程序
(10) 图 1-10　配置 IDLE 并增加清屏菜单和快捷键
(11) 图 1-11　Python 内存管理模式
(12) 图 1-12　Python 内置帮助系统
(13) 图 1-13　从 Python 安装文件夹的 scripts 文件夹进入命令提示符环境
(14) 图 1-14　展开 Python 启动程序并右击后选择"属性"
(15) 图 1-15　选择"打开文件位置"按钮进入 Python 安装文件夹
(16) 图 2-1　Python 序列分类示意图
(17) 图 2-2　双向索引示意图
(18) 图 2-3　reduce()函数执行过程示意图
(19) 图 3-1　逻辑运算符与几种电路的类比关系
(20) 图 3-2　单分支选择结构
(21) 图 3-3　双分支选择结构
(22) 图 3-4　代码层次与隶属关系
(23) 图 3-5　函数示意图
(24) 图 3-6　使用注释来为用户提示函数使用说明
(25) 图 3-7　几个好玩的表情
(26) 图 3-8　一维序列卷积计算原理示意图
(27) 图 3-9　把卷积结果转换为数字
(28) 图 3-10　函数递归调用示意图
(29) 图 4-1　列出对象公开成员

(30) 图 4-2　列出对象所有成员
(31) 图 4-3　二叉树
(32) 图 5-1　字符串格式化
(33) 图 5-2　字符串加密与解密结果
(34) 图 6-1　二进制文件无法使用文本编辑器直接查看
(35) 图 6-2　使用 Winhex 十六进制编辑器打开可执行文件
(36) 图 7-1　doctest 测试过程示意图
(37) 图 7-2　IDLE 调试器窗口
(38) 图 7-3　程序调试截图(一)
(39) 图 7-4　程序调试截图(二)
(40) 图 7-5　程序调试截图(三)
(41) 图 7-6　程序调试截图(四)
(42) 图 7-7　运行程序自动进行 pdb 调试模式
(43) 图 7-8　在命令提示符环境运行程序
(44) 图 7-9　使用命令行调试程序
(45) 图 8-1　运行结果(一)
(46) 图 8-2　运行结果(二)
(47) 图 9-1　使用 ipconfig 命令查看本机 IP 地址和网卡物理地址
(48) 图 9-2　UDP 通信程序运行结果
(49) 图 9-3　TCP 通信程序运行结果
(50) 图 9-4　在 IIS 中创建网站
(51) 图 9-5　配置 IIS 的程序映射
(52) 图 9-6　网站运行效果(一)
(53) 图 9-7　网站运行效果(二)
(54) 图 9-8　网站启动界面(三)
(55) 图 9-9　网站运行效果(四)
(56) 图 9-10　启动 django 网站
(57) 图 9-11　素数判断
(58) 图 9-12　使用网页模板
(59) 图 10-1　WPS 进程中某个线程的属性
(60) 图 10-2　在 IDLE 环境中运行
(61) 图 10-3　在命令提示符环境中运行
(62) 图 10-4　WPS 创建的部分线程
(63) 图 10-5　系统中每个进程的线程数量
(64) 图 10-6　使用 Condition 实现线程同步
(65) 图 10-7　使用 Event 对象实现线程同步
(66) 图 11-1　MapReduce 流程
(67) 图 11-2　文件切分结果

(68) 图 11-3　Map 结果
(69) 图 11-4　Reduce 结果
(70) 图 11-5　pyspark 开发界面
(71) 图 11-6　执行 Python 程序
(72) 图 12-1　使用 OpenGL 绘制简单图形
(73) 图 12-2　纹理映射
(74) 图 12-3　光照模型
(75) 图 12-4　原始 lena 图像
(76) 图 12-5　部分区域被旋转 180°以后的 lena 图像
(77) 图 12-6　比例分配图
(78) 图 12-7　验证码图片
(79) 图 13-1　lena 图像处理结果
(80) 图 13-2　lena 图像模糊处理结果
(81) 图 13-3　原始图像
(82) 图 13-4　高斯滤波结果
(83) 图 13-5　边缘锐化结果
(84) 图 13-6　中值滤波结果
(85) 图 13-7　原始随机图像
(86) 图 13-8　开运算结果
(87) 图 13-9　膨胀运算结果
(88) 图 13-10　闭运算结果
(89) 图 13-11　绘制曲线图结果
(90) 图 13-12　绘制柱状图结果
(91) 图 13-13　水平柱状图绘制结果
(92) 图 13-14　正弦曲线
(93) 图 13-15　余弦散点图
(94) 图 13-16　散点图
(95) 图 13-17　绘制饼状图
(96) 图 13-18　中文标签和图例
(97) 图 13-19　标签中带公式的图
(98) 图 13-20　多图形同时显示
(99) 图 13-21　绘制三维参数曲线
(100) 图 13-22　绘制三维图形(一)
(101) 图 13-23　绘制三维图形(二)
(102) 图 13-24　使用指令绘制任意图形
(103) 图 13-25　把 matplotlib 绘图结果嵌入 tkinter 程序中
(104) 图 13-26　matplotlib 组件的应用
(105) 图 13-27　根据最新数据动态更新图形

(106）图 13-28 使用 Slider 组件调整曲线参数
(107）图 14-1 维吉尼亚密码替换表
(108）图 14-2 计算文件的 MD5 值
(109）图 15-1 用户登录界面
(110）图 15-2 密码正确
(111）图 15-3 密码错误
(112）图 15-4 程序运行效果
(113）图 15-5 简单文本编辑器
(114）图 15-6 简单画图程序
(115）图 15-7 电子时钟运行截图
(116）图 15-8 动画截图
(117）图 15-9 turtle 画图示例
(118）图 15-10 MP3 播放器界面
(119）图 16-1 教师端主界面
(120）图 16-2 学生端主界面
(121）图 16-3 离线点名与加分功能
(122）图 16-4 随机提问界面

附录 C 本书中表格清单

(1) 表 1-1　IDLE 中的常用快捷键
(2) 表 1-2　Python 内置对象
(3) 表 1-3　Python 常用内置函数
(4) 表 1-4　Python 运算符
(5) 表 1-5　常用 pip 命令使用方法
(6) 表 2-1　常用的列表对象方法
(7) 表 4-1　Python 类特殊方法
(8) 表 5-1　常见的转义字符
(9) 表 5-2　格式字符
(10) 表 5-3　常用的正则表达式元字符
(11) 表 5-4　常用子模式扩展语法
(12) 表 5-5　re 模块常用方法
(13) 表 6-1　文件打开模式
(14) 表 6-2　文件对象的常用属性
(15) 表 6-3　文件对象的常用方法
(16) 表 6-4　os 模块常用成员
(17) 表 6-5　os.path 模块常用成员
(18) 表 6-6　shutil 模块常用成员
(19) 表 7-1　TestCase 类的常用方法
(20) 表 7-2　常用 pdb 调试命令
(21) 表 8-1　Connection 对象的主要方法
(22) 表 10-1　threading 模块常用方法与类
(23) 表 10-2　Thread 对象成员
(24) 表 12-1　mode 取值
(25) 表 13-1　scipy 工具包的主要模块
(26) 表 15-1　tkinter 的常用组件
(27) 表 15-2　pygame 的主要模块
(28) 表 15-3　mixer 模块的主要方法
(29) 表 16-1　表结构

附录 D 本书中拓展知识摘要清单

(1) 拓展知识：如果有读者想尝试一下在安卓手机上编写 Python 程序，可以安装支持 Python 3.x 的 QPython3 或者支持 Python 2.x 的 QPython，关于 SL4A 和安卓类库调用的相关知识可以查阅相关资料。

(2) 拓展知识：自定义 IDLE 清屏快捷键。

(3) 拓展知识：Python 标准库 fractions 中的 Fraction 对象支持分数运算。

(4) 拓展知识：Python 字符串对象提供了一个方法 isidentifier()可以用来判断指定字符串是否可以作为变量名、函数名、类名等标识符。

(5) 拓展知识：Python 之禅。

(6) 拓展知识：位运算规则为 1&1=1、1&0=0&1=0&0=0，1|1=1|0=0|1=1、0|0=0，1^1=0^0=0、1^0=0^1=1，左移位时右侧补 0，右移位时左侧补 0。

(7) 拓展知识：复合赋值运算符。

(8) 拓展知识：Python 标准库 sys 还提供了 read()和 readline()方法用来从键盘接收指定数量的字符。

(9) 拓展知识：Python 标准库 pprint 还提供了更加友好的输出函数(pretty printer) pprint()，可以更好地控制输出格式，如果要输出的内容多于一行则会自动添加换行和缩进来更好地展示内容的结构。

(10) 拓展知识：Python 支持创建多个虚拟环境，每个虚拟环境都是包含 Python 和相应扩展库的一个目录，多个虚拟环境(文件夹)之间互相不干扰。

(11) 拓展知识：重新导入模块。

(12) 拓展知识：导入模块时文件的搜索顺序。

(13) 拓展知识：垃圾回收机制。一般来说，使用 del 删除对象之后 Python 会在恰当的时机调用垃圾回收机制来释放内存，我们也可以在必要的时候导入 Python 标准库 gc 之后调用 gc.collect()函数立刻启动垃圾回收机制来释放内存。

(14) 拓展知识：排序方法的 key 参数。

(15) 拓展知识：使用列表模拟向量运算。

(16) 拓展知识：生成器对象。

(17) 拓展知识：内置函数 globals()和 locals()分别返回包含当前作用域内所有全局变量和局部变量的名称及值的字典。

(18) 拓展知识：有序字典。

(19) 拓展知识：内置函数 sorted() 可以对字典元素进行排序并返回新列表，充分利用 key 参数可以实现丰富的排序功能。

(20) 拓展知识：Python 支持字典推导式快速生成符合特定条件的字典。

(21) 拓展知识：字典和集合的 in 操作比列表快很多。

(22) 拓展知识：自定义枚举类型。

(23) 拓展知识：集合中的元素不允许重复，Python 集合的内部实现为此做了大量相应的优化，判断集合中是否包含某元素时比列表速度快很多。

(24) 拓展知识：Python 也支持集合推导式。

(25) 拓展知识：逻辑运算符与常见电路连接方式的相似之处。

(26) 拓展知识：Python 还提供了一个三元运算符，可以实现与选择结构相似的效果。

(27) 拓展知识：Python 标准库 datetime。

(28) 拓展知识：标准库 calendar 也提供了一些与日期操作有关的方法。

(29) 拓展知识：也可以自己编写代码模拟 Python 标准库 calendar 中查看日历的方法。

(30) 拓展知识：math 是用于数学计算的标准库。

(31) 拓展知识：有时候换个角度来思考和解决问题，或许会更加有效和快捷。

(32) 拓展知识：也可以直接使用 Python 标准库 itertools 提供的函数来解决组合数计算的问题。

(33) 拓展知识：函数属于可调用对象。

(34) 拓展知识：局部变量的空间是在栈上分配的，而栈空间是由操作系统维护的，每当调用一个函数时，操作系统会为其分配一个栈帧，函数调用结束后立刻释放这个栈帧。

(35) 拓展知识：除了局部变量和全局变量，Python 还支持使用 nonlocal 关键字定义一种介于两者之间的变量。

(36) 拓展知识：例 3-18 描述的实际上是将列表循环左移 k 位的算法，下面的代码使用了更加直接的方法，但对于长列表来说效率不如上面的代码高。

(37) 拓展知识：在 Python 3.5 版本中，标准库 math 也提供了计算最大公约数的函数 gcd()。

(38) 拓展知识：例 3-24 给出的算法是快速排序算法中非常重要的一个步骤，当然也可以使用下面更加简洁的代码来实现。

(39) 拓展知识：函数递归调用。

(40) 拓展知识：利用类数据成员的共享性，可以实时获得该类的对象数量，并且可以控制该类可以创建的对象最大数量。

(41) 拓展知识：在 Python 中，函数和方法是有区别的。

(42) 拓展知识：所谓多态，是指基类的同一个方法在不同派生类对象中具有不同的表现和行为。

(43) 拓展知识：Python 标准库 queue 提供了 LILO 队列类 Queue、LIFO 队列类

LifoQueue、优先级队列类 PriorityQueue，标准库 collections 提供了双端队列。

（44）拓展知识：堆也是一种很重要的数据结构，在进行排序时使用较多，优先队列也是堆结构的一个重要应用。

（45）拓展知识：转义字符。

（46）拓展知识：在字符串格式化方法 format() 中常用的格式字符。

（47）拓展知识：Python 标准库 string 还提供了用于字符串格式化的模板类 Template。

（48）拓展知识：实际开发时应优先考虑使用 Python 内置函数和内置对象的方法，运行速度快，并且运行稳定。

（49）拓展知识：timeit 模块还支持下面代码演示的用法，从运行结果可以看出，当需要对大量数据进行类型转换时，内置函数 map() 可以提供非常高的效率。

（50）拓展知识：Python 标准库中的 string 提供了英文字母大小写、数字字符、标点符号等常量，可以直接使用，下面的代码实现了随机密码生成功能。

（51）拓展知识：在 Python 中，字符串属于不可变对象，不支持原地修改，如果需要修改其中的值，只能重新创建一个新的字符串对象。

（52）拓展知识：Python 标准库 unicodedata 提供了不同形式数字字符到十进制数字的转换方法。

（53）拓展知识：Python 标准库 textwrap 提供了更加友好的排版函数。

（54）拓展知识：Python 扩展库 jieba 和 snownlp 很好地支持了中文分词，可以使用 pip 命令进行安装。

（55）拓展知识：Python 扩展库 pypinyin 支持汉字到拼音的转换，并且可以和分词扩展库配合使用。

（56）拓展知识：文件操作一般都要遵循"打开文件→读写文件→关闭文件"的标准套路，但是如果文件读写操作代码引发了异常，很难保证文件能够被正确关闭，使用上下文管理关键字 with 可以避免这个问题。

（57）拓展知识：在交互模式下使用文件对象的 write() 方法写入文件时，会显示成功写入的字符数量。

（58）拓展知识：JSON(JavaScript Object Notation)是一种轻量级的数据交换格式，易于阅读和编写，同时也易于机器解析和生成（一般用于提升网络传输速率），是一种比较理想的编码与解码格式。

（59）拓展知识：CSV(Comma Separated Values)格式的文件常用于电子表格和数据库中内容的导入和导出。

（60）拓展知识：除了用于文件操作和文件夹操作的方法之外，os 模块还提供了大量其他方法。

（61）拓展知识：CRC 又称为循环冗余检验码，常用于数据存储和通信领域，具有极强的检错能力。

（62）拓展知识：Pywin32 模块需要单独安装，这是一个功能非常强大的模块，提供了 Windows 底层 API 函数的封装，使得可以在 Python 中直接调用 Windows API 函数，支持大量的 Windows 底层操作。

（63）拓展知识：Python 标准库 ctypes 提供了访问 DLL 动态链接库的功能，很好地支持了与 C/C++ 等语言混合编程的需求，也可以调用系统底层 API 函数。

（64）拓展知识：系统运维涵盖的内容非常多，还包括电力系统维护、数据库维护、磁盘配额、用户账号与权限、网络设备与带宽分配、病毒防护与入侵检测、系统资源分配等。

（65）拓展知识：正如前面所说，系统运维涉及面非常广，也包括系统中进程的创建与结束、系统服务状态等。

（66）拓展知识：Python 程序编译与打包。

（67）拓展知识：回调函数原理。

（68）拓展知识：断言语句 assert 也是一种比较常用的技术，常用来在程序的某个位置确认指定条件必须满足，常和异常处理结构一起使用。

（69）拓展知识：在工程界，不管是安全专家还是恶意攻击者，最常使用的漏洞发现和挖掘方法是 Fuzz，属于"灰"盒测试技术，也可以说是一种特殊的黑盒测试技术。

（70）拓展知识：有时候可能需要把代码执行过程中的一些调试信息、出错信息或其他信息记录下来而不影响正常的输出，这时可以使用 Python 标准库 logging 提供的功能。

（71）拓展知识：软件性能测试。

（72）拓展知识：SQL 注入式攻击与防范。

（73）拓展知识：如果想知道某个 IP 地址的详细信息，如国家、城市、经纬度等信息，可以使用 Python 扩展库 pygeoip 配合数据库 GeoLiteCity.dat 来获取这些信息。

（74）拓展知识：使用 Python 查看本机的 IP 地址与网卡的物理地址。

（75）拓展知识：发送数据时，如果目标 IP 地址中最后一组数字是 255，表示广播地址，也就是说局域网内的所有主机都会收到信息。

（76）拓展知识：Python 标准库 socket 除了支持 UDP 和 TCP 编程之外，还提供了用来获取本地主机名的 gethostname()、根据主机名获取 IP 地址的 gethostbyname()、根据 IP 地址获取主机名的 gethostbyaddr()、根据端口号获取对应服务名称的 getservbyport()、根据服务名称获取对应端口号的 getservbyname()等方法。

（77）拓展知识：sniffer pro 是 NAI 公司出品的一款一流的便携式网管和应用故障诊断分析软件，拥有强大的网络抓包和协议分析能力，软件能够完美支持全系统 Windows 平台，性能优越，是网络管理员必备的一款网络协议分析软件。

（78）拓展知识：scapy 是一款功能非常强大的交互式包处理程序，可以伪造或解码很多种网络协议的数据包，可以发送和捕获数据包，可以对请求数据包和回复数据包进行匹配，可以处理扫描、路由跟踪、探测、单元测试、攻击、网络发现等任务，还具有很多其他工具所不具有的功能。

（79）拓展知识：运行例 9-4 的代码会发现，虽然扫描效果不错，但是速度非常慢，远不如 xscan 快。

（80）拓展知识：Python 扩展库 netaddr 提供了大量可以处理网络地址的类和对象。

（81）拓展知识：在例 9-4 的代码中是使用 IP 地址来表示目标主机的，但是很多网站为了防止黑客攻击或者进行负载均衡,会经常变换主机,这样同一个域名在不同时间可能会对应不同的 IP 地址,在这种情况下可以通过 socket 模块的 gethostbyname()函数来实时获取目标主机的 IP 地址。

(82) 拓展知识:Nmap 是一款非常棒的网络扫描工具,首先下载并安装 Nmap 工具,把安装路径添加到系统 Path 环境变量,然后使用 pip 安装 python-nmap,就可以使用了。

(83) 拓展知识:Python 标准库 ftplib 提供了 FTP 客户端的主要功能。

(84) 拓展知识:如果你仍然不舍得放弃 Python 2.x,或许可以试试 Python 扩展库 machanize,这也是一款不错的网页内容读取工具。

(85) 拓展知识:以创建并启动线程的方式来执行一个函数可以实现多个函数或功能代码并发或同时运行,而直接调用函数的话会阻塞当前线程,直到函数执行结束返回后才能继续执行当前线程的代码。

(86) 拓展知识:前面我们已经多次用过 Python 标准库 time 中的 sleep()函数,它的功能是暂停(或者说阻塞当前线程)指定时间(单位是秒)。

(87) 拓展知识:多线程技术的提出并不仅仅是为了提高处理速度和硬件资源利用率,还有就是为了增加系统的可扩展性(采用多线程技术编写的代码移植到多处理器平台上继续不需要改写就能立刻适应新的平台)和提高用户体验。

(88) 拓展知识:在 Windows 7 系统中,单击"开始"→"计算机"→"管理"命令,然后展开"事件查看器",选择感兴趣的事件类别并右击,选择"将所有事件另存为",就可以把系统日志保存为记事本文档。

(89) 拓展知识:对于贝塞尔曲线而言,其特点在于第一个控制点恰好是曲线的起点,最后一个控制点是曲线的终点,其他控制点并不在曲线上,而是起到控制曲线形状的作用。

(90) 拓展知识:MD5 算法属于单向变换算法,不存在反函数,暴力测试几乎成为唯一可能的 MD5 破解方法。

(91) 拓展知识:也可以使用 ssdeep 工具来计算文件的模糊哈希值或分段哈希值,或者编写 Python 程序调用 ssdeep 提供的 API 函数来计算文件的模糊哈希值,模糊哈希值可以用来比较两个文件的相似百分比。

(92) 拓展知识:Python 标准库 turtle 也提供了很多绘图功能。

(93) 拓展知识:对于例 15-10 的程序有几点需要补充说明一下:①通过 socket 进行网络通信时收发数据包的数量和大小最好是一致的,要不然会导致混乱而无法正常监控;②如果需要进行屏幕广播,需要使用 UDP,可以参考本书第 16 章介绍的服务器发现功能代码;③在上面程序的实现中,发送端把屏幕截图转换为字节串后发送给接收端,而接收端接收完整个屏幕截图后还原为图像进行显示,这些操作都是在内存中进行的,如果同时监控多个客户端,可能会占用监控端大量的内存,可以接收截图的内容后不存储在内存中而是写入磁盘文件来缓解监控端的内存压力;④每次收发数据包的大小不能太大或太小,如果太大可能会因为出错引起的频繁重传而降低效率,如果太小则会因为额外开销(在网络体系结构中,发送端数据包每往下一层就会加一层"皮",而接收端每往上一层就会为数据包去掉一层"皮",这些"皮"是有开销的)太大而降低效率。

参 考 文 献

[1] 董付国. Python 程序设计基础[M]. 北京：清华大学出版社，2015.
[2] 董付国. Python 程序设计[M]. 北京：清华大学出版社，2015.
[3] 董付国. Python 程序设计[M]. 2 版. 北京：清华大学出版社，2016.
[4] 张颖，赖勇浩. 编写高质量代码——改善 Python 程序的 91 个建议[M]. 北京：机械工业出版社，2014.
[5] 杨佩璐，宋强，等. Python 宝典[M]. 北京：电子工业出版社，2014.
[6] 张若愚. Python 科学计算[M]. 北京：清华大学出版社，2012.
[7] 赵家刚，狄光智，吕丹桔，等. 计算机编程导论——Python 程序设计[M]. 北京：人民邮电出版社，2013.
[8] TJ. O'Connor. Python 绝技——运用 Python 成为顶级黑客[M]. 崔孝晨，武晓音，等译. 北京：电子工业出版社，2016.